CONCEPTS AND METHODS OF 2D INFRARED SPECTROSCOPY

2D infrared (IR) spectroscopy is a cutting-edge technique, with applications in subjects as diverse as the energy sciences, biophysics and physical chemistry. This book introduces the essential concepts of 2D IR spectroscopy step-by-step to build an intuitive and in-depth understanding of the method.

Taking a unique approach, this book outlines the mathematical formalism in a simple manner, examines the design considerations for implementing the methods in the laboratory, and contains working computer code to simulate 2D IR spectra and exercises to illustrate the concepts involved. Readers will learn how to accurately interpret 2D IR spectra, design their own spectrometer and invent their own pulse sequences. It is an excellent starting point for graduate students and researchers new to this exciting field. Computer codes and answers to the exercises can be downloaded from the authors' website, available at www.cambridge.org/9781107000056.

PETER HAMM is a Professor at the Institute of Physical Chemistry, University of Zurich.

MARTIN ZANNI is Meloche-Bascom Professor in the Department of Chemistry, University of Wisconsin-Madison.

They specialize in using 2D IR spectroscopy to study molecular structures and dynamics.

CONCEPTS AND METHODS OF 2D INFRARED SPECTROSCOPY

PETER HAMM
University of Zurich

and

MARTIN ZANNI
University of Wisconsin-Madison

CAMBRIDGE UNIVERSITY PRESS
Cambridge, New York, Melbourne, Madrid, Cape Town,
Singapore, São Paulo, Delhi, Mexico City

Cambridge University Press
The Edinburgh Building, Cambridge CB2 8RU, UK

Published in the United States of America by Cambridge University Press, New York

www.cambridge.org
Information on this title: www.cambridge.org/9781107000056

© P. Hamm and M. Zanni 2011

This publication is in copyright. Subject to statutory exception
and to the provisions of relevant collective licensing agreements,
no reproduction of any part may take place without the written
permission of Cambridge University Press.

First published 2011

A catalogue record for this publication is available from the British Library

Library of Congress Cataloging in Publication data
Hamm, Peter, 1966–
Concepts and methods of 2d infrared spectroscopy / Peter Hamm
and Martin T. Zanni.
p. cm.
Includes bibliographical references and index.
ISBN 978-1-107-00005-6
1. Infrared spectra. I. Zanni, Martin T. II. Title.
QC457.H35 2011
543.57–dc22
2010041962

ISBN 978-1-107-00005-6 Hardback

Additional resources for this publication at www.cambridge.org/9781107000056

Cambridge University Press has no responsibility for the persistence or
accuracy of URLs for external or third-party internet websites referred to in
this publication, and does not guarantee that any content on such websites is,
or will remain, accurate or appropriate. Information regarding prices, travel
timetables, and other factual information given in this work is correct at
the time of first printing but Cambridge University Press does not guarantee
the accuracy of such information thereafter.

Dedicated to Robin M. Hochstrasser.
We appreciate the help of our students, postdoctoral researchers,
colleagues, mentors and families.

Contents

			page
1	Introduction		1
	1.1	Studying molecular structure with 2D IR spectroscopy	3
	1.2	Structural distributions and inhomogeneous broadening	10
	1.3	Studying structural dynamics with 2D IR spectroscopy	12
	1.4	Time domain 2D IR spectroscopy	14
	Exercises		16
2	Designing multiple pulse experiments		18
	2.1	Eigenstates, coherences and the emitted field	18
	2.2	Bloch vectors and molecular ensembles	23
	2.3	Bloch vectors are a graphical representation of the density matrix	27
	2.4	Multiple pathways visualized with Feynman diagrams	31
	2.5	What is absorption?	37
	2.6	Designing multi-pulse experiments	38
	2.7	Selecting pathways by phase matching	42
	2.8	Selecting pathways by phase cycling	44
	2.9	Double sided Feynman diagrams: Rules	46
	Exercises		47
3	Mukamelian *or* perturbative expansion of the density matrix		48
	3.1	Density matrix	48
	3.2	Time dependent perturbation theory	52
	Exercises		60
4	Basics of 2D IR spectroscopy		61
	4.1	Linear spectroscopy	61
	4.2	Third-order response functions	65
	4.3	Time domain 2D IR spectroscopy	69

	4.4	Frequency domain 2D IR spectroscopy	82
	4.5	Transient pump–probe spectroscopy	84
		Exercises	86
5	Polarization control		88
	5.1	Using polarization to manipulate the molecular response	88
	5.2	Diagonal peak, no rotations	92
	5.3	Cross-peaks and orientations of coupled transition dipoles	93
	5.4	Combining pulse polarizations: Eliminating diagonal peaks	99
	5.5	Including (or excluding) rotational motions	100
	5.6	Polarization conditions for higher-order pulse sequences	106
		Exercises	108
6	Molecular couplings		109
	6.1	Vibrational excitons	109
	6.2	Spectroscopy of a coupled dimer	114
	6.3	Extended excitons in regular structures	120
	6.4	Isotope labeling	128
	6.5	Local mode transition dipoles	133
	6.6	Calculation of coupling constants	134
	6.7	Local versus normal modes	137
	6.8	Fermi resonance	140
		Exercises	142
7	2D IR lineshapes		145
	7.1	Microscopic theory of dephasing	145
	7.2	Correlation functions	149
	7.3	Homogeneous and inhomogeneous dynamics	152
	7.4	Nonlinear response	155
	7.5	Photon echo peak shift experiments	161
		Exercises	164
8	Dynamic cross-peaks		166
	8.1	Population transfer	166
	8.2	Dynamic response functions	172
	8.3	Chemical exchange	174
9	Experimental designs, data collection and processing		176
	9.1	Frequency domain spectrometer designs	176
	9.2	Experimental considerations for impulsive spectrometer designs	180
	9.3	Capabilities made possible by phase control	191

9.4	Phase control devices	197
9.5	Data collection and data workup	201
9.6	Experimental issues common to all methods	214
Exercises		216
10	Simple simulation strategies	217
10.1	2D lineshapes: Spectral diffusion of water	217
10.2	Molecular couplings by *ab initio* calculations	226
10.3	2D spectra using an exciton approach	229
Exercises		232
11	Pulse sequence design: Some examples	233
11.1	Two-quantum pulse sequence	233
11.2	Rephased 2Q pulse sequence: Fifth-order spectroscopy	236
11.3	3D IR spectroscopy	239
11.4	Transient 2D IR spectroscopy	243
11.5	Enhancement of 2D IR spectra through coherent control	245
11.6	Mixed IR–Vis spectroscopies	247
11.7	Some of our dream experiments	249
Exercises		252
Appendix A	Fourier transformation	254
A.1	Sampling theorem, aliasing and under-sampling	256
A.2	Discrete Fourier transformation	257
Appendix B	The ladder operator formalism	260
Appendix C	Units and physical constants	262
C.1	Physical constants	262
C.2	Units of common physical quantities	262
C.3	Emitted field $E_{sig}^{(3)}$	263
Appendix D	Legendre polynomials and spherical harmonics	265
Appendix E	Recommended reading	267
References		269
Index		281

1
Introduction

Scientific questions encompassing both the structure and dynamics of molecular systems are difficult to address. Take the case of a folding protein, a fluctuating solvent environment or a transferring electron. In each case, one wants to know the reaction pathway, which requires time-resolving the structure. But the range of time-scales can easily span from femtoseconds to hours, depending on the system. If time-scales are slow, then exquisite structural information can be obtained with nuclear magnetic resonance (NMR) spectroscopy. If time-scales are fast, then fluorescence or absorption spectroscopy can be used to probe the dynamics with a corresponding tradeoff in structural resolution. In between, there is an experimental gap in time- and structure-resolution. The gap is even broader when the dynamics takes place in a confined environment like a membrane, which makes it especially difficult to apply many standard structural techniques.

2D IR spectroscopy is being used to fill this gap because it provides bond-specific structural resolution and can be applied to all relevant time-scales (see these Special Issues [96, 143, 144] and review articles [19, 26, 27, 56, 63, 67, 80, 87, 103, 108, 142, 165, 191, 200, 208]). It has the fast time-resolution to follow electron transfer and solvent dynamics, for instance, or can be applied in a "snapshot" mode to study kinetics to arbitrarily long time-scales. Moreover, it can be applied to any type of sample, including dilute solutions, solid-state systems, or membranes. Its structural sensitivity stems from couplings between vibrational modes that give rise to characteristic infrared bands and cross-peaks. Structures can also be probed through hydrogen bonding and electric field effects that generate dynamic 2D lineshapes. Moreover, 2D IR spectra can be quantitatively computed from molecular dynamics simulations, which provides a direct comparison to all-atom models.

Constructing a 2D IR spectrometer, collecting the data, and interpreting the spectra requires a very broad skill set. Of course, one can qualitatively use 2D IR spectroscopy as an analytical tool, but a little bit of knowledge about nonlinear optics, vibrational potentials and lineshape theory, enables a much deeper interpretation of 2D IR spectra and a broader range of applications. These topics are explained in various textbooks (see Appendix E) and research articles, but there is no single source that contains all of the fundamental concepts that pertain to 2D IR spectroscopy from which students and researchers new to the field can easily draw upon. This book is intended to foster the ease at which new graduate students and experienced researchers that are moving into the field can learn about the mathematical formalism and technical challenges of 2D IR spectroscopy. Many of the topics also pertain to 2D visible spectroscopy that probes electronic transitions.

But the interpretation of 2D IR spectroscopy is not the sole motivating force for writing this book. Rather, it is our belief that 2D IR spectroscopy will evolve into 3D and higher dimensions. If 2D was the highest-order spectroscopy possible, then one could just memorize the relatively few types of possible 2D pulse sequences (there are really only three) and apply the best one to the problem at hand without intimate knowledge of the chemistry or physics. But with 3D IR spectroscopy there is the potential to develop more sophisticated pulse sequences that are specifically tailored to the problem at hand. Since most of these pulse sequences are yet to be developed, their design and application will require a deeper understanding of the technique, which this book is intended to facilitate.

Only two types of 3D IR experiments have been explored so far, examples of which are shown in Fig. 1.1. The first example is a 3D IR spectrum of a metal carbonyl compound. This spectrum and the accompanying work demonstrated the feasibility of collecting 3D IR spectra, outlined a novel two-quantum pulse sequence, and showed that cascading processes that are major problems in other

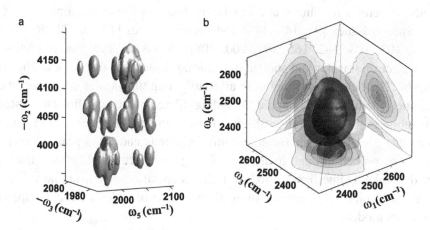

Figure 1.1 3D IR spectra. (a) 3D IR spectrum of a metal dicarbonyl using a two-quantum pulse sequence (adapted from Ref. [43] with permission). (b) Absorptive 3D IR spectrum of the OD stretch of HOD in H_2O [64].

nonlinear spectroscopies are not an issue in multidimensional IR spectroscopy [43–45, 59, 61]. The second experiment is the 3D IR spectrum of water (more precisely, it is the OD stretch of HOD in H_2O) [64]. This spectrum is of interest not only because of what it reveals about the structural dynamics of water, but also because it demonstrates the feasibility of collecting 3D IR spectra even on weakly absorbing chromophores (as compared to the metal carbonyls). These two experiments suggest that many new and exciting 3D and higher-order spectroscopies are possible for a wide range of samples. Designing the pulse sequences to extract the interesting information in these experiments requires the methods contained in this book. We return to 3D IR spectroscopy in the final chapter.

1.1 Studying molecular structure with 2D IR spectroscopy

When one thinks of 2D IR spectroscopy, cross-peaks usually come to mind. Cross-peaks are the hallmark of multidimensional spectroscopy. They are a measure of the coupling between molecular vibrations and thus contain information on the molecular structure. To illustrate the concept, consider two carbonyl stretches, such as from two acetone molecules shown in Fig. 1.2(a). The molecules are made of negative electrons and positive nuclei, which together create the electronic structure of the molecules. The electronic structure, that is the molecular orbitals, dictate the bond lengths and thus the vibrational frequencies [125]. Moreover, the charge distributions of the electrons and nuclei create an electrostatic potential that surrounds the molecule. If the two acetone molecules are close enough, they will feel one another's potentials, which will slightly alter their molecular orbitals, thereby leading to a shift in the vibrational frequencies. When this perturbation occurs, we say that the vibrational modes are *coupled*. Thus, if we can measure the vibrational coupling and understand its distance and angular dependence, we can determine the distances and orientations of the two molecules with respect to one another. 2D IR spectroscopy provides the measurement, through the cross-peaks, and models provide the structure dependence of the coupling.

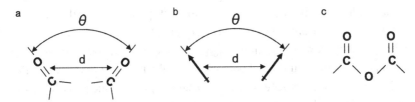

Figure 1.2 Two coupled acetone molecules. (a) The coupling strength will depend on the distance and orientations. (b) Representing the carbonyl stretches with transition dipoles. (c) A molecule in which both mechanical and electrostatic couplings are probably important.

Let us give an example of a coupling model and what it would predict for a 2D IR spectrum. The electrostatic potential around each molecule has a complex distance and angular dependence, at least at short radii, but at distances larger than the carbonyl length, the electrostatic potential can be described by the potential of a dipole [74, 99]. It is actually the *transition dipole* that we are interested in, not the dipole itself, since it is the transition dipole that couples the modes. The transition dipole is the change in the charge distribution of a molecule when it is vibrationally excited [131, 174]. Since acetone is a symmetric molecule, its transition dipole for the carbonyl stretch lies along the carbonyl bond. Thus, when the two acetone molecules are sufficiently far apart, they can each be represented as a transition dipole, which is shown in Fig. 1.2(b). The coupling between two dipoles is given by

$$\beta_{ij} = \frac{1}{4\pi\epsilon_0} \left[\frac{\vec{\mu}_i \cdot \vec{\mu}_j}{r_{ij}^3} - 3\frac{(\vec{r}_{ij} \cdot \vec{\mu}_i)(\vec{r}_{ij} \cdot \vec{\mu}_j)}{r_{ij}^5} \right] \quad (1.1)$$

where $\vec{\mu}_i$ are the directions of the transition dipoles and $\vec{r}_{ij}$ are the vectors connecting the two sites. The coupling β_{ij} scales as $1/r^3$ and depends on the orientation. Of course this formula for transition dipole–dipole coupling breaks down at close distances and for complicated molecular vibrations. Moreover, if the two vibrational modes share common atoms, like two carbonyl stretches located on the same molecule, then the carbonyl modes may be mechanically coupled as well (i.e. one stretch influences the other because of the intervening molecular bonds, see Fig. 1.2c), in which case a more sophisticated relationship between coupling and structure is needed [179]. Nonetheless, if it is understood how the molecular potential depends on the structure, then one can quantitatively interpret 2D IR spectra.

Shown in Fig. 1.3 are simulated 2D IR spectra for two coupled acetone molecules oriented at 45° with respect to one another. In the first spectrum (Fig. 1.3a), the acetone molecules are only separated by a few angstroms so that they are strongly coupled (10 cm^{-1}, see Eq. 1.1). In the second spectrum (Fig. 1.3b) they are farther apart so that the coupling is smaller (4 cm^{-1}). The frequency of one acetone molecule is simulated as if it were an isotope labeled with ^{13}C so that the two molecules have different vibrational frequencies even if they are not coupled. We label the two axes as ω_{pump} and ω_{probe}, for reasons that will become apparent soon. Each acetone molecule creates a pair of peaks near the diagonal of the spectrum, which we collectively refer to as the *diagonal peaks*. One peak lies exactly on the diagonal and the other is shifted off the diagonal to a different ω_{probe} frequency. The on-diagonal peak lies at the fundamental frequency, ω_{01}, along both axes (i.e. $\omega_{\text{pump}} = \omega_{\text{probe}} = \omega_{01}$). In the convention of this book, this peak is negative. The other peak is shifted because of the anharmonicity of the carbonyl stretch so that

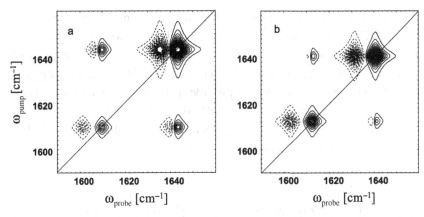

Figure 1.3 Simulated 2D IR spectra of two acetone molecules with a relative orientation of 45° and a transition dipole coupling of (a) 10 cm^{-1} and (b) 4 cm^{-1}, respectively.

the difference in frequency between the two peaks is what is known as the *diagonal anharmonic shift*. Since the two acetone molecules are coupled to one another, cross-peaks appear. The cross-peaks also appear in 180° phase-shifted pairs. On the upper half of the spectrum, one cross-peak will appear at $\omega_{pump} = \omega_{01}^{CO_2}$ and $\omega_{probe} = \omega_{01}^{CO_1}$ where the superscripts are labels for the two acetone carbonyl groups. This cross-peak most often has a negative intensity but some polarized 2D IR pulse sequences will generate a positive peak instead, depending on the orientation of the carbonyl transition dipoles, which gives additional structural information. The other cross-peak in the pair has the opposite sign and a different ω_{probe} frequency. The frequency difference between the two is the *off-diagonal* anharmonic shift, which is related to the coupling. Another pair of cross-peaks lies on the bottom half of the 2D IR spectrum. Notice that the anharmonic shifts make 2D IR spectra intrinsically nonsymmetric. As a result, in congested spectra with broad lineshapes and/or cross-peaks that are partially obscured by the diagonal peaks, the cross-peaks in the upper and lower halves of the spectrum may appear different, but in well-resolved spectra they should be identical.

In Fig. 1.3(b), where the coupling is weak, the off-diagonal anharmonic shift is smaller, leading to smaller cross-peak separation. In the limit that there is no coupling, the negative cross-peak will sit on top of the positive cross-peak so that they entirely cancel. One may notice upon careful inspection that the coupling not only creates cross-peaks but also changes the diagonal peaks. The diagonal peak frequencies, anharmonic shifts and intensities change because the coupling creates a multidimensional potential energy surface that has a slightly different curvature than each isolated molecule. In fact, one can extract the coupling strength and orientation of the molecules without using 2D spectroscopy by measuring the frequency shifts and intensity change of each fundamental transitions with standard linear (FTIR) spectroscopy and isotope labeling. However, in practice, 2D IR spectroscopy does a much better job of measuring the coupling with much less work (although isotope labeling is still very useful in 2D IR spectroscopy).

These simulations are intended to provide a qualitative understanding of how coupling alters the curvature of the molecular potential energy surface which results in cross-peaks. In the following section, we expand on how 2D IR spectroscopy probes the molecular potential energy surface.

1.1.1 2D IR spectrum of a single vibrational mode

Before we explain the origin of the cross-peaks, let us describe a simple way of collecting a 2D IR spectrum and what it will look like for a single vibrational mode, such as the carbonyl stretch of an acetone molecule. All we need to construct a 2D IR spectrum are the eigenstates and transition dipoles for the vibrational modes of the molecule that we are interested in. We represent the potential energy curve of the carbonyl stretch by a Morse oscillator (Fig. 1.4a):

$$V(r) = D(1 - e^{-ar})^2 \quad (1.2)$$

where r is the carbonyl bond length, D is the well depth, and a gives the curvature of the potential. The vibrational eigenstates generated from the Hamiltonian with this potential are

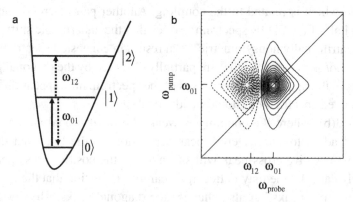

Figure 1.4 (a) Level scheme of an anharmonic oscillator with the dipole-allowed transitions depicted. The solid arrow represents the pump process, the dotted arrow the probe process. (b) Resulting 2D IR spectrum. Solid contour lines represent negative response (bleach and stimulated emission), dotted contour lines positive response (excited state absorption).

$$E_n = \hbar\omega\left(n+\frac{1}{2}\right) - x\left(n+\frac{1}{2}\right)^2 \tag{1.3}$$

where ω is the harmonic frequency of the oscillator, x is the *anharmonicity*, and n is the quantum number [93, 131, 132, 174]. This potential will produce the 2D IR spectrum simulated in Fig. 1.4(b).

The 2D IR spectrum can be measured in either the time or frequency domain. We begin our discussion in the frequency domain in which the 2D IR spectrum can be generated by a simple pump–probe experiment. Imagine that we scan the frequency of a pump pulse across the resonance frequency of the vibrator and plot its absorption as the y-axis of a 2D graph.[1] Whenever resonance with a dipole-allowed 0–1 transition is achieved, $\omega_{\text{pump}} = \hbar\omega - 2x \equiv \omega_{01}$, a certain fraction of molecules in the laser focus will be excited from their ground state $|0\rangle$ into their first vibrationally excited state $|1\rangle$ (solid arrow in Fig. 1.4a). Following the pump pulse, we scan the frequency of a probe pulse for the x-axis. The probe pulse will now measure two possible transitions from the excited state, i.e. the stimulated emission back into the ground state and the excited state absorption into the second excited state $|2\rangle$ (dotted arrows in Fig. 1.4a). In addition, since there are now fewer molecules in the ground state, the probe pulse will not be absorbed as much as it is when there is no pump pulse. Since we typically measure difference spectra (i.e. the difference of absorption between pump pulse switched on minus pump pulse switched off), the difference spectrum will be negative, which is an effect that is called a *bleach*. Both bleach and stimulated emission occur at the original ω_{01} with identical signs. The two contributions result in less absorption or gain, respectively, and by convention we give the signal a negative sign. In contrast, the excited state absorption will be positive because it is a new absorption induced by the pump and its frequency, $\omega_{12} = \hbar\omega - 4x$ is red-shifted from ω_{01} because of the anharmonicity of the potential. The shift is equal to $\omega_{12} - \omega_{01} = 2x \equiv \Delta$, which is the diagonal anharmonic shift.

Thus, by plotting the absorption as a function of the pump and probe frequencies, we will see a doublet of peaks in the 2D IR spectrum with opposite signs. There will be an on-diagonal peak at $\omega_{\text{pump}} = \omega_{\text{probe}} = \omega_{01}$ due to the bleach and stimulated emission signals, whereas the excited state absorption signal appears at $\omega_{\text{pump}} = \omega_{01}$ and $\omega_{\text{probe}} = \omega_{12}$. Even though there are two signals contributing to the on-diagonal peak and only one to the off-diagonal peak, both peaks will have roughly the same intensities because the 1–2 excited state absorption is twice as strong as a 0–1 transition (the 1–2 transition dipoles for a close-to

---

[1] There is currently no agreement in the community as to whether the x-axis should be the pump or the probe frequency axis. We use the convention of NMR spectroscopy with the probe frequency axis being the x-axis. Moreover, not all research groups follow the same convention for positive and negative signals.

harmonic oscillator scale as $\mu_{12}^2 = 2\mu_{01}^2$). If both peaks are well separated, then the anharmonic shift of the oscillator can be directly read off from a 2D IR spectrum. This condition is true only if the anharmonic shift is larger than the bandwidth of the transition. If it does not hold, then the 0–1 and 1–2 transitions overlap and partially cancel, as in Fig. 1.4(b), in which case the anharmonic shift has to be determined by deconvolution or peak fitting.

1.1.2 2D IR spectrum of two coupled vibrational modes

Using the same procedure as above, we can construct the 2D IR spectrum of two coupled oscillators from their vibrational energy levels and transition dipoles. For two oscillators, we have a 2D potential, which we write as

$$V(r_1, r_2) = V_1(r_1) + V_2(r_2) + \beta_{12} r_1 r_2 \qquad (1.4)$$

where $V_n(r_n)$ are the 1D potentials of each carbonyl stretch given by Eq. 1.2 and β_{12} is the coupling given by Eq. 1.1 if transition dipole–dipole coupling is adequate. We refer to the individual carbonyl groups and their parameters as *local modes* (e.g. local mode frequency). To get the eigenstates of the 2D potential, one must diagonalize $H(r_1, r_2)$ generated from $V(r_1, r_2)$. In Chapter 6, we solve this Hamiltonian explicitly, but even without doing so here, one can see that the coupling will shift the observed frequencies because they are no longer pure local modes. Moreover, it will also shift the ω_{12} transitions (the sequence transitions) and create a combination band.[2] Now, we need anharmonic shifts not only for the diagonal peaks, which we call Δ_{ii} for oscillator i, but we also need to describe the shift of the combination band, which we call Δ_{ij} and is the off-diagonal anharmonic shift. The eigenstates before and after diagonalization are shown in Fig. 1.5(a). The anharmonic constants Δ_{ij} describe the deviation of the energy of overtones and combination modes from just being a simple sum of the harmonic energies. That is, if there is no coupling, then $\omega_{0i} + \omega_{0j} = \omega_{0,i+j}$ and $\Delta_{ij} = 0$.

Figure 1.5(a) gives our nomenclature for labeling the eigenstates of two coupled oscillators. $|kl\rangle$ represents a state with k quanta of excitation in the first mode and l quanta of excitation in the second mode. If anharmonicity is small, which typically is the case, then the selection rules of harmonic oscillators still apply, i.e. only one oscillator can be changed by one quantum at a time. And the strength of the transition is determined by the transition dipole of that oscillator. For example,

---

[2] Oftentimes in the 2D IR literature, ω_{12} is referred to as the overtone transition, which is incorrect. It is actually a sequence band. An overtone transition would give the frequency ω_{02}. Nonetheless, we use these terms interchangeably in this book.

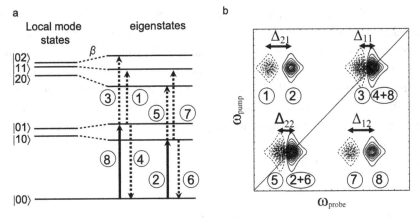

Figure 1.5 (a) Level scheme of two coupled oscillators before coupling (local modes) and after coupling (eigenstates). The dipole-allowed transitions are depicted. The solid arrows represent the pump process, the dotted arrows the probe process. (b) Resulting 2D IR spectrum. Solid contour lines represent negative response (bleach and stimulated emission), dotted contour lines positive response (excited state absorption). The labels (1)–(8) relate each peak in the 2D IR spectrum to the corresponding transition in the level scheme.

transitions $|10\rangle \to |20\rangle$ and $|10\rangle \to |11\rangle$ are *dipole allowed*, whereas $|10\rangle \to |02\rangle$ is *forbidden*. The arrows in Fig. 1.5(a) show all possible allowed transitions for two coupled oscillators.

With these rules in mind, we can now construct the 2D IR spectrum by imagining a pump–probe experiment. That is, we scan the pump frequency across the resonances of the two oscillators. When the pump frequency comes into resonance with an eigenstate, we mark that frequency along the y-axis. For example, when the pump is resonant with the higher-frequency oscillator, we will excite state $|01\rangle$. The subsequent probe pulse now has three possible transitions labeled (1), (3) and (4) in Fig. 1.5(a). In addition, the probe pulse will observe a bleach of *both* oscillators, giving rise to transitions (8) and (2), since the number of molecules in the common ground state $|00\rangle$ is diminished. Transitions (8), (4) and (3) are the same as for a single oscillator (Fig. 1.4), i.e. bleach, stimulated emission and excited state absorption, respectively, of the higher-frequency oscillator, whereas transitions (1) and (2) are new. Transition (1) excites the lower oscillator by one quantum from its ground state to its first excited state when there is already one quantum of excitation in the higher-frequency oscillator: $|01\rangle \to |11\rangle$. If the two oscillators were not coupled, then the excitation frequency of the second oscillator would not depend on the number of quanta of the first oscillator, and we would have exactly the same frequency for the $|00\rangle \to |10\rangle$ and the $|01\rangle \to |11\rangle$ transitions. In that case, peaks (1) and (2) would exactly coincide and cancel each other due to their identical transition strength but opposite sign. On the other hand, if the off-diagonal anharmonicity Δ_{12} is nonzero, then the two peaks do not cancel, and we obtain a doublet in the off-diagonal region, which we call a *cross-peak*. The existence of a cross-peak in a 2D IR spectrum is a direct manifestation of the coupling between both oscillators. In this context coupling means that the transition frequency of the one oscillator depends on the excitation level of the other oscillator. The off-diagonal anharmonicity Δ_{12} can directly be read off from a 2D IR spectrum, as depicted in Fig. 1.5(b). We will discuss in Chapter 6 how such cross-peaks are related to molecular structure.

1.2 Structural distributions and inhomogeneous broadening

The above sections pertain to an ensemble of identical molecules, but usually there are differences in the structure, hydrogen bonding, and environments of the molecules in the ensemble. Consider, for instance, the OH stretch vibration of a water molecule in liquid water. Each water molecule will sit in a different hydrogen bond environment (Fig. 1.6). Hydrogen bonding deforms the stretch potential such that the vibrational frequency is lowered. Hence, at each instant of time, each water molecule will have a different stretch frequency so that all the molecules together create a distribution of frequencies. If the molecules do not move on the time-scale of the 2D IR pulse sequence, then we say that there is an *inhomogeneous* distribution of frequencies. Each molecule also has an intrinsic linewidth that cannot be narrower than dictated by its vibrational lifetime, which we call the *homogeneous* linewidth. The overall 2D IR spectrum is a superposition of the 2D IR spectra for each individual molecule (Fig. 1.7a). The overall 2D IR spectrum will be broader than the homogeneous linewidths of the individual molecules, especially if the inhomogeneous distribution of the center frequencies is larger than the homogeneous linewidth.

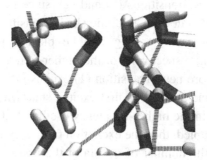

Figure 1.6 Snapshot from a molecular dynamics (MD) simulation of water with the hydrogen bonds indicated.

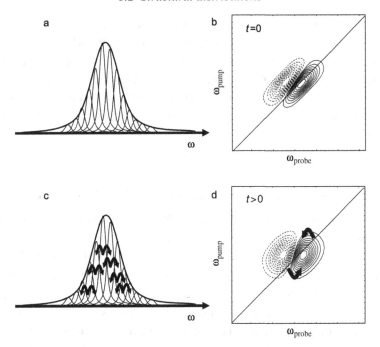

Figure 1.7 (a) An inhomogeneously broadened vibrational transition and (b) resulting 2D IR spectrum. Panels (c) and (d) show the same at a later delay time, when spectral diffusion has occurred. Solid contour lines represent negative response (bleach and stimulated emission), dotted contour lines positive response (excited state absorption).

Now imagine measuring a 2D IR spectrum of an inhomogeneous distribution of molecules. If the spectral width of the pump pulse is smaller than or equal to the homogeneous linewidth, then the pump pulse will be tuned into the particular subensemble of molecules whose spectral width is on-resonance with the center wavelength of the pump pulse. As a result, only that subensemble will be excited, creating a 2D IR response similar to Fig. 1.4 with the characteristic doublet of bands. The subensemble will be significantly narrower than the overall lineshape if the inhomogeneous distribution is much larger than the homogeneous linewidth. Scanning the pump frequency across the inhomogeneous ensemble, one will then observe a 2D IR spectrum that is elongated along the diagonal (Fig. 1.7b). In certain limits, the antidiagonal linewidth provides the homogeneous linewidth whereas the diagonal width represents the total linewidth (i.e. inhomogeneous width convoluted with the homogeneous width). This process is called *hole-burning*, and so the pump–probe process for measuring 2D IR spectra that we have described above is often referred to as hole-burning 2D IR spectroscopy. Cross-peaks also have 2D lineshapes which contain information on the frequency correlation between modes (see Problem 1.2) [41, 68].

2D lineshapes are another way to investigate molecular structure. To this end, one often employs a localized molecular reporter group, which can be the OD vibration of HOD in liquid H_2O or an isotope labeled amino acid in a protein, and investigates its inhomogeneous broadening. Depending on the direct environment of this reporter group (e.g. whether the amino acid is inside a protein or is exposed to the solvent water), the inhomogeneous broadening will vary in a systematic way [194].

1.3 Studying structural dynamics with 2D IR spectroscopy

One of the most powerful capabilities of 2D IR spectroscopy is its ability to monitor chemical and structural dynamics on all relevant time-scales, including those as short as femtoseconds. It has an intrinsic time resolution of about 50 fs, which is commensurate with the amount of time it takes a chemical bond to break. Thus, it can monitor even the fastest dynamics in equilibrium and nonequilibrium systems. Or it can be used to take nearly instantaneous snapshots of structures that evolve on much longer time-scales. We elaborate on some of these methods below.

1.3.1 Spectral diffusion

Going back to the example of Fig. 1.6, we note that the environment of a molecular probe in solution-phase systems is not always static. In glasses or in the interior of proteins, dynamics can be longer than seconds. In liquids, it is usually on the order of picoseconds. Our window to look into such diffusion processes is through the vibrational frequency of the reporter group. So the term "inhomogeneous broadening" really depends of the time-scale on which we investigate the molecular system. If the molecules in the vicinity of the reporter group move, then the vibrational frequency of the reporter group will change and alter the observed distribution of frequencies (Fig. 1.7c). This process is called *spectral diffusion*. Spectral diffusion can be measured with 2D IR spectroscopy. To do so, we must vary the time delay between the pump and probe pulses. At small enough time delays, the molecules will not have enough time to move and so we measure a seemingly static inhomogeneous distribution of frequencies, as discussed above (Fig. 1.7b). As we increase the time delay, we give the molecules time to move and thus change their frequencies (Fig. 1.7c). In the 2D IR spectrum, this results in a shift of intensity away from the diagonal, as indicated by the arrows in Fig. 1.7(d). Consequently, the 2D IR lineshape becomes more and more round with increasing delay time. The time-scale on which that happens reflects directly the time-scale on which the environment of the reporter group changes.

1.3.2 Chemical exchange

In the previous example, we had considered a continuous distribution of vibrational frequencies created by a large diversity in environments, for example. But distributions are often bimodal rather than continuous (Fig. 1.8b). For example, a hydrogen bond either exists or it does not. A molecule is either in a *cis* or *trans* conformation. For each of these examples, the two states are typically separated by a reaction barrier, so the probability of finding molecules in the transition state is very low. In the 2D IR spectrum of a bimodal distribution, each structure creates a pair of diagonal peaks. At early pump–probe delay times, there are no cross-peaks because one molecule cannot exist in both states (see Fig. 1.8a). However, if a hydrogen bond breaks during the delay between the pump and the probe pulse, then the molecule that was initially pumped with a hydrogen bond is now probed at the frequency without one. The same is true if a molecule converts from a *cis* to a *trans* conformation. Either way the dynamics creates a cross-peak on one side of the diagonal (depending on which species is higher frequency). Likewise, the reverse reaction will create a cross-peak on the other side. This process is called *chemical exchange*. What is unique about 2D IR spectroscopy as compared to NMR or other methods that also monitor chemical exchange is that it is sensitive to exchange rates ranging from femtoseconds to the vibrational lifetime of the chromophore. Thus, phenomena can now be explored on time-scales that were previously not possible [107, 192, 207].

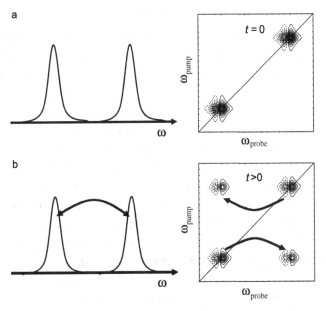

Figure 1.8 2D IR spectrum of a system undergoing chemical exchange with (a) no pump–probe delay and (b) a delay that is roughly that of the exchange.

1.3.3 Transient 2D IR spectroscopy

Chemical exchange is an equilibrium process, so that the dynamics are set by the forward and backward rate constants. A second mode of measuring chemical or structural dynamics is often referred to as *transient* 2D IR spectroscopy, in which some sort of nonequilibrium dynamics are initiated (e.g. by a temperature jump) and then the resulting kinetics are probed with 2D IR spectroscopy. Consider again the example of Fig. 1.8 for a *cis–trans* isomerization, but imagine that at equilibrium the system prefers the *cis* state so that there is only one set of diagonal peaks and no cross-peaks. With an appropriate actinic pump pulse, we might be able to excite the molecule to an electronic state that then relaxes to the *trans* conformation. In doing so, the *trans* diagonal peaks would appear as would cross-peaks on one side of the diagonal, and by monitoring their intensities as a function of time delay between the actinic pump and the 2D IR probe, we will measure the time-scale for this chemical reaction. The details of the transient 2D IR spectra one gets will depend greatly on the system at hand, but the approach can be used to study bond breakage, electron transfer and protein structural changes. In any case, to the extent equilibrium 2D IR spectroscopy reports on the structure of a molecular system, transient 2D IR spectroscopy reports on the kinetics of structural change. Thus, by triggering a chemical reaction or a conformational change, 2D IR spectroscopy can monitor the resulting kinetics over all relevant time-scales and down to 100 fs.

1.4 Time domain 2D IR spectroscopy

2D IR spectra can be collected in either the frequency or time domains, just as there are two methods of measuring the tone of a wine glass. One method is to rub a finger around the lip of a wine glass, gradually increasing the velocity until a tone is heard, at which point the frequency of the finger matches a harmonic of the natural resonance frequency of the glass. This method is a frequency domain approach, because the speed of the finger is scanned into resonance with the wine glass. A second method is to instead rap the wine glass with a finger nail and listen. As the wine glass reverberates, it emits a sound wave that our ear Fourier transforms into a frequency (a tone). This second method is a time domain approach, because our finger swiftly induces the vibrations (it is *impulsive*) and the detector (our ear) converts a time domain response into the frequency domain. If there are many wine glasses and lots of synchronized fingers (as done by a skilled glass harmonica player), then in the frequency domain one or more wine glasses will come into resonance for a particular finger frequency, while the others remain quiet. In the time domain, all of the wine glasses are rapped simultaneously, so that they all emit, irrespective of their frequencies. Thus, the resulting sound is quite complicated in the time domain, but by performing a Fourier transform, we get a spectrum that is (in principle) identical to the one actually measured in the frequency domain.

What we explained here for the wine glass can be done in an analogous manner for the vibrations of a molecule. That is, we can either scan the center frequency of spectrally narrow IR light to find the vibrational modes by absorption or we may hit the molecule with an ultrashort femtosecond IR pulse and excite all of the vibrational modes simultaneously. In the second case, all vibrational modes will vibrate and thereby emit light since the vibrations create moving charges. Just as our ear takes a Fourier transform to disentangle the sounds of multiple ringing wine glasses, the vibration of the various modes can be disentangled by a Fourier transformation of the emitted light field. And, as for a wine glass, one gets the same absorption spectrum whether one measures the molecular vibrations in the frequency or time domains.

How does one measure a 2D IR spectrum in the time domain? In the frequency domain, we scanned the center frequency of a narrowband IR pump pulse (Fig. 1.9a) and used a probe pulse to measure which modes the pump pulse came

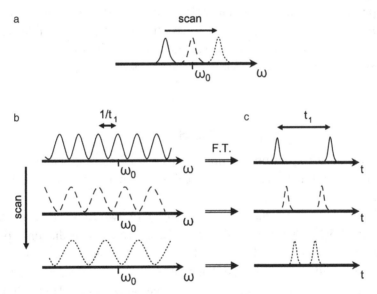

Figure 1.9 (a) Narrowband spectrum of the pump pulse in a frequency domain 2D IR experiment. The center frequency of the pump pulse is scanned to come in and out of resonance with the vibrational modes, ω_0, which translates directly into the ω_{pump}-axis of 2D IR spectra. (b) Alternatively, one may imagine a pump-pulse spectrum with a sinusoidal shape, the periodicity of which is scanned. (c) The inverse Fourier transform (FT) of the sinusoidal pump spectrum gives the pulse shape in the time domain, which is two pulses separated by a time delay. Thus, scanning the sinusoidal periodicity in the frequency domain is equivalent to scanning the delay in the time domain.

into resonance with. Now consider simultaneously using many narrow pump pulses spaced such that they span the frequency range of interest like a sinusoidal wave (Fig. 1.9b). Vibrational modes that are resonant with one of the peaks of the sinusoidal pump spectrum (such as at ω_0) will be excited while other modes whose frequencies lie in a dip will not. Therefore, the response measured by the probe will be a *linear superposition* of all excited absorption lines. How does one deconvolute these overlapping responses? We scan the periodicity of the pump spectrum, driving the molecular vibrations by bringing each vibrational mode in and out of resonance with the pump. And then, as for linear spectroscopy, we take a Fourier transform to separate the frequency components. Thus, the probe spectrum gives us one dimension and its modulation by the pump pulse, after Fourier transformation, gives us the second dimension.

How do we generate the necessary sinusoidal pump spectrum? The easiest way is by using two femtosecond pulses. The Fourier transform of two pulses in the time domain is a sinusoidally shaped spectrum in the frequency domain whose period is inversely proportional to the time delay between the two pulses (Fig. 1.9c, see analogous Eq. A.11 in Appendix A). Thus, to scan the periodicity of the sinusoidally shaped pump spectrum, we increment the time separation of two femtosecond pump pulses. Therefore, to collect 2D IR spectra in the time domain instead of the frequency domain, we substitute our narrowband pump pulse with two femtosecond probe pulses and scan their relative delays rather than their center frequencies.

Exercises

1.1 Draw a 2D IR spectrum of three coupled oscillators in which the coupling is quite strong between 1 and 2, weak between 1 and 3, and weak between 2 and 3.

1.2 Schematically draw the 2D IR spectrum of two coupled oscillators like in Fig. 1.5(b), except consider the case in which both diagonal peaks are inhomogeneously broadened. (a) Draw one 2D spectrum assuming that the frequency fluctuations of the diagonal modes are correlated. That is, when one mode is at higher frequency, so is the other mode. (Hint: Think of the ensemble 2D IR spectrum as the addition of many 2D IR spectra of individual molecules at different frequencies.) (b) Draw a second 2D IR spectrum in which the two diagonal modes are anticorrelated. That is, when one mode is at a higher frequency, the other is at a lower one. (c) Are the shapes of the cross-peaks different in these two situations? [68]. (d) Describe two physical process that could create these types of correlations between coupled modes.

1.3 Plot β_{ij} as a function of angle for the two acetone molecules shown in Fig. 1.2. Do this again, but fix the orientation of one acetone molecule perpendicular to the normal between the two, and rotate the other one. (The magnitude and sign of the coupling is very important for determining the 2D spectrum, which we will see in the following chapters.)

1.4 Plot the intensities of the upper and lower cross-peaks as a function of pump–probe time delay for (a) $A \rightleftharpoons B$, (b) $A \rightarrow B$ and (c) $B \rightarrow A$.

1.5 In Fig. 1.4(b), only the peaks arising from allowed transitions are drawn. Draw a 2D IR spectrum that also includes the two missing forbidden transitions.

2
Designing multiple pulse experiments

Researchers that are new to the field of 2D IR spectroscopy will find an enormous literature on the mathematical formalism behind the technique. To fully understand the capabilities of 2D IR spectroscopy, one needs to know nonlinear optics, lineshape theory, quantum mechanics and density matrices, to name a few topics (see Appendix E for recommended reading material). It can take years to learn all of these topics, but for many applications such a detailed understanding is not necessary. On a day-to-day basis, researchers in the field do not dwell on these topics, but instead rely on a few methods to design and interpret experiments. In later chapters, we focus on many aspects of the detailed mathematical formalism. In this chapter, we outline a view of 2D IR spectroscopy that we think provides intuition for the interpretation and design of 2D IR experiments based on physical phenomena. We will end up with *double sided Feynman diagrams* that are a useful tool for designing multiple pulse experiments.

2.1 Eigenstates, coherences and the emitted field

We begin in the same way that one would do an experiment; we shine light on a molecule. Consider a molecule like the one shown in Fig. 2.1(a). It has many vibrational modes and can be oriented in any direction in the laboratory frame. Describing the vibrational modes of this molecule is the subject of Chapter 6 and calculating the signal strength for an isotropically distributed sample is the subject of Chapter 5. For now, let us consider just one vibrational mode (like a carbonyl stretch) and have that mode oriented along the z-axis. Furthermore, let us consider a gas-phase molecule that is completely isolated from other molecules. As for the real-valued electric field of the light that we shine on the molecule (Fig. 2.1b), it can be mathematically written as

$$\vec{E}(t) = \vec{E}'(t) \cos(\vec{k} \cdot \vec{r} - \omega t + \phi) \qquad (2.1)$$

2.1 Eigenstates, coherences and the emitted field

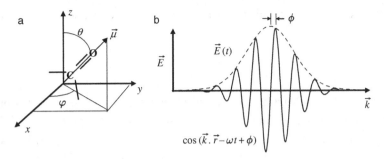

Figure 2.1 (a) Molecule oriented in the laboratory frame. (b) Electric field of laser pulse.

where $\vec{k}$ is a *wavevector* that describes the direction of light propagation, ω is the frequency, ϕ is the phase, and $\vec{E}'(t)$ is the pulse envelope that is a vectorial property which includes the polarization of the pulse. We will utilize all of these properties in the coming chapters to design particular 2D IR pulse sequences, but for now let us consider that our pulse is simply

$$E(t) = E'(t)\cos(\omega t), \qquad (2.2)$$

and polarized along the z-axis so that it is parallel to the molecule.

We will treat the light–molecule interaction semiclassically by considering the time dependent electric field classically, and the vibrational states of the molecule quantum mechanically. With this treatment, the energy of interaction between the molecule's dipole $\vec{\mu}$ and an external electrostatic field $\vec{E}$ is

$$\hat{W}(t) = -\hat{\mu} E(t) \qquad (2.3)$$

where, for the time being, we treat the electric field and the dipole operator as scalars and use hats to identify operators. The total Hamiltonian then is

$$\hat{H} = \hat{H}_0 + \hat{W}(t), \qquad (2.4)$$

where $\hat{H}_0$ is the Hamiltonian for the isolated molecule and $\hat{W}$ is the quantum mechanical operator for the interaction energy between the laser pulse and the molecule. The molecular eigenstates $|n\rangle$ of $\hat{H}_0$ are found by solving the time independent Schrödinger equation

$$\hat{H}_0|n\rangle = E_n|n\rangle. \qquad (2.5)$$

The time dependence of a wavefunction $|\Psi\rangle$ is governed by the time dependent Schrödinger equation

$$i\hbar\frac{\partial}{\partial t}|\Psi\rangle = \hat{H}|\Psi\rangle. \qquad (2.6)$$

In the absence of a laser pulse, $\hat{H}_0$ is time independent, and so the solution is simply:

$$|\Psi\rangle = \sum_n c_n e^{-iE_n t/\hbar} |n\rangle. \tag{2.7}$$

Before the laser pulse arrives, the molecule is probably in its lowest vibrational state $|0\rangle$, although it does not have to be. After the laser pulse, it may be in a linear combination of eigenstates dictated by the coefficients c_n. Either way, the time dependence of the wavefunctions is given by Eq. 2.7. However, when the laser pulse is interacting with the molecule, the coefficients c_n themselves are time dependent because the laser field is coupling the molecular eigenstates. Their time dependence can be solved by substituting Eq. 2.7 into Eq. 2.6, which gives (see Problem 2.1)

$$\frac{\partial}{\partial t} c_m(t) = -\frac{i}{\hbar} \sum_n c_n(t) e^{-i(E_n - E_m)t/\hbar} \langle m|\hat{W}(t)|n\rangle. \tag{2.8}$$

Equation 2.8 is a set of coupled equations that can be used to calculate the wavefunction after a laser pulse of duration t. For example, if we are only concerned with a two-level system, such as the ground and first excited states, then by substituting Eq. 2.3 and defining $\omega_{01} \equiv (E_1 - E_0)/\hbar$, we get the two coupled equations

$$\frac{\partial}{\partial t} c_1(t) = +\frac{i}{\hbar} c_0(t) e^{-i\omega_{01} t} \langle 1|\hat{\mu}|0\rangle E(t)$$

$$\frac{\partial}{\partial t} c_0(t) = +\frac{i}{\hbar} c_1(t) e^{+i\omega_{01} t} \langle 0|\hat{\mu}|1\rangle E(t), \tag{2.9}$$

where $\langle n|\hat{\mu}|m\rangle$ is known as the *transition dipole moment* which we refer to as $\hat{\mu}_{nm}$. In Chapter 6 we describe the transition dipole in more detail, but for now one just needs to know that it has two terms

$$\langle n|\hat{\mu}|m\rangle = \frac{d\mu}{dx} \langle n|\hat{x}|m\rangle \tag{2.10}$$

where x is the coordinate of the vibrating bond, and $d\mu/dx$ is the change of the static dipole of the molecule as the bond is stretched or compressed. The term $d\mu/dx$ is the *transition dipole strength*, which scales the intensity of the corresponding peak in the IR spectrum and $\langle n|\hat{x}|m\rangle$ gives rise to the *vibrational selection rules* of $\Delta v = \pm 1$.

In general, these equations must be solved numerically if the laser pulse has a complicated shape, but their solution is exact (in so far as the Hamiltonian is exact). Thus, if we start in the ground vibrational state $|0\rangle$, for instance, we can use Eq. 2.9 to calculate the c_0 and c_1 coefficients of $|0\rangle$ and $|1\rangle$ of the wavefunction after the laser pulse has altered the molecule. Many textbooks cover methods for solving these equations [31]. We are not so much concerned about their exact solution, because in most femtosecond mid-IR experiments the laser pulses cannot be easily adjusted to tailor the coupling and thus choose the magnitude of the coefficients. Most often, if one starts in the state $|0\rangle$, then after the laser pulse it is typically the case that $c_1 \ll c_0$. For now, what is important is that after the laser pulse, the molecule is in a linear combination of eigenstates of $|0\rangle$ and $|1\rangle$

$$|\Psi(t)\rangle = c_0 e^{-iE_0 t/\hbar}|0\rangle + ic_1 e^{-iE_1 t/\hbar}|1\rangle \qquad (2.11)$$

where c_n includes the transition dipole moment and other factors. The way we write the equations, the coefficients c_0 and c_1 are real, positive numbers. They are also no longer time dependent because the laser pulse is off. We explicitly write the i for c_1, which originates from Eq. 2.9, since it determines the phase of the rotating wavefunction (however, since $c_1 \ll c_0$, we still have $c_0 \approx 1$ and we ignore the imaginary part of c_0). Thus, the laser pulse has created a coherent linear superposition of states, or a *wavepacket*. The time dependence of this wavepacket is just the intrinsic time dependence of the isolated molecular Hamiltonian Eq. 2.7, which is called the *molecular response*, $R(t)$. This is equivalent to the electric field of the laser pulse pushing and pulling the charges to get the molecule vibrating. The molecular vibration is synchronized to the phase of the laser pulse. If there were more than one molecule in the laser beam, then the molecules would be vibrating in phase. Thus, our laser pulse has created a nonequilibrium charge distribution in the sample that we refer to as a *macroscopic polarization*, $P(t)$, which then evolves according to the molecular response (wavefunction) in Eq. 2.11. It contains information on the structure of the molecule according to the energies of the eigenstates as well as the dynamics of the wavefunction which we will learn gives information on the solvent environment, energy flow, and other properties. It is the objective of 2D IR spectroscopy to measure the macroscopic polarization, $P(t)$, and extract the molecular response $R(t)$. In general, $P(t)$ does not equal $R(t)$ because the laser pulse blurs the measurement.

The macroscopic polarization is measured by detecting the emission field that oscillating changes create according to Maxwell's equations. Since the laser field is coupled to the molecule through the interaction term $\hat{W}(t) = -\hat{\mu}E(t)$, it stands to reason that the same coupling term is responsible for the emission of the field as well. Thus, we calculate the macroscopic polarization as the expectation value of the transition dipole

$$\begin{aligned}
P(t) = \langle \mu \rangle &= \langle \Psi(t)|\hat{\mu}|\Psi(t)\rangle \\
&= \left(c_0 e^{iE_0 t/\hbar}\langle 0| - ic_1 e^{iE_1 t/\hbar}\langle 1|\right) \hat{\mu} \left(c_0 e^{-iE_0 t/\hbar}|0\rangle + ic_1 e^{-iE_1 t/\hbar}|1\rangle\right) \\
&= c_0 c_1 \langle 0|\hat{\mu}|1\rangle \sin(\omega_{01} t) + c_0^2 \langle 0|\hat{\mu}|0\rangle + c_1^2 \langle 1|\hat{\mu}|1\rangle. \qquad (2.12)
\end{aligned}$$

This equation can be simplified by realizing that the terms with $\langle 0|\hat{\mu}|0\rangle$ and $\langle 1|\hat{\mu}|1\rangle$ are related to the static dipoles of the molecule in its ground or first excited states. These terms are time independent, so they do not emit a field and can be neglected. Moreover, the coefficient c_1 is proportional to their transition dipoles (Eq. 2.9 and 2.11) and $c_0 \approx 1$, so that Eq. 2.13 becomes[1]

$$P(t) \equiv c_0 c_1 \mu_{01} \sin(\omega_{01} t)$$
$$\propto +\mu_{01}^2 \sin(\omega_{01} t). \tag{2.13}$$

We see that it is the term dependent on the transition dipole $\langle 0|\hat{\mu}|1\rangle$, which is related to the product of the c_0 and c_1 coefficients, that is responsible for the time dependent vibrational dipole. Since it is oscillating at the difference in frequency of the two eigenstates, it will emit an electric field at the fundamental frequency of the vibrator.

According to Maxwell's equations [74, 99], oscillating charges create an electromagnetic wave that is 90° phase shifted with respect to the macroscopic polarization. Thus, since $P(t) = +\mu_{01}^2 \sin(\omega_{01} t)$, it will create an emitted electric field that is proportional to $-\mu_{01}^2 \cos(\omega_{01} t)$. Classically, it is the charges oscillating at the vibrational frequency that create the emitted signal field, $E_{\text{sig}}(t)$. Quantum mechanically, it is the coherent superposition of eigenstates that creates the macroscopic polarization, as we have just seen. Later in the book we outline ways of measuring $E_{\text{sig}}(t)$ and hence the macroscopic polarization, but for now the important point is that the laser pulse prepares a vibrational *coherence* by creating a linear superposition of eigenstates that then radiates, as drawn in Fig. 2.2(a). In this case, the emitted field gives the fundamental frequency between $|0\rangle$ and $|1\rangle$. Thus, if we measure the emitted field and then calculate its Fourier transform, we get the *absorption spectrum*, which we will refer to as the *linear spectrum* throughout the book.

If one wanted to measure the frequency of the overtone state $|2\rangle$, then one could use a second pulse that couples $|1\rangle$ and $|2\rangle$ to create $|\Psi\rangle = |0\rangle + |2\rangle$, as drawn in Fig. 2.2(b). This is not the optimum pulse sequence for measuring a $|0\rangle\langle 2|$ coherence, for a number of reasons (one being that the vibrational selection rules $\Delta v = \pm 1$ would make $E_{\text{sig}}(t)$ very weak), but it provides the basic idea of how a pulse sequence can be used to manipulate the coefficients c_n in order to measure molecular properties. The sample will also emit light in between the two pulses. However, we are not usually interested in this field and can discriminate against it in a number of ways that will be explained. Thus, by using tailored sequences of pulses, we can create linear superpositions of eigenstates which will radiate an electric field from which we obtain information about the system.

---

[1] We ignore a factor of 2 in most equations in this chapter that arises from the equality between a cos or a sin and two exponentials.

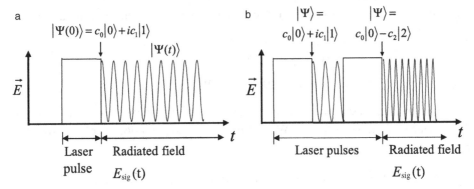

Figure 2.2 Schematic diagrams illustrating how a pulse sequence creates a wavefunction that is the source of a radiated field. (a) A laser pulse that creates a coherent superposition between $|0\rangle$ and $|1\rangle$ quantum states that subsequently radiates a signal field $E_{\text{sig}}(t)$. (b) A pulse sequence that creates a coherence between the ground and second vibrational state that might be used to measure overtone and combination levels.

2.2 Bloch vectors and molecular ensembles

The quantum mechanical description in the preceding section provides an overview of how the laser pulses start a vibrational coherence that gives rise to an emitted signal field. The formalism is perfectly well suited for describing isolated molecules such as low-density gases, but is usually insufficient to describe ensembles of condensed-phase molecules. In the condensed phase, the environment surrounding each molecule is typically different enough that each molecule has a slightly different vibrational frequency, or the molecules themselves have different structures and thereby different vibrational frequencies (such as in a protein). In this case, quantum mechanics alone is not sufficient. We must also utilize statistical mechanics to calculate how the polarizations from the molecules interfere with each other to create the macroscopic polarization. For example, if we had two molecules that had slightly different frequencies, the polarization would be

$$P(t) = c_0^{(1)} c_1^{(1)} \mu_{01} \sin(\omega_{01}^{(1)} t) + c_0^{(2)} c_1^{(2)} \mu_{01} \sin(\omega_{01}^{(2)} t). \tag{2.14}$$

Immediately after the laser pulse the polarization created by each molecule would constructively interfere, but after a while they would become out of phase and destructively interfere, so that $P(t)$ would have a beating pattern. For an ensemble of molecules, recurrences are unlikely and so the signal irreversibly decays.

To better understand how the individual polarizations from an ensemble of molecules interfere to create the macroscopic polarization, we visualize the quantum mechanical coherences using a vector diagram. We start simply, by plotting the vector of a single molecule in the coherent superposition presented above in Eq. 2.11, namely

$$|\Psi\rangle = c_0 e^{-iE_0 t/\hbar}|0\rangle + ic_1 e^{-iE_1 t/\hbar}|1\rangle$$
$$\equiv c_0(t)|0\rangle + ic_1(t)|1\rangle. \quad (2.15)$$

In the following, when we use a coefficient c_i with time dependence, it is the time dependence of the eigenstates. We define the three components of the so-called *Bloch vector*

$$B_z(t) = c_0(t)c_0^*(t) - c_1(t)c_1^*(t)$$
$$B_x(t) = i\left(c_0(t)c_1(t)^* - c_0^*(t)c_1(t)\right) = c_0 c_1 \sin(\omega_{01} t)$$
$$B_y(t) = c_0(t)c_1^*(t) + c_0^*(t)c_1(t) = c_0 c_1 \cos(\omega_{01} t) \quad (2.16)$$

which gives the vector shown in Fig. 2.3. By construct, the length of the vector is unity for the time being (but can change, as we will see). Before the laser pulse, when the system is in its ground state with $c_0 = 1$ and $c_1 = 0$, the Bloch vector is upright with $B_z = +1$. Right after the laser pulse, at $t = 0$, it lies in the (z, y)-plane at an angle defined by the strength of the pulse, i.e. by $c_0(0)$ and $c_1(0)$. As time evolves, it precesses around the z-axis due to the time evolution of the coefficients $c_0(t)c_1(t)^*$ and $c_1(t)c_0(t)^*$. In this vectorial picture, it is this rotation around the z-axis that is responsible for the emitted signal field. The macroscopic polarization can be calculated from the Bloch vector using $P(t) \propto B_x$, which is equivalent to the wavefunction notation in Eq. 2.13. We refer to B_x and B_y as the coherence of the state, and to the deviation of B_z from +1 as its population.

According to B_x and Eq. 2.13, the largest signal is created by the superposition of states with $c_0 = c_1 = 1/\sqrt{2}$. This superposition corresponds to a vector rotating in the (x, y)-plane. A laser pulse that would be intense enough to achieve this perfect superposition is called a $\pi/2$-*pulse*. However, it is currently very difficult to create such a large coherence with existing mid-infrared laser sources. We typically can

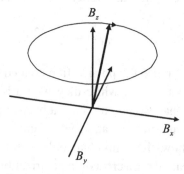

Figure 2.3 Vector diagram of the wavefunction from Eqs. 2.15 and 2.16.

only create coefficients representing ≈10% excitation, so that the *flipping* or *tilt* angle is quite small in nonlinear IR spectroscopies.

The Bloch vector representation is not really necessary to describe the wavefunction of a single molecule since the equations themselves are simple, but it is quite useful for describing an ensemble as we now demonstrate. Consider that our sample consists of an ensemble of molecules with (slightly) different frequencies. At the end of the laser pulse, the wavefunction of each molecule s will be in a coherent superposition of states, $|\Psi^s\rangle$, giving rise to an individual Bloch vector. The magnitude of their coefficients c_0^s and c_1^s may differ depending on the action of the laser pulse, and so could have different tilt angles, but will nonetheless constructively interfere at $t = 0$ (see Fig. 2.4a). However, each will precess with a different frequency, and so the macroscopic polarization $P(t)$ will be the sum of the projections B_x

$$\begin{aligned} P(t) = \langle \mu \rangle &= \langle B_x \rangle \\ &= \sum_s i p_s \left(c_0^s(t) c_1^{s*}(t) - c_0^{s*}(t) c_1^s(t) \right) \\ &= i \left(\langle c_0(t) c_1^*(t) \rangle - \langle c_0^*(t) c_1(t) \rangle \right) \end{aligned} \quad (2.17)$$

where the brackets $\langle ... \rangle$ specify an ensemble average and p_s weights the contribution of each wavefunction. At early times, the electric fields generated by the molecules constructively interfere, but as time progresses at some point their vectors will have opposite projections onto the B_x-axis and thereby destructively interfere, giving rise to a net loss of the macroscopic polarization (unless a photon echo pulse sequence is used, which we describe later).

For more than a few molecules, plotting individual vectors is tedious. Instead, we plot the averaged vector that is responsible for the macroscopic polarization:

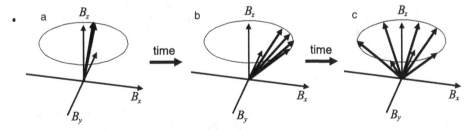

Figure 2.4 Vector diagrams for an ensemble of molecules that all have the same c_0 and c_1 coefficients at $t = 0$ but different frequencies. (a) At $t = 0$, all vectors lie in the (z, y)-plane and so constructively interfere. (b) As time progresses, the vectors have different projections onto the B_x-axis and so begin to destructively interfere. (c) Eventually, the projections onto B_x and B_y will sum to zero, at least for large ensembles.

$$\langle B_z \rangle = \langle c_0(t)c_0^*(t) \rangle - \langle c_1^*(t)c_1(t) \rangle$$
$$\langle B_x \rangle = i \left(\langle c_0(t)c_1^*(t) \rangle - \langle c_0^*(t)c_1(t) \rangle \right)$$
$$\langle B_y \rangle = \langle c_0(t)c_1^*(t) \rangle + \langle c_0^*(t)c_1(t) \rangle. \tag{2.18}$$

Note that the length of such a Bloch vector is no longer necessarily unity. Using this averaged vector, let us consider two common cases of destructive interference that are often observed in infrared spectroscopy. The first case is when the ensemble of molecules in the sample have a distribution of frequencies and that the frequency of each molecule does not change with time. In such a case, we say that there is an inhomogeneous distribution of frequencies. An example would be the antisymmetric stretch mode of the azide ion trapped in an ionic glass [59]. The electrostatic forces between the azide and glass ions are very strong, which changes the vibrational frequency of the azide antisymmetric stretch by perturbing its potential energy surface (see Fig. 2.5). Since the glass and azide ions are randomly mixed, each azide ion feels a different electrostatic field, which results in a nearly Gaussian distribution of azide antisymmetric stretch frequencies. Since neither the glass nor the azide can rearrange any significant amount, the frequency of each azide ion does not change.

The second case is when the vibrational frequencies of the molecular ensemble are dynamic (Fig. 2.5). Depending on the magnitude and time-scale of the frequency fluctuations, this leads to dephasing. The difference from the static inhomogenous case is that the potentials of the individual molecules change with time.

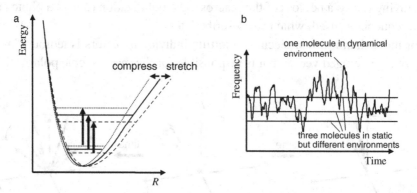

Figure 2.5 (a) A Morse potential describing a chemical bond with coordinate R, at which the solvent is pushing and pulling. (b) If different molecules have different solvent environments, but the environments do not change, then each molecule will have a different frequency that is constant in time. The horizontal lines represent three different molecules in such a scenario, the ensemble of which creates an inhomogeneous distribution. If the environments change with time, then the frequency will as well, which is the source of dephasing. The jagged line is for a single fluctuating molecule.

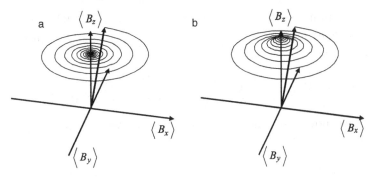

Figure 2.6 (a) Trajectory of a Bloch vector when including dephasing. (b) Trajectory of a Bloch vector when including both dephasing and population relaxation.

For example, the asymmetric stretch vibration of the azide ion in water is close to homogeneously broadened since its environment, i.e. individual water molecules, keeps changing on a fast time-scale [82]. Both homogeneous and inhomogeneous dynamics lead to a decay in the macroscopic polarization with time. In a Bloch vector diagram, dephasing causes the vector to spiral towards $\langle B_z \rangle$ as its projections on the $\langle B_x \rangle$ and $\langle B_y \rangle$ axes decay (Fig. 2.6a). The decrease in the projection along $\langle B_x \rangle$ corresponds to a decay in the macroscopic polarization $P(t)$. In principle, the decay of the macroscopic polarization is different for a homogeneous versus an inhomogeneous sample, but in practice it is very hard to distinguish between the two by linear spectroscopy. 2D IR spectroscopy is a powerful means of quantifying dephasing mechanisms through a photon echo pulse sequence, which we describe later.

In addition, we may have *population relaxation*. Population relaxation results when the wavefunction coefficients of the individual molecules change. For example, if the laser pulse creates the state $|\Psi\rangle = c_0|0\rangle + ic_1|1\rangle$, then as time evolves energy transfer will cause the wavefunction to collapse to the ground state $|\Psi\rangle = |0\rangle$. In the Bloch vector picture, population relaxation causes the z-component to increase until it finally reaches $B_z = +1$ (Fig. 2.6b).

2.3 Bloch vectors are a graphical representation of the density matrix

In the previous section, we found that the coefficients $c_0^s(t)$ and $c_1^s(t)$ from the wavefunctions of individual molecules could be averaged and the result plotted to give a graphic description of quantum mechanical coherences and dephasing. It turns out that these quantities are directly related to the elements in the *density matrix*, ρ. We will introduce the density matrix on a more formal level in a later chapter, but first let us qualitatively link what we have learned about coherences and dephasing to the density matrix.

For the vibrational system we have been discussing in which only the $|0\rangle$ and $|1\rangle$ quantum states are being accessed by the laser pulse, the density matrix is defined as

$$\rho = \begin{pmatrix} \rho_{00} & \rho_{01} \\ \rho_{10} & \rho_{11} \end{pmatrix} = \begin{pmatrix} \langle c'_0(t) c'^*_0(t) \rangle & \langle c'_0(t) c'^*_1(t) \rangle \\ \langle c'_1(t) c'^*_0(t) \rangle & \langle c'_1(t) c'^*_1(t) \rangle \end{pmatrix} \quad (2.19)$$

into which we now insert the coefficients of our wavefunction, $c'_0 = c_0$ and $c'_1 = ic_1$:

$$\rho = \begin{pmatrix} \langle c_0(t) c^*_0(t) \rangle & -i\langle c_0(t) c^*_1(t) \rangle \\ i\langle c_1(t) c^*_0(t) \rangle & \langle c_1(t) c^*_1(t) \rangle \end{pmatrix}. \quad (2.20)$$

Comparing Eq. 2.20 to Eq. 2.18, we find that $\langle B_z \rangle$, $\langle B_x \rangle$ and $\langle B_y \rangle$ can be written as

$$\langle B_z \rangle = \rho_{00} - \rho_{11}$$
$$\langle B_x \rangle = -(\rho_{01} + \rho_{10})$$
$$\langle B_y \rangle = i(\rho_{01} - \rho_{10}). \quad (2.21)$$

Thus, it is the difference in the diagonal elements of the density matrix, ρ_{00} and ρ_{11}, that determines the z-component of the Bloch vector. Therefore, the diagonal terms are the *populations*. In addition, it is clear from Eq. 2.21 that it is the off-diagonal elements, ρ_{01} and ρ_{10}, of the density matrix which are the source of the quantum mechanical *coherences* that cause the vector in the Bloch diagrams to rotate and are ultimately the source of the emitted electric field, $E_{sig}(t)$. In the absence of dephasing and population relaxation, the time evolution of the density matrix elements is:

$$\rho_{00}(t) = c_0^2 = \text{const.}$$
$$\rho_{11}(t) = c_1^2 = \text{const.}$$
$$\rho_{01}(t) = -ic_0 c_1 e^{+i\omega_{01} t}$$
$$\rho_{10}(t) = ic_0 c_1 e^{-i\omega_{01} t} \quad (2.22)$$

which can be seen when plugging in the time dependence of the coefficients c_0 and c_1 from Eq. 2.11.

Homogeneous dephasing is a result of the off-diagonal elements, ρ_{10} and ρ_{01}, decaying with time, which we describe phenomenologically using a homogeneous dephasing time T_2:

$$\rho_{01}(t) = -ic_0 c_1 e^{+i\omega_{01} t} e^{-t/T_2}$$
$$\rho_{10}(t) = ic_0 c_1 e^{-i\omega_{01} t} e^{-t/T_2}. \quad (2.23)$$

In addition, population relaxation results in a decay of the ρ_{11} diagonal element with a time constant T_1 together with a refilling of the ground state ρ_{00}:

2.3 Graphical representation of the density matrix

$$\rho_{11}(t) = \rho_{11}(0)e^{-t/T_1}$$
$$\rho_{00}(t) = 1 - \rho_{11}(t). \tag{2.24}$$

Homogeneous dephasing T_2 and population relaxation T_1 are related by

$$\frac{1}{T_2} = \frac{1}{2T_1} + \frac{1}{T_2^*} \tag{2.25}$$

where the *pure dephasing* T_2^* is caused by fluctuations of the environment (Fig. 2.5). This follows since population relaxation T_1 necessarily causes a decay of the corresponding off-diagonal elements, i.e. gives rise to homogeneous dephasing even in the absence of pure dephasing. To see that, consider a weakly excited state with $\rho_{11} \ll 1$ and $\rho_{00} \approx 1$. When

$$\rho_{11}(t) = c_1^2 e^{-t/T_1} \tag{2.26}$$

then

$$c_1 \propto e^{-t/2T_1} \tag{2.27}$$

but c_0 stays ≈ 1 at all times. Consequently, the off-diagonal element

$$\rho_{01}(t) = -ic_0c_1e^{+i\omega_{01}t}e^{-t/2T_1} \tag{2.28}$$

decays with at least $T_2 = 2T_1$. In Section 7.1 we will introduce a more sophisticated description of dephasing.

With this correspondence between the coefficients of the wavefunctions, the Bloch vectors and the density matrix, let us summarize the typical evolution of these three quantities in an experiment with a laser pulse, which is illustrated in Fig. 2.7. Before the laser interacts with the sample, molecules in room temperature liquids are usually in their ground vibrational states because the frequency of most vibrations is larger than $k_B T$. In this case, the wavefunction of all molecules is $|\Psi^s\rangle = |0\rangle$, which gives the density matrix

$$\rho(-\infty) = \begin{pmatrix} 1 & 0 \\ 0 & 0 \end{pmatrix} \tag{2.29}$$

and a Bloch vector, which points up the z-axis. After the laser pulse, each molecule is in a coherent superposition of states, $|\Psi^s\rangle = c_0^s|0\rangle + ic_1^s|1\rangle$. We have to sum the polarizations from all the molecules to get the macroscopic $P(t)$, but immediately after the laser pulse is off (defined as $t = 0$), all the molecules are in phase and the density matrix is

$$\rho(t=0) = \begin{pmatrix} 1/2 & -i/2 \\ +i/2 & 1/2 \end{pmatrix} \tag{2.30}$$

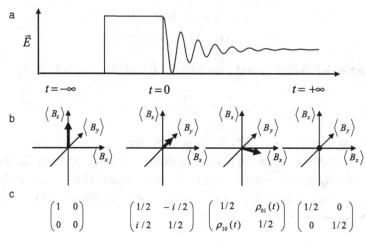

Figure 2.7 Illustration of the relationship between (a) the laser pulse and the corresponding polarization, (b) Bloch vectors, and (c) the density matrices for a system that dephases (but does not undergo population relaxation).

for the case when the laser pulse was able to maximize the signal strength. This density matrix corresponds to a vector along $\langle B_y \rangle$. As time passes, the density matrix evolves according to Eq. 2.23, which results in an oscillatory coherence that decays as the individual molecules destructively interfere:

$$\rho(t) = \begin{pmatrix} 1/2 & -(i/2)e^{+i\omega_{01}t - t/T_2} \\ (i/2)e^{-i\omega_{01}t - t/T_2} & 1/2 \end{pmatrix}. \tag{2.31}$$

In the Bloch vector picture, the vector rotates in the (x, y)-plane like in Fig. 2.7(b) (we are ignoring population relaxation). Eventually, the system will be completely dephased, at which time the density matrix will be

$$\rho(t = +\infty) = \begin{pmatrix} 1/2 & 0 \\ 0 & 1/2 \end{pmatrix} \tag{2.32}$$

and the Bloch vector disappears.

There are three interesting points to be made about this last density matrix. First, this density matrix cannot be made by a laser pulse directly, because a laser pulse creates coherences through linear combinations of eigenstates. This density matrix represents a macroscopically incoherent sample. Second, there is no wavefunction of a single oscillator that can result in this density matrix. The only way a single molecule can be in both the $|0\rangle$ and $|1\rangle$ states is through a linear superposition, which will also create off-diagonal terms in the density matrix. Thus, at least two molecules are necessary to create this matrix, one in $|0\rangle$ and one in $|1\rangle$. Third, the averaged number of excited molecules, and hence the energy of the state, is the same as in the state described by Eq. 2.30. We call Eq. 2.30 a *coherent state* (which is still a pure state although it is not an eigenstate of the Hamiltonian), and Eq. 2.32 an *incoherent state*. A pure state is a state whose Bloch vector has length one. Equation 2.31 is a state in between those two extremes; it is partially coherent since the off-diagonal elements have not yet decayed to zero. To make the distinction between these two types of excitation, quantum mechanics is insufficient and we need statistical mechanics.

We can think of the density matrix also in terms of a quantum mechanical operator, $\rho = |\psi\rangle\langle\psi|$. Its matrix elements are found by expanding the wavefunction ψ^s in an eigenstate basis

$$|\psi^s\rangle = \sum_n c_n^s |n\rangle \qquad (2.33)$$

or, for its conjugate complex:

$$\langle\psi^s| = \sum_n c_n^{s*} \langle n| \qquad (2.34)$$

and then performing the ensemble average by summing over s:

$$\hat{\rho}_{nm} = \langle c_n c_m^* \rangle |n\rangle\langle m|. \qquad (2.35)$$

Hence, the matrix elements $\langle c_n c_m^* \rangle$ of the density operator $\hat{\rho}$ are related to a coherence $|n\rangle\langle m|$ between states $|n\rangle$ and $|m\rangle$. We will refer to the left or right of the density matrix as its *bra* or *ket*, respectively. We also note that the polarization is calculated by

$$\langle \mu \rangle = \sum_{nm} \langle c_n c_m^* \rangle \mu_{mn} = \sum_{nm} \rho_{nm} \mu_{mn}. \qquad (2.36)$$

The right side of this expression is called the trace of $\rho\mu$:

$$\langle \mu \rangle \equiv \mathrm{Tr}\,(\rho\mu) \equiv \langle \rho\mu \rangle. \qquad (2.37)$$

The mathematical formalism behind the density matrix is presented in more detail in Chapter 3.

2.4 Multiple pathways visualized with Feynman diagrams

2.4.1 Manipulating the density matrix

We have motivated why we use density matrices (because we need a statistical average of the wavefunctions) and illustrated their time evolution using Bloch vectors. Now, to set the stage for multi-pulse experiments, we need to add laser pulses, which interact with the molecule through the transition dipole operator $\hat{W}(t) = -\hat{\mu}E(t)$ given in Eq. 2.3. The transition dipole operator is off-diagonal:

$$\mu = \begin{pmatrix} 0 & \mu_{01} \\ \mu_{01} & 0 \end{pmatrix} \tag{2.38}$$

in order to generate the linear combinations of the $|0\rangle$ and $|1\rangle$ eigenstates (we do not distinguish between μ_{01} and μ_{10}). To properly account for the effect of $\hat{W}(t)$ on the ensemble of wavefunctions, we really need to time-propagate the density matrix under the influence of the time dependent Hamiltonian $\hat{H} = \hat{H}_0 + \hat{W}(t)$ during the laser pulse. This is done through a perturbative expansion of the so-called Liouville–von Neumann equation, which we will derive in Chapter 3. For now, all we need to know is that when a laser pulse is interacting with the molecules, we multiply ρ by μ, and we must do so from each side (e.g. the bra and ket sides of ρ). For the interaction with one laser pulse, this procedure gives (see Eq. 3.63 from Chapter 3):

$$\rho^{(1)} = i\left(\mu(0)\rho(-\infty) - \rho(-\infty)\mu(0)\right) \tag{2.39}$$

where we have skipped writing the electric field $E(t)$ for the moment. The macroscopic polarization evolves according to the molecular response, which is obtained by taking the trace $\langle \mu(t_1)\rho^{(1)}\rangle$ (Eq. 2.37), so that we obtain for the so-called *linear response function*, or *first-order response function*:

$$R^{(1)}(t_1) = i\langle \mu(t_1)\mu(0)\rho(-\infty)\rangle - i\langle \mu(t_1)\rho(-\infty)\mu(0)\rangle$$
$$= i\langle \mu(t_1)\mu(0)\rho(-\infty)\rangle - i\langle \rho(-\infty)\mu(0)\mu(t_1)\rangle \tag{2.40}$$

where we have made use of the invariance of the trace under cyclic permutation in order to write the two terms in a symmetric manner (by convention). The superscript in $R^{(1)}$ indicates that this is the first term in a perturbative expansion, which also is why there is an i (see Chapter 3). In certain limits the response function equals the polarization as we shall see.

The terms appearing in the linear response function are often visualized with the help of so-called *Feynman diagrams*, which we now introduce. To that end, it is instructive to calculate the response function for a two-level system in a step-by-step manner. We start with the density matrix in its ground state:

$$\rho(-\infty) = \begin{pmatrix} 1 & 0 \\ 0 & 0 \end{pmatrix} \tag{2.41}$$

and act with the dipole operator (which we write here unit-less for simplicity):

$$\mu = \begin{pmatrix} 0 & 1 \\ 1 & 0 \end{pmatrix} \tag{2.42}$$

from the left (i.e. the first term in Eq. 2.40):

$$i\mu(0)\rho(-\infty) = i\begin{pmatrix} 0 & 1 \\ 1 & 0 \end{pmatrix}\begin{pmatrix} 1 & 0 \\ 0 & 0 \end{pmatrix} = \begin{pmatrix} 0 & 0 \\ i & 0 \end{pmatrix}. \tag{2.43}$$

2.4 Multiple pathways visualized with Feynman diagrams

Since the (10) matrix element of the density matrix is nonzero after the first interaction with the dipole operator, we say that we have generated a (10) coherence state. Starting from there, the system propagates freely under the influence of the unperturbed molecular Hamiltonian. Thus, the time propagation is given by the difference of energies of the ground and excited states (Eq. 2.23):

$$ie^{-i\omega_{01}t_1}\mu(0)\rho(-\infty) = \begin{pmatrix} 0 & 0 \\ ie^{-i\omega_{01}t_1} & 0 \end{pmatrix}. \tag{2.44}$$

In the second term of Eq. 2.40, the dipole operator is operating from the right:

$$\begin{pmatrix} 1 & 0 \\ 0 & 0 \end{pmatrix} \xrightarrow{-i\rho(-\infty)\mu(0)} \begin{pmatrix} 0 & -i \\ 0 & 0 \end{pmatrix} \xrightarrow{-i\rho(-\infty)\mu(0)e^{+i\omega_{01}t_1}} \begin{pmatrix} 0 & -ie^{+i\omega_{01}t_1} \\ 0 & 0 \end{pmatrix}$$
$$\tag{2.45}$$

and we obtain a (01) coherence after the first interaction. We see that the two terms are just the conjugate complex of each other, which can also be proven on very general grounds:

$$\langle \rho(-\infty)\mu(0)\mu(t_1) \rangle = \langle \rho^\dagger(-\infty)\mu^\dagger(0)\mu^\dagger(t_1) \rangle$$
$$= \langle \mu(t_1)\mu(0)\rho(-\infty) \rangle^\dagger \tag{2.46}$$

where we have used the fact that all operators are Hermitian.

Taking both terms together, we obtain for the density matrix after the first laser pulse interaction:

$$\rho^{(1)} = \begin{pmatrix} 0 & -ie^{+i\omega_{01}t_1} \\ ie^{-i\omega_{01}t_1} & 0 \end{pmatrix} \tag{2.47}$$

which, according to Eq. 2.21, corresponds to a Bloch vector:

$$B_z = 0$$
$$B_x = \sin(\omega_{01}t_1)$$
$$B_y = \cos(\omega_{01}t_1) \tag{2.48}$$

rotating in the (xy)-plane (Fig. 2.8a). However, to plot the complete Bloch vector, we need to keep in mind that $\rho^{(1)}$ is just the first-order term in a power expansion of the density matrix. When considering the total density matrix including the zeroth-order term (which is the unperturbed density matrix):

$$\rho^{(0)} + \epsilon\rho^{(1)} = \begin{pmatrix} 1 & 0 \\ 0 & 0 \end{pmatrix} + \epsilon \begin{pmatrix} 0 & -ie^{+i\omega_{01}t_1} \\ ie^{-i\omega_{01}t_1} & 0 \end{pmatrix} \tag{2.49}$$

with the smallness parameter ϵ, we obtain a Bloch vector shown in Fig. 2.8(b). But for the practical purpose of understanding the time dependence, we only need to consider $\rho^{(1)}$.

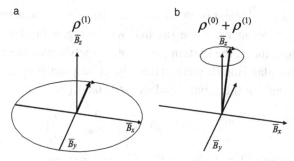

Figure 2.8 (a) Bloch vector representation of the first-order density matrix $\rho^{(1)}$. (b) Bloch vector representation of the total density matrix $\rho^{(0)} + \epsilon\rho^{(1)}$.

We have seen in Section 2.2 that an oscillating Bloch vector will emit a macroscopic polarization. This is described by the second interaction of the density matrix with the dipole operator at time t_1 (Eq. 2.40), which switches the system back into a (00) state, i.e. the ground state:

$$i\mu(t_1)e^{-i\omega_{01}t_1}\mu(0)\rho(-\infty) = \begin{pmatrix} 0 & 1 \\ 1 & 0 \end{pmatrix} \begin{pmatrix} 0 & 0 \\ ie^{-i\omega_{01}t_1} & 0 \end{pmatrix}$$

$$= \begin{pmatrix} ie^{-i\omega_{01}t_1} & 0 \\ 0 & 0 \end{pmatrix}. \quad (2.50)$$

Finally, we take the trace and obtain for the linear response function:

$$i\langle\mu(t_1)e^{-i\omega_{01}t_1}\mu(0)\rho(-\infty)\rangle = \left\langle\begin{pmatrix} ie^{-i\omega_{01}t_1} & 0 \\ 0 & 0 \end{pmatrix}\right\rangle = ie^{-i\omega_{01}t_1}. \quad (2.51)$$

The second, conjugate complex term of Eq. 2.40 evolves in an analogous manner:

$$-i\langle\rho(-\infty)\mu(0)e^{+i\omega_{01}t_1}\mu(t_1)\rangle = -ie^{+i\omega_{01}t_1} \quad (2.52)$$

and we obtain for the total response function:

$$R^{(1)}(t_1) \propto \sin(\omega_{01}t_1). \quad (2.53)$$

This is the same result as Eq. 2.13, in the limit when the response function equals the polarization.

Thus, we have the two pathways (often called *Liouville pathways* or *Feynman pathways*)

$$(\rho_{00} \rightarrow \rho_{01} \rightarrow \rho_{00}) \equiv (|0\rangle\langle 0| \rightarrow |0\rangle\langle 1| \rightarrow |0\rangle\langle 0|)$$
$$(\rho_{00} \rightarrow \rho_{10} \rightarrow \rho_{00}) \equiv (|0\rangle\langle 0| \rightarrow |1\rangle\langle 0| \rightarrow |0\rangle\langle 0|). \quad (2.54)$$

2.4.2 Rotating wave approximation

So far we have ignored the role of the electrical field $E(t)$ of the laser pulse in creating the macroscopic polarization $P(t)$. $E(t)$ enters the equations through the perturbation terms $\hat{W}$ of the Hamiltonian. The most precise way to include $E(t)$ is to solve the time dependent Schrödinger equation (Eq. 2.9), which is difficult for an ensemble of different molecules. Instead, we use what is known as *linear response theory*, which relies on the assumption that our laser pulse is weak, so that the macroscopic polarization scales linearly with the electric field strength. If the envelope of the electric field were a δ-function in time, $E'(t) = \delta(t)$, then the macroscopic polarization would exactly reproduce the molecular response

$$P^{(1)}(t) \propto R^{(1)}(t) \tag{2.55}$$

like in Eq. 2.53.

If linear response theory holds, then we can treat a finite width pulse as a sum of δ-function pulses that together give the pulse envelope $E'(t)$. In other words, to calculate the macroscopic polarization, we convolute the first-order response function with the laser pulse:

$$P^{(1)}(t) = \int_0^\infty dt_1 E'(t - t_1) R^{(1)}(t_1). \tag{2.56}$$

In Chapter 3 we rigorously derive the convolution on which linear response is based (see Eq. 3.62).

Now we recognize that the electric fields are real valued in an actual experiment (Eq. 2.1), but can formally be written as two terms with positive and negative frequencies:

$$2E'(t)\cos(\omega t) = E'(t)\left(e^{-i\omega t} + e^{+i\omega t}\right) = E(t) + E^*(t). \tag{2.57}$$

It is useful to write the electric field this way because these terms cause different transitions in the molecule, which we see when we substitute into Eq. 2.56, which gives

$$P^{(1)}(t) = \int_0^\infty dt_1 \left(E(t - t_1) + E^*(t - t_1)\right) R^{(1)}(t_1). \tag{2.58}$$

To continue, we need the molecular response, which contains two terms (Eq. 2.40). In the first term, the dipole operator acts twice on the ket side of the density matrix and twice on the bra side in the second term. Thus, there are four combinations of electric fields and molecular responses. We start by considering the ket side interaction with homogeneous dephasing added:

$$R^{(1)}(t_1) = i \langle \mu(t_1)\mu(0)\rho(-\infty) \rangle = ie^{-i\omega_{01}t_1}e^{-t_1/T_2}. \quad (2.59)$$

Assuming that the laser field is resonant with the transition $\omega = \omega_{01}$, we get:

$$P^{(1)}(t) \propto ie^{-i\omega t}\int_0^\infty dt_1 E'(t-t_1)e^{-t_1/T_2}$$

$$+ie^{+i\omega t}\int_0^\infty dt_1 E'(t-t_1)e^{-t_1/T_2}e^{-i2\omega t_1}. \quad (2.60)$$

The integrand in the first term is slowly varying as a function of time t_1, while that of the second is highly oscillating. The second integral therefore is much smaller than the first. When the smaller term is neglected, we are making the so-called *rotating wave approximation* (RWA). Under this approximation, we find that when μ operates on the *ket* (left) it is $E(t)$ that creates the coherence, and we obtain for the linear polarization one term only:

$$P^{(1)}(t) = \int_0^\infty dt_1 E(t-t_1)R^{(1)}(t_1) \quad (2.61)$$

while the corresponding term with $E^*(t-t_1)$ vanishes. Repeating the same convolution for the right term of Eq. 2.40:

$$R^{*(1)}(t_1) = -i\langle \rho(-\infty)\mu(0)\mu(t_1)\rangle = -ie^{+i\omega_{01}t_1}e^{-t_1/T_2} \quad (2.62)$$

we find that it is $E^*(t)$ that excites the *bra* to create a coherence. The rotating wave approximation is extremely useful. In fact, we often unconsciously make this approximation by only considering transitions caused by resonant laser fields.

Thus, to calculate the macroscopic polarization and its consequential emitted electric field, we must keep track of:

- the time ordering of field interactions,
- whether the laser field operates on the *bra* or the *ket* of the density matrix,
- the states (coherences *versus* population states) through which the system runs,
- and whether it interacts with $E(t) \propto e^{-i\omega t}$ or $E^*(t) \propto e^{+i\omega t}$,

which can be tedious using algebraic notation, but is quite elegant when done graphically with so-called *double sided Feynman diagrams* (Fig. 2.9). In these diagrams, two vertical lines represent the time evolution of the *ket* (left) and the *bra* (right) of the density matrix. Time is running from the bottom to the top. Interactions with the dipole operator at a given time are represented by arrows. After each field interaction, we keep track of the particular coherence or population state

2.5 What is absorption?

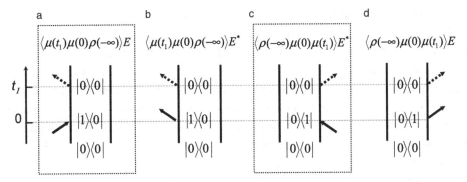

Figure 2.9 The four possible double sided Feynman diagrams of linear response. The diagrams in the dashed boxes survive the RWA.

by noting the corresponding matrix element of the density matrix in between two vertical lines.

For the linear response there are two pathways (Eq. 2.54). Each pathway has two possibilities to interact with the electric field with positive or negative frequencies, $E(t) \propto e^{-i\omega t}$ or $E^*(t) \propto e^{+i\omega t}$, which creates the four Feynman diagrams shown in Fig. 2.9. We indicate the two terms using the directions of the arrows in the Feynman diagram, i.e. a term $E(t) \propto e^{-i\omega t}$ by an arrow pointing to the right, and $E^*(t) \propto e^{+i\omega t}$ by an arrow pointing to the left. For example, the two terms in Eq. 2.60 correspond to the two diagrams shown in Fig. 2.9(a,b), however, Fig. 2.9(b) does not survive the rotating wave approximation. This leads to an intuitive physical interpretation: In Fig. 2.9(a), the ground state density matrix $|0\rangle\langle 0|$ is excited on the left side (light is going in) to yield a $|1\rangle\langle 0|$ coherence. In contrast, Fig. 2.9(b) would try to de-excite the ground state (light is going out), which is of course not possible. If we had started at $t = -\infty$ with a density matrix that is already in an excited state, then a term $E(t) \propto e^{+i\omega t}$ could act on the *ket* and de-excite it (see Problem 2.4). Very generally speaking, a term $E(t) \propto e^{-i\omega t}$ excites the *ket* of the density whereas $E^*(t) \propto e^{+i\omega t}$ de-excites it, and vice versa for the *bra*.

2.5 What is absorption?

The 90° phase shift of the emitted field relative to the macroscopic polarization can be written as:

$$E_{\text{sig}}(t) = iP(t). \tag{2.63}$$

When we combine this with the i from one of our response functions, such as $R^{(1)}(t)$ from Eq. 2.59, we find that the emitted light field has an opposite sign relative to the incident light field $E(t)$, hence they interfere destructively. In other words, the amount of light after the sample is less than before the sample, which indicates that light has been *absorbed* by the sample. It is worth comparing this result with our intuitive picture of absorption. In fact, at the end of the Feynman diagram, the system is back in the ground state (Fig. 2.9), and not in a vibrationally excited state, as one might expect. In other words, our formalism seems to violate energy conservation, since energy is missing in the light field (due to the destructive interference), but the energy, seemingly, does not appear in the molecule. There are several ways to resolve this paradox; one goes as follows. The paradox is the result of considering the density matrix only up to first order in the field:

$$\rho = \rho^{(0)} + \rho^{(1)} + \cdots = \begin{pmatrix} 1 & 0 \\ 0 & 0 \end{pmatrix} + \begin{pmatrix} 0 & \epsilon \\ \epsilon & 0 \end{pmatrix} + \cdots \qquad (2.64)$$

where $\epsilon \ll 1$ is the small (01) and (10) coherence generated by the laser pulse (see Eq. 2.49). Since $|\rho_{00}| \equiv |c_0|^2 = 1$, we know that $|\rho_{01}| = |c_0 c_1| = |c_1|$. Hence $|\rho_{11}| = |c_1|^2$. Using these coefficients, we know the density matrix to second order, without having to explicitly calculate it:

$$\rho = \rho^{(0)} + \rho^{(1)} + \rho^{(2)} + \cdots = \begin{pmatrix} 1 & \epsilon \\ \epsilon & \epsilon^2 \end{pmatrix}. \qquad (2.65)$$

Thus, a small off-diagonal density matrix element of order ϵ requires a corresponding population of the first excited state. The off-diagonal density matrix element scales like $\epsilon \propto \mu E$, hence the population of the first excited state scales with the square of the transition dipole moment, and with the intensity of the laser pulse ($\propto \mu^2 E^2$), as one would expect. So the paradox is resolved by the second-order density matrix which indeed shows that the missing energy ends up in the excited state.

2.6 Designing multi-pulse experiments

The method we have outlined above for linear spectroscopies can be extended in a straightforward manner to design multiple pulse experiments. We concentrate on third-order spectroscopies which are the typical methods used to collect 2D IR spectra (the second-order response vanishes for isotropic media; nevertheless, the following approach could be used for those as well). Shown in Fig. 2.10 is the basic 2D IR pulse sequence that uses three infrared pulses. Once again, since μ can act on either the *bra* or the *ket* of the density matrix, there are many possible combinations of states that can be accessed, which we call pathways. We obtain for

2.6 Designing multi-pulse experiments

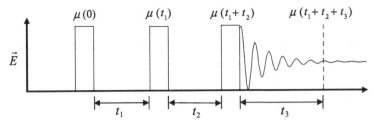

Figure 2.10 Generic pulse sequence for 2D IR experiments.

the *third-order nonlinear response function*, this time written in a more compact way with nested commutators (see Eq. 3.65 from Chapter 3):[2]

$$R^{(3)}(t_3, t_2, t_1) \propto i \langle \mu(t_3 + t_2 + t_1) [\mu(t_2 + t_1), [\mu(t_1), [\mu(0), \rho(-\infty)]]] \rangle$$
$$= i \langle \mu_3 [\mu_2, [\mu_1, [\mu_0, \rho(-\infty)]]] \rangle \qquad (2.66)$$

where the second line is just a short notation. When expanding the three nested commutators, we obtain eight terms:

$$i \langle \mu_3 [\mu_2, [\mu_1, [\mu_0, \rho(-\infty)]]] \rangle = \qquad (2.67)$$

$$\begin{aligned}
&i \langle \mu_3 \mu_1 \rho(-\infty) \mu_0 \mu_2 \rangle - i \langle \mu_2 \mu_0 \rho(-\infty) \mu_1 \mu_3 \rangle && \Rightarrow R_1 + R_1^* \\
&i \langle \mu_3 \mu_2 \rho(-\infty) \mu_0 \mu_1 \rangle - i \langle \mu_1 \mu_0 \rho(-\infty) \mu_2 \mu_3 \rangle && \Rightarrow R_2 + R_2^* \\
&i \langle \mu_3 \mu_0 \rho(-\infty) \mu_1 \mu_2 \rangle - i \langle \mu_2 \mu_1 \rho(-\infty) \mu_0 \mu_3 \rangle && \Rightarrow R_4 + R_4^* \\
&i \langle \mu_3 \mu_2 \mu_1 \mu_0 \rho(-\infty) \rangle - i \langle \rho(-\infty) \mu_0 \mu_1 \mu_2 \mu_3 \rangle && \Rightarrow R_5 + R_5^*.
\end{aligned}$$

(We will introduce the missing terms R_3 and R_6 a little later.) In sorting these terms, we again made use of the invariance of the trace under cyclic rotation. By convention, we rotate the terms such that the last interaction with the dipole operator appears at the very left in R_1, R_2, R_4 and R_5, and the very right in R_1^*, R_2^*, R_4^* and R_5^*. As before, one finds that the terms R_1^*, R_2^*, R_4^* and R_5^* are just the complex conjugates of R_1, R_2, R_4 and R_5.

We exemplify the evolution of the density matrix by illustrating the pathway R_4:

$$\begin{pmatrix} 1 & 0 \\ 0 & 0 \end{pmatrix} \xrightarrow{i\mu_0\rho} \begin{pmatrix} 0 & 0 \\ i & 0 \end{pmatrix} \xrightarrow{t_1} \begin{pmatrix} 0 & 0 \\ ie^{-i\omega_{01}t_1} & 0 \end{pmatrix} \xrightarrow{i\mu_0\rho\mu_1}$$
$$\begin{pmatrix} 0 & 0 \\ 0 & ie^{-i\omega_{01}t_1} \end{pmatrix} \xrightarrow{i\mu_0\rho\mu_1\mu_2} \begin{pmatrix} 0 & 0 \\ ie^{-i\omega_{01}t_1} & 0 \end{pmatrix} \xrightarrow{t_3}$$
$$\begin{pmatrix} 0 & 0 \\ ie^{-i\omega_{01}(t_3+t_1)} & 0 \end{pmatrix} \xrightarrow{i \langle \mu_3 \mu_0 \rho \mu_1 \mu_2 \rangle} ie^{-i\omega_{01}(t_3+t_1)} \qquad (2.68)$$

---

[2] There is an additional minus sign in all third-order response functions, which originates from the $(i/\hbar)^3$ term in the perturbative expansion of the density matrix (see Eq. 3.61 below). We skip that minus sign throughout the book, since this is commonly done so in the literature. Due to that term, the response functions of first- and third-order have opposite signs.

or, if we combine it with its conjugate complex $R_4 + R_4^*$:

$$\begin{pmatrix} 1 & 0 \\ 0 & 0 \end{pmatrix} \xrightarrow{\text{pulse}} \begin{pmatrix} 0 & -i \\ +i & 0 \end{pmatrix} \xrightarrow{t_1} \begin{pmatrix} 0 & -ie^{+i\omega_{01}t_1} \\ +ie^{-i\omega_{01}t_1} & 0 \end{pmatrix} \xrightarrow{\text{pulse}}$$

$$\begin{pmatrix} 0 & 0 \\ 0 & \sin(\omega_{01}t_1) \end{pmatrix} \xrightarrow{\text{pulse}} \begin{pmatrix} 0 & -ie^{+i\omega_{01}t_1} \\ +ie^{-i\omega_{01}t_1} & 0 \end{pmatrix} \xrightarrow{t_3}$$

$$\begin{pmatrix} 0 & -ie^{+i\omega_{01}(t_1+t_3)} \\ +ie^{-i\omega_{01}(t_3+t_1)} & 0 \end{pmatrix} \xrightarrow{\text{emit}} \sin(\omega_{01}(t_3+t_1)). \qquad (2.69)$$

For simplicity, we have assumed here that the density matrix does not evolve during time period t_2 (which will be the case if we neglect T_1 relaxation).

Doing the same for R_1 reveals:

$$\begin{pmatrix} 1 & 0 \\ 0 & 0 \end{pmatrix} \xrightarrow{i\rho\mu_0} \begin{pmatrix} 0 & i \\ 0 & 0 \end{pmatrix} \xrightarrow{t_1} \begin{pmatrix} 0 & ie^{+i\omega_{01}t_1} \\ 0 & 0 \end{pmatrix} \xrightarrow{i\mu_1\rho\mu_0}$$

$$\begin{pmatrix} 0 & 0 \\ 0 & ie^{+i\omega_{01}t_1} \end{pmatrix} \xrightarrow{i\mu_1\rho\mu_0\mu_2} \begin{pmatrix} 0 & 0 \\ ie^{+i\omega_{01}t_1} & 0 \end{pmatrix} \xrightarrow{t_3}$$

$$\begin{pmatrix} 0 & 0 \\ ie^{-i\omega_{01}(t_3-t_1)} & 0 \end{pmatrix} \xrightarrow{i\langle\mu_3\mu_1\rho\mu_0\mu_2\rangle} ie^{-i\omega_{01}(t_3-t_1)} \qquad (2.70)$$

or, if we combine it with its conjugate complex $R_1 + R_1^*$:

$$\begin{pmatrix} 1 & 0 \\ 0 & 0 \end{pmatrix} \xrightarrow{\text{pulse}} \begin{pmatrix} 0 & +i \\ -i & 0 \end{pmatrix} \xrightarrow{t_1} \begin{pmatrix} 0 & +ie^{+i\omega_{01}t_1} \\ -ie^{-i\omega_{01}t_1} & 0 \end{pmatrix} \xrightarrow{\text{pulse}}$$

$$\begin{pmatrix} 0 & 0 \\ 0 & -\sin(\omega_{01}t_1) \end{pmatrix} \xrightarrow{\text{pulse}} \begin{pmatrix} 0 & -ie^{-i\omega_{01}t_1} \\ +ie^{+i\omega_{01}t_1} & 0 \end{pmatrix} \xrightarrow{t_3}$$

$$\begin{pmatrix} 0 & -ie^{+i\omega_{01}(t_3-t_1)} \\ +ie^{-i\omega_{01}(t_3-t_1)} & 0 \end{pmatrix} \xrightarrow{\text{emit}} \sin(\omega_{01}(t_3-t_1)). \qquad (2.71)$$

Figure 2.11 shows a Bloch vector representation of $R_1 + R_1^*$ and $R_4 + R_4^*$. In both cases, the density matrix is in a coherent state after the first field interaction. In $R_4 + R_4^*$ the coherent state continues unaltered after the second and third field interactions, whereas in $R_1 + R_1^*$ it is changed into its complex conjugate. As a result, in $R_1 + R_1^*$ time t_3 counteracts the evolution during time t_1 by effectively inverting this time.

Now consider the difference of R_1 and R_4 if we have an inhomogeneous ensemble of molecules. In this case, the ability of R_1 to invert time t_1 has important consequences, as illustrated in Fig. 2.12. After the first pulse, the Bloch vectors of the various molecules start to oscillate with their individual frequencies, and run out of phase. Those molecules which are slower during period t_1 will lag behind. As a result of the second and third pulse, however, each of these Bloch vectors is mirrored on the B_y-axis, and those which lagged behind will now be in front of the

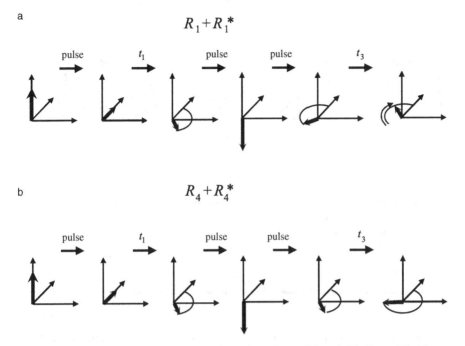

Figure 2.11 Bloch vector representation of (a) $R_1 + R_1^*$ and (b) $R_4 + R_4^*$, illustrating that $R_1 + R_1^*$ flips the vector to the reverse coherence whereas $R_4 + R_4^*$ does not.

others. They will continue to rotate slower during time period t_3, so the individual vectors will rephase at time $t_1 = t_3$. This reappearance of a macroscopic polarization is called a *photon echo*, which occurs whenever we have inhomogeneous broadening. Rephasing allows one to disentangle inhomogeneous from homogeneous dephasing. We call R_1 a *photon echo pulse sequence* or a *rephasing pulse sequence*. Diagram R_4, in contrast, does not have this ability to rephase the distribution since the Bloch vectors just continue to rotate unaltered during period t_3, so we call it a *non-rephasing pulse sequence*. One can show that R_2 is rephasing as well, whereas R_5 is non-rephasing.

In the presence of inhomogeneous broadening, the response functions in Eq. 2.68 and Eq. 2.70 are convoluted with a Gaussian distribution for the transition frequency ω_{01}:

$$R_1 \rightarrow i \int d\omega_{01} e^{-\frac{(\omega_{01} - \overline{\omega_{01}})^2}{2\Delta\omega^2}} e^{-i\omega_{01}(t_3 - t_1)}$$

$$R_4 \rightarrow i \int d\omega_{01} e^{-\frac{(\omega_{01} - \overline{\omega_{01}})^2}{2\Delta\omega^2}} e^{-i\omega_{01}(t_3 + t_1)} \quad (2.72)$$

where $\overline{\omega_{01}}$ is the center frequency of this distribution, and $\Delta\omega$ its width. Using the convolution theorem of the Fourier transformation (Appendix A), one obtains

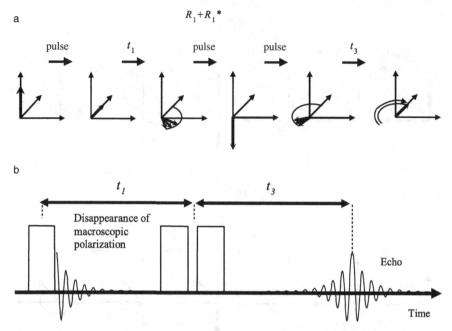

Figure 2.12 (a) Rephasing of an inhomogeneous ensemble, plotting individual Bloch vectors. (b) Pulse sequence of three pulses (with the separation between the second and third pulse practically zero) generating a photon echo.

$$R_1 \propto i e^{-i\omega_{01}(t_3-t_1)} e^{-\Delta\omega^2(t_1-t_3)^2/2}$$
$$R_4 \propto i e^{-i\omega_{01}(t_3+t_1)} e^{-\Delta\omega^2(t_1+t_3)^2/2}. \quad (2.73)$$

The second term in the rephasing diagram R_1, $e^{-\Delta\omega^2(t_1-t_3)^2/2}$, masks the emitted light field in such a manner that it peaks at $t_3 = t_1$. That is the appearance of the photon echo. The same is not happening in the non-rephasing diagram R_4 that has $e^{-\Delta\omega^2(t_1+t_3)^2/2}$. We will return to this topic many times in the book.

2.7 Selecting pathways by phase matching

There are many third-order Feynman diagrams (Fig. 2.13). By carefully designing the experimental setup, we can discriminate some of them against others. The most common experimental trick is to use *phase matching*, which we now explain.

Analogous to linear spectroscopy (Eq. 2.58), the third-order polarization is a convolution of the third-order nonlinear response function with the three laser fields (see Eq. 3.64 from Chapter 3):

$$P^{(3)}(t) \propto \int_0^\infty dt_3 \int_0^\infty dt_2 \int_0^\infty dt_1 E_3'(t-t_3) E_2'(t-t_3-t_2)$$
$$\cdot E_1'(t-t_3-t_2-t_1) R^{(3)}(t_3, t_2, t_1) \quad (2.74)$$

2.7 Selecting pathways by phase matching

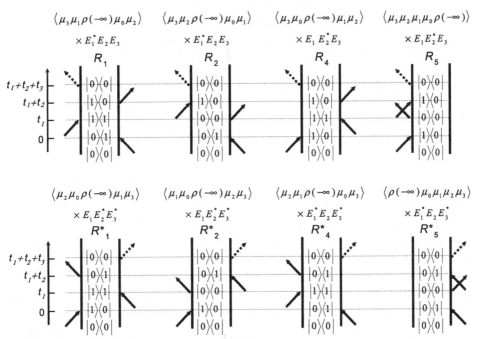

Figure 2.13 The third-order Feynman diagrams that survive the rotating wave approximation.

where E'_1, E'_2 and E'_3 are the real-valued electrical fields of the three laser pulses. However, if the rotating wave approximation applies, particular Feynman diagrams are created by either $E(t) \propto e^{-i\omega t}$ or $E^*(t) \propto e^{+i\omega t}$ only for each of the three laser pulses, as indicated by the directionality of the field interactions in the Feynman diagrams (Fig. 2.13). For example, when we consider diagram R_1 we get:

$$P_1^{(3)}(t) \propto \int_0^\infty dt_3 \int_0^\infty dt_2 \int_0^\infty dt_1 E_3(t - t_3)$$
$$\cdot E_2(t - t_3 - t_2) E_1^*(t - t_3 - t_2 - t_1) R_1(t_3, t_2, t_1) \qquad (2.75)$$

whereas we obtain for diagram R_4:

$$P_4^{(3)}(t) \propto \int_0^\infty dt_3 \int_0^\infty dt_2 \int_0^\infty dt_1 E_3(t - t_3)$$
$$\cdot E_2^*(t - t_3 - t_2) E_1(t - t_3 - t_2 - t_1) R_4(t_3, t_2, t_1). \qquad (2.76)$$

Remember that the electric field also includes the wavevector and phase:

$$E_n(t) = E'_n(t) \cos(\vec{k} \cdot \vec{r} - \omega t + \phi), \qquad (2.77)$$

so that the rotating wave approximation not only selects the frequency $e^{\mp i\omega t}$, but in doing so, also dictates the phase and wavevector via $E(t) \propto e^{-i\omega t + i\phi + i\vec{k} \cdot \vec{r}}$ or

$E^*(t) \propto e^{+i\omega t - i\phi - i\vec{k}\cdot\vec{r}}$ for each of the laser pulses. We then obtain for R_1:

$$P_1^{(3)}(t) \propto e^{i(-\vec{k}_1+\vec{k}_2+\vec{k}_3)\cdot\vec{r}} e^{i(-\phi_1+\phi_2+\phi_3)} \int_0^\infty dt_3 \int_0^\infty dt_2 \int_0^\infty dt_1 \, E_3''(t-t_3)$$
$$\cdot E_2''(t-t_3-t_2) E_1''^*(t-t_3-t_2-t_1) R_1(t_3, t_2, t_1) \quad (2.78)$$

and for R_4:

$$P_4^{(3)}(t) \propto e^{i(+\vec{k}_1-\vec{k}_2+\vec{k}_3)\vec{r}} e^{i(+\phi_1-\phi_2+\phi_3)} \int_0^\infty dt_3 \int_0^\infty dt_2 \int_0^\infty dt_1 \, E_3''(t-t_3)$$
$$\cdot E_2''^*(t-t_3-t_2) E_1''(t-t_3-t_2-t_1) R_4(t_3, t_2, t_1) \quad (2.79)$$

where E_n'' represents the envelope and time dependence of the electric fields. The prefactors can be used to select pathways. For example, by placing a detector in the direction $-\vec{k}_1 + \vec{k}_2 + \vec{k}_3$, one can discriminate R_1 versus R_4, respectively. More generally speaking, one can discriminate rephasing from non-rephasing diagrams.

The response functions in Eq. 2.74 are *single sided*, which means that $R^{(3)}(t_1, t_2, t_3) \neq 0$ only when $t_1 \geq 0$, $t_2 \geq 0$ and $t_3 \geq 0$. When the input laser pulses E_1', E_2' and E_3' do not overlap in time, we say that we have strict time ordering, and only one set of Feynman diagrams is measured in a given phase matching direction. However, when pulses do overlap in time, then the time ordering changes during the convolution in Eq. 2.74. As a result, one needs to switch sets of Feynman diagrams during the integration. Moreover, the rephasing and non-rephasing diagrams will each be emitted in both phase matching directions when this occurs. For example, when $t_1 = 0$, then the $-\vec{k}_1 + \vec{k}_2 + \vec{k}_3$ and $+\vec{k}_1 - \vec{k}_2 + \vec{k}_3$ directions both contain the same information.

The concept of phase matching is illustrated in Fig. 2.14. Three input beams enter the sample with directions $\vec{k}_1$, $\vec{k}_2$ and $\vec{k}_3$ to excite the sample, and various nonlinear signals are emitted in different directions (Fig. 2.14a). Most commonly, only the beams in the $-\vec{k}_1 + \vec{k}_2 + \vec{k}_3$ and the $+\vec{k}_1 - \vec{k}_2 + \vec{k}_3$ directions are used, but Fig. 2.14(b) illustrates that signals are emitted in many other directions as well. Less intense fifth-order signals are also seen. Each beam is characterized by a particular phase matching condition, which in turn can be related to a certain subset of Feynman diagrams. We will see in the following chapters that different diagrams carry different information.

2.8 Selecting pathways by phase cycling

Alternatively, one can use *phase cycling* to discriminate pathways, which relies on manipulating the sum of the pulse phases [38, 165, 176, 185]. Phase cycling is necessary when the chosen beam geometry does not fully discriminate between two or more pathways. For example, consider rephasing R_1 and non-rephasing R_4

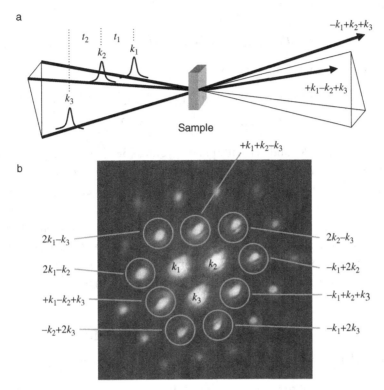

Figure 2.14 (a) A typical phase matching geometry used in 2D spectroscopy. Three input beams with wavevectors $\vec{k}_1$, $\vec{k}_2$ and $\vec{k}_3$ excite the sample, and various nonlinear fields are emitted (in most cases, one uses only the emitted field in the $-\vec{k}_1+\vec{k}_2+\vec{k}_3$ and the $+\vec{k}_1-\vec{k}_2+\vec{k}_3$ directions). (b) Photograph of the emitted fields. Besides all possibilities of phase matching for third-order responses (encircled), also weaker beams from fifth-order responses can be seen. This photograph originates from the Raman response of CS_2, rather than from an IR experiment, but the concepts of phase matching are the same. Picture courtesy of Tõnu Pullerits.

diagrams when $\vec{k}_1 = \vec{k}_2$, (which occurs when 2D IR spectra are collected in a pump–probe beam geometry.) In this situation, both the diagrams will be emitted collinearly, according to Eqs. 2.78 and 2.79, and we observe the sum of both. Nonetheless, the phase dependence of R_1 is still $e^{i(-\phi_1+\phi_2+\phi_3)}$ whereas for R_4 it is still $e^{i(+\phi_1-\phi_2+\phi_3)}$. This allows one to separate R_1 from R_4. To understand how this is done, we define $\phi_{12} = \phi_1 - \phi_2$ and ignore ϕ_3 for the time being. Thus, the polarization that creates the emitted electric field will be

$$P^{(3)}_{\text{sum}}(\phi_{12}) \propto R_1 e^{-i\phi_{12}} + R_4 e^{+i\phi_{12}}. \tag{2.80}$$

To obtain just one pathway from the sum, we make two measurements, one with $\phi_{12} = 0$ and another with $\phi_{12} = \pi/2$,

$$P^{(3)}_{\text{sum}}(\phi_{12} = 0) \propto R_1 + R_4$$
$$P^{(3)}_{\text{sum}}(\phi_{12} = \pi/2) \propto -i(R_1 - R_4) \tag{2.81}$$

and then take their linear combinations

$$P^{(3)}_{\text{sum}}(\phi_{12} = 0) + i P^{(3)}_{\text{sum}}(\phi_{12} = \pi/2) = R_1 \tag{2.82}$$
$$P^{(3)}_{\text{sum}}(\phi_{12} = 0) - i P^{(3)}_{\text{sum}}(\phi_{12} = \pi/2) = R_4. \tag{2.83}$$

Thus, we can regain R_1 and R_4 with phase cycling [38, 176]. The procedure outlined above demonstrates the concept of phase cycling: two or more signals are summed to isolate one or more desired pathways. In practice, the procedure above is too simple, because it requires the measurement of both the real and imaginary components of R_1 and R_4, whereas we usually can measure only one. In the next chapter, we introduce the local oscillator, which enables us to measure both components. In Section 9.3.2, we outline the exact method of using phase cycling to extract R_1 and R_4 from a pump–probe style 2D IR experiment.

2.9 Double sided Feynman diagrams: Rules

We end this chapter by summarizing the rules for drawing Feynman diagrams.

1. The left and right vertical lines represent the time evolution of the *ket* and *bra*, respectively, of the density matrix. Time is running from the bottom to the top.
2. Interactions with the light field are represented by arrows. The last interaction, which originates from the trace $P^{(n)}(t) = \langle \mu \rho^{(n)}(t) \rangle$, is emission and hence is often indicated using a different arrow. By convention, we plot only diagrams with the emission from the *ket* (left); the corresponding diagrams with the emission from the *bra* are just the conjugate complex and do not carry any additional information.
3. Each diagram has a sign $(-1)^n$, where n is the number of interactions on the right (*bra*). This is because each time an interaction is from the right in the commutator it carries a minus sign. Since the last interaction is not part of the commutator, it is *not* counted in this sign rule.[3]
4. An arrow pointing to the right represents an electric field with $e^{-i\omega t + i\vec{k}\cdot\vec{r} + i\phi}$, while an arrow pointing to the left represents an electric field with $e^{+i\omega t - i\vec{k}\cdot\vec{r} - i\phi}$. This rule expresses the fact that the real electric field $E(t) = 2E'(t) \cdot \cos(\omega t - \vec{k} \cdot \vec{r} - \phi)$ can be separated into positive and negative frequencies $E(t) = E'(t) \cdot \left(e^{-i\omega t + i\vec{k}\cdot\vec{r} + i\phi} + e^{+i\omega t - i\vec{k}\cdot\vec{r} - i\phi} \right)$. The emitted light,

---

[3] We have not used this rule yet because we considered only two-level systems so far.

i.e. the last interaction, has a frequency and wavevector which is the sum of the input frequencies and wavevectors (considering the appropriate signs).
5. An arrow pointing towards the system represents an up-climbing of the *bra* or *ket* of the density matrix, while an arrow pointing away represents a de-excitation. This rule is a consequence of the rotating wave approximation. Since the last interaction corresponds to emission of light, it always points away from the diagram.
6. The last interaction must end in a population state. In linear spectroscopy, this will be the ground state $|0\rangle\langle 0|$, but in nonlinear spectroscopy, this can also be higher excited states $|n\rangle\langle n|$.

Exercises

2.1 Verify Eq. 2.8.
2.2 With Eq. 2.37, show that the macroscopic polarization of state 2.30 is maximal, and that of an incoherent state 2.32 is zero.
2.3 Calculate the evolution of the density matrix for R_2 and R_5 in a manner that is analogous to Eq. 2.68. In what aspect are they different from R_1 and R_4?
2.4 Repeat the calculation of the propagation of the density matrix analogous to Eqs. 2.43, 2.44 and 2.50 with the first field interaction acting from the left, however, now starting out from an excited state with:

$$\rho(-\infty) = \begin{pmatrix} 0 & 0 \\ 0 & 1 \end{pmatrix}. \tag{2.84}$$

Prove that $E^*(t) \propto e^{+i\omega_{01}t}$ now survives the rotating wave approximation, whereas the term related to $E(t) \propto e^{-i\omega_{01}t}$ vanishes. Draw the corresponding Feynman diagram.
2.5 The trace is invariant under cyclic permutation, so $\langle \mu(t_1)\mu(0)\rho(-\infty)\rangle = \langle \mu(0)\rho(-\infty)\mu(t_1)\rangle$. By convention we choose the left term, but the right one is mathematically identical. Plot the corresponding Feynman diagram for the right term, taking into account the rotating wave approximation, and discuss how it might be interpreted.
2.6 Draw a Feynman diagram that emits in the $+\vec{k}_1 + \vec{k}_2 - \vec{k}_3$ direction.
2.7 Draw a Feynman diagram that emits in the $+\vec{k}_1 + \vec{k}_2 + \vec{k}_3$ direction. At what frequency will it emit?
2.8 In Fig. 2.14, find the phase matching conditions for the fifth-order beams. In each case, think of a corresponding Feynman diagram, assuming that you can apply the rotating wave approximation. Hint: In some cases, you will need more than a two-level system, e.g. a slightly anharmonic oscillator with a set of almost equidistant quantum states.

3
Mukamelian *or* perturbative expansion of the density matrix

In the present chapter, we derive the two essential equations 2.66 and 2.74 that we used in the previous chapter to develop the Feynman diagrams. This chapter is intended for people who want to more rigorously understand the formalism of 2D IR spectroscopy. A comprehensive description is given by Mukamel [141], Boyd [16], and Cho [27].

3.1 Density matrix

3.1.1 Density matrix of a pure state

We call a *pure state* a quantum mechanical state of a single molecule that can be described by a single wavefunction. Let the total Hamiltonian be the sum of the time independent molecular Hamiltonian $\hat{H}_0$ and an interaction with a time dependent electric field:

$$\hat{H}(t) = \hat{H}_0 + \hat{W}(t) \tag{3.1}$$

with

$$\hat{W}(t) = -\hat{\mu}E(t). \tag{3.2}$$

The molecular wavefunction evolves according to the time dependent Schrödinger equation:

$$\frac{\partial}{\partial t}|\psi(t)\rangle = -\frac{i}{\hbar}\hat{H}(t)|\psi(t)\rangle. \tag{3.3}$$

We expand the wavefunction in an eigenstate basis of the molecular Hamiltonian $\hat{H}_0$:

$$|\psi(t)\rangle = \sum_n c_n(t)|n\rangle \tag{3.4}$$

with the probability amplitudes c_n and

$$H_0|n\rangle = E_n|n\rangle. \quad (3.5)$$

We substitute this into Eq. 3.3, and get:

$$\frac{dc_m(t)}{dt} = -\frac{i}{\hbar} \sum_n H_{mn} c_n(t). \quad (3.6)$$

The expectation value of any operator $\hat{A}$ (which for most cases will be the dipole operator $\hat{\mu}$ in our case) is defined as:

$$\begin{aligned}\langle \hat{A} \rangle &\equiv \langle \psi(t)|\hat{A}|\psi(t)\rangle \\ &= \sum_m c_m^* \sum_n c_n \langle m|\hat{A}|n\rangle \\ &= \sum_{mn} c_m^* c_n A_{mn}. \end{aligned} \quad (3.7)$$

It is convenient to introduce the density matrix:

$$\rho_{nm} = c_n c_m^* \quad (3.8)$$

with which the expectation value of $\hat{A}$ can be written as:

$$\langle \hat{A} \rangle = \sum_{mn} \rho_{nm} A_{mn} \equiv \text{Tr}(\hat{\rho}\hat{A}) \equiv \langle \hat{\rho}\hat{A} \rangle. \quad (3.9)$$

We care about the time dependence of the density matrix which is obtained by applying the chain rule to Eq. 3.8

$$\dot{\rho}_{nm} = \left(\frac{dc_n}{dt}\right) c_m^* + c_n \left(\frac{dc_m^*}{dt}\right). \quad (3.10)$$

When plugging Eq. 3.6 into this equation, we obtain:

$$\begin{aligned}\frac{d}{dt}\rho &= -\frac{i}{\hbar}\hat{H}\rho + \frac{i}{\hbar}\rho\hat{H} \\ &= -\frac{i}{\hbar}[\hat{H}, \rho] \end{aligned} \quad (3.11)$$

with the commutator $[., .]$. This is the *Liouville–von Neumann equation*, which describes the time evolution of the density matrix. Notice that the Hamiltonian acts from the *left* and the *right* on the density matrix. We have already seen in Eq. 2.22 how the density matrix elements time-evolve in the absence of a laser pulse (i.e. with $\hat{H} = \hat{H}_0$):

$$\begin{aligned}\rho_{nn}(t) &= \rho_{nn}(0) = \text{const.} \\ \rho_{nm}(t) &= \rho_{nm}(0)e^{-i\omega_{nm}t}. \end{aligned} \quad (3.12)$$

An alternative derivation of the Liouville–von Neumann equation starts with ρ in a basis-free representation:

$$\rho \equiv |\psi\rangle \langle\psi| \qquad (3.13)$$

which can be seen when expanding ψ

$$|\psi\rangle = \sum_n c_n |n\rangle \qquad (3.14)$$

or, its complex conjugate:

$$\langle\psi| = \sum_n c_n^* \langle n| \qquad (3.15)$$

or both together:

$$\rho \equiv |\psi\rangle \langle\psi| = \sum_{n,m} c_n c_m^* |n\rangle \langle m|. \qquad (3.16)$$

For the time evolution of the density matrix:

$$\frac{d}{dt}\rho = \frac{d}{dt}(|\psi\rangle\langle\psi|) = \left(\frac{d}{dt}|\psi\rangle\right)\cdot\langle\psi| + |\psi\rangle\cdot\left(\frac{d}{dt}\langle\psi|\right) \qquad (3.17)$$

we take the Schrödinger equation, which describes the time evolution of $|\psi\rangle$:

$$\frac{d}{dt}|\psi\rangle = -\frac{i}{\hbar}\hat{H}|\psi\rangle \qquad (3.18)$$

or, for its complex conjugate $\langle\psi|$:

$$\frac{d}{dt}\langle\psi| = +\frac{i}{\hbar}\langle\psi|\hat{H}. \qquad (3.19)$$

Plugging this into Eq. 3.17, we again obtain the Liouville–von Neumann equation:

$$\begin{aligned}\frac{d}{dt}\rho &= -\frac{i}{\hbar}\hat{H}|\psi\rangle\langle\psi| + \frac{i}{\hbar}|\psi\rangle\langle\psi|\hat{H} \\ &= -\frac{i}{\hbar}\hat{H}\rho + \frac{i}{\hbar}\rho\hat{H} \\ &= -\frac{i}{\hbar}\left[\hat{H},\rho\right].\end{aligned} \qquad (3.20)$$

3.1.2 Density matrix of a statistical average

So far, we have discussed the density matrix of a pure state, described by a single wavefunction ψ, in which case the density matrix is $\rho = |\psi\rangle\langle\psi|$. As long as this is the case, both equations

$$\frac{d}{dt}|\psi\rangle = -\frac{i}{\hbar}H|\psi\rangle \quad\Leftrightarrow\quad \frac{d}{dt}\rho = -\frac{i}{\hbar}[H,\rho] \qquad (3.21)$$

are identical, and the density matrix does not add any additional physics.

However, in condensed-phase systems, we in general have to deal with statistical ensembles of molecules, rather than pure states. There is no way to write a wavefunction of a statistical average (see Section 2.3), but we can write the density matrix of a statistical average. Let p_s be the probability of a system being in a state $|\psi_s\rangle$. Then the density matrix of the ensemble is defined as:

$$\rho = \sum_s p_s |\psi_s\rangle \langle\psi_s| \tag{3.22}$$

with $p_s \geq 0$ and $\sum_s p_s = 1$ (normalization). Since Eq. 3.22 is linear, we still get for the expectation value of an operator $\hat{A}$:

$$\langle \hat{A} \rangle = \text{Tr}(\hat{\rho}\hat{A}) \equiv \langle \hat{\rho}\hat{A} \rangle. \tag{3.23}$$

Note that ρ is by no means equivalent to a wavefunction of the form

$$\Psi \stackrel{?}{=} \sum_s p_s |\psi_s\rangle \tag{3.24}$$

which would still be a coherent superposition of states.

In a basis representation, Eq. 3.22 reads:

$$\rho_{nm} = \sum_s p_s c_m^{s*} c_n^s = \langle c_m^{s*} c_n^s \rangle. \tag{3.25}$$

The time dependence of the density matrix of a statistical average (using the chain rule):

$$\dot{\rho}_{nm} = \sum_s \frac{dp_s}{dt} c_n c_m^* + \sum_s p_s \left(\frac{dc_n}{dt} c_m^* + c_n \frac{dc_m^*}{dt} \right)$$

$$= \sum_s \frac{dp_s}{dt} c_n c_m^* - \frac{i}{\hbar} [\hat{H}, \rho]_{nm} \tag{3.26}$$

and contains two terms. The second term is related to quantum mechanics and leads to the Liouville–von Neumann equation (Eq. 3.20), as we have already seen. The first term is related to statistical mechanics, and leads to dephasing and population relaxation. A rigorous treatment of that term is quite difficult, and is beyond the scope of this book. Homogeneous dephasing and population relaxation have been introduced phenomenologically in Section 2.3, and will also be discussed on a semiclassical level in Sections 7.1 and 8.1.

We close this section by summarizing the rules for the trace:

- the trace of a matrix A is defined as the sum over its diagonal elements: $\text{Tr}(A) \equiv \sum_n A_{nn}$;
- The trace is invariant to cyclic permutation: $\text{Tr}(ABC) = \text{Tr}(CAB) = \text{Tr}(BCA)$;

- from which it follows that the trace of a commutator vanishes: $\text{Tr}([A, B]) = \text{Tr}(AB - BA) = \text{Tr}(AB) - \text{Tr}(BA) = 0$;
- and that the trace is invariant to a unitary transformation (i.e. is invariant to the basis): $\text{Tr}(Q^{-1}AQ) = \text{Tr}(QQ^{-1}A) = \text{Tr}(A)$;

and with the properties of the density matrix:

- the density matrix is Hermitian: $\rho_{nm} = \rho_{mn}^*$;
- the diagonal elements of the density matrix are nonnegative: $\rho_{nn} \geq 0$ (the diagonal elements of a density matrix ρ_{nn} are interpreted as the probability of the system to be found in state $|n\rangle$);
- $\text{Tr}(\rho) = 1$ (normalization);
- $\text{Tr}(\rho^2) \leq 1$ (in general);
- $\text{Tr}(\rho^2) = 1$ (if and only if it is a pure state);
- $\rho = \rho^2$ (if and only if it is a pure state).

3.2 Time dependent perturbation theory

Now that we have the density matrix and the Liouville–von Neumann equation defined, let us use them to calculate the linear response. We start from

$$\hat{H}(t) = \hat{H}_0 + \hat{W}(t) \qquad (3.27)$$

but now consider the fact that the electric field of the laser pulse is much weaker than the molecule's internal fields $\hat{W}(t) \ll \hat{H}_0$. In other words, the laser field will alter the coefficients of the wavefunctions, c_n, but not the eigenstates themselves. That is, we may use the eigenstates of the system Hamiltonian:

$$\hat{H}_0|n\rangle = E_n|n\rangle \qquad (3.28)$$

whose time evolution is trivial to calculate, and treat the influence of the laser electric field perturbatively. To that end, we write the Liouville–von Neumann equation and phenomenologically include dephasing for the first term in Eq. 3.26

$$\dot{\rho}_{nm} = -\frac{i}{\hbar}[\hat{H}_0, \rho]_{nm} - \frac{i}{\hbar}[\hat{W}_0, \rho]_{nm} - \rho_{nm}/T_2. \qquad (3.29)$$

The term related to $\hat{H}_0$ can be solved trivially since the density matrix is expanded in an eigenstate basis of $\hat{H}_0$:

$$[\hat{H}_0, |n\rangle\langle m|] = \hat{H}_0|n\rangle\langle m| - |n\rangle\langle m|\hat{H}_0$$
$$= (E_n - E_m)|n\rangle\langle m| \equiv \omega_{mn}|n\rangle\langle m|, \qquad (3.30)$$

hence

$$\dot{\rho}_{nm} = -\left(i\omega_{mn} + \frac{1}{T_2}\right)\rho_{nm} - \frac{i}{\hbar}[\hat{W}, \rho]_{nm}. \quad (3.31)$$

We now use perturbation theory to solve this equation. Since it is the laser pulse that initiates the dynamics, the zeroth-order density matrix is $\rho_{nm}^{(0)}=0$ for $n \neq m$, which we can plug into Eq. 3.31 to get the first-order dynamics

$$\dot{\rho}_{nm}^{(1)}(\tau) = -\left(i\omega_{mn} + \frac{1}{T_2}\right)\rho_{nm}^{(1)}(\tau) - \frac{i}{\hbar}[\hat{W}(\tau), \rho^{(0)}]_{nm} \quad (3.32)$$

where τ is the absolute time. Integrating this equation and converting to relative time delays produces

$$\rho_{nm}^{(1)}(t) = \frac{i}{\hbar}\int_0^\infty [\hat{W}(t-t_1), \rho^{(0)}]_{nm} e^{-\left(i\omega_{mn}+\frac{1}{T_2}\right)t_1} dt_1 \quad (3.33)$$

(see Problem 3.7). We now expand the commutator by substituting for $\hat{W}(t)$, remembering from Eq. 2.57 that the electric field can be written with two terms

$$\hat{W}(t) = \hat{\mu} E'(t) \cos \omega t \propto \hat{\mu} E'(t)\left(e^{-i\omega t} + e^{i\omega t}\right) = \hat{\mu}\left(E(t) + E^*(t)\right) \quad (3.34)$$

which gives

$$\left[\hat{W}(t), \rho^{(0)}\right] = \left(\mu\rho^{(0)} - \rho^{(0)}\mu\right)_{nm}\left(E(t) + E^*(t)\right)$$
$$= -\left(\rho_{mm}^{(0)} - \rho_{nn}^{(0)}\right)\mu_{nm}\left(E(t) + E^*(t)\right) \quad (3.35)$$

since $\rho_{nm}^{(0)} = 0$ for $n \neq m$. Assuming that the laser field is resonant with the laser transition $\omega_{nm} = \omega$, the first-order coherences become

$$\rho_{nm}^{(1)}(t) = \frac{i}{\hbar}\left(\rho_{mm}^{(0)} - \rho_{nn}^{(0)}\right)\mu_{nm}\left\{e^{-i\omega t}\int_0^\infty E'(t-t_1)e^{-t_1/T_2}e^{-2i\omega t_1}dt_1\right.$$
$$\left. + e^{i\omega t}\int_0^\infty E'(t-t_1)e^{-t_1/T_2}dt_1\right\}. \quad (3.36)$$

Except for the prefactors, this equation is the same as Eq. 2.60. And, like we did there, we make the rotating wave approximation by neglecting the first term which is highly oscillatory. Thus, only the second term survives, which is due to the interaction with $E(t)$. If we calculate the off-diagonal element on the opposite corner of the density matrix, $\rho_{mn}^{(1)}(t)$, we would find that it is the $E^*(t)$ term that survives because the coherences evolve as the complex conjugate of those here. Either way, the signal is given by $\langle \mu(t) \rangle = \text{Tr}(\rho^{(1)}(t)\mu)$.

To go beyond first order, we could follow the same procedure, and iteratively plug the solution for $\rho^{(n)}$ back into Eq. 3.31 to get the next-highest order of the perturbative expansion. In fact, this is the approach we take below, but written in the current formalism it is quite tedious. Thus, let us first rewrite the Hamiltonian in the *interaction picture*.

3.2.1 Interaction picture

The interaction picture is helpful in spectroscopy because it allows us to separate out dynamics caused by the laser pulses from those that occur in between the laser pulses and thus are intrinsic to the molecule itself. To that end, we first define the wavefunction in the interaction picture (denoted by the subscript I):

$$|\psi(t)\rangle \equiv e^{-\frac{i}{\hbar}\hat{H}_0(t-t_0)}|\psi_I(t)\rangle \tag{3.37}$$

with some reference time point t_0. $|\psi(t)\rangle$ is the wavefunction under the full Hamiltonian $\hat{H}(t)$, whereas $e^{-\frac{i}{\hbar}\hat{H}_0(t-t_0)}$ describes the time evolution with respect to the system Hamiltonian $\hat{H}_0$ only. Hence, the time dependence of $|\psi_I(t)\rangle$ is caused by the difference between $\hat{H}(t)$ and $\hat{H}_0$, i.e. the weak perturbation $\hat{W}(t)$. If $\hat{W}(t)$ is zero, $|\psi_I(t)\rangle$ will be constant in time:

$$|\psi_I(t)\rangle = |\psi(t_0)\rangle. \tag{3.38}$$

When introducing Eq. 3.37 into the Schrödinger equation:

$$\frac{d}{dt}|\psi(t)\rangle = -\frac{i}{\hbar}\hat{H}|\psi(t)\rangle \tag{3.39}$$

we obtain, after a little bit of algebra:

$$\frac{d}{dt}|\psi_I(t)\rangle = -\frac{i}{\hbar}\hat{W}_I(t)|\psi_I(t)\rangle \tag{3.40}$$

where the perturbation $\hat{W}_I(t)$ in the interaction picture is defined as:

$$\hat{W}_I(t) = e^{\frac{i}{\hbar}\hat{H}_0(t-t_0)}\hat{W}(t)e^{-\frac{i}{\hbar}\hat{H}_0(t-t_0)}. \tag{3.41}$$

3.2.2 Perturbative expansion of the wavefunction

Equation 3.40 is formally equivalent to the Schrödinger equation which contains the interaction part only, albeit in the interaction picture. We formally integrate Eq. 3.40:

$$|\psi_I(t)\rangle = |\psi_I(t_0)\rangle - \frac{i}{\hbar}\int_{t_0}^{t} d\tau \hat{W}_I(\tau)|\psi_I(\tau)\rangle \tag{3.42}$$

and solve it iteratively (like we did in the previous section) by plugging it into itself:

$$|\psi_I(t)\rangle = |\psi_I(t_0)\rangle - \frac{i}{\hbar} \int_{t_0}^{t} d\tau \hat{W}_I(\tau) |\psi_I(t_0)\rangle \qquad (3.43)$$

$$+ \left(-\frac{i}{\hbar}\right)^2 \int_{t_0}^{t} d\tau_2 \int_{t_0}^{\tau_2} d\tau_1 \hat{W}_I(\tau_2) \hat{W}_I(\tau_1) |\psi_I(\tau_1)\rangle$$

and so on:

$$|\psi_I(t)\rangle = |\psi_I(t_0)\rangle + \sum_{n=1}^{\infty} \left(-\frac{i}{\hbar}\right)^n \int_{t_0}^{t} d\tau_n \int_{t_0}^{\tau_n} d\tau_{n-1} \cdots \int_{t_0}^{\tau_2} d\tau_1 \qquad (3.44)$$

$$\hat{W}_I(\tau_n) \hat{W}_I(\tau_{n-1}) \cdots \hat{W}_I(\tau_1) |\psi_I(t_0)\rangle .$$

This is a power expansion in terms of the small interaction term $\hat{W}(t)$.

Going back to the Schrödinger picture, we obtain:

$$|\psi(t)\rangle = |\psi^{(0)}(t)\rangle + \sum_{n=1}^{\infty} \left(-\frac{i}{\hbar}\right)^n \int_{t_0}^{t} d\tau_n \int_{t_0}^{\tau_n} d\tau_{n-1} \cdots \int_{t_0}^{\tau_2} d\tau_1$$

$$e^{-\frac{i}{\hbar}\hat{H}_0(t-\tau_n)} \hat{W}(\tau_n) e^{-\frac{i}{\hbar}\hat{H}_0(\tau_n-\tau_{n-1})} \hat{W}(\tau_{n-1}) \cdots$$

$$\cdots e^{-\frac{i}{\hbar}\hat{H}_0(\tau_2-\tau_1)} \hat{W}(\tau_1) e^{-\frac{i}{\hbar}\hat{H}_0(\tau_1-t_0)} |\psi(t_0)\rangle. \qquad (3.45)$$

The first term:

$$|\psi^{(0)}(t)\rangle \equiv e^{-\frac{i}{\hbar}\hat{H}_0(t-t_0)} |\psi(t_0)\rangle \qquad (3.46)$$

is the zeroth-order wavefunction, i.e. the time propagation of the wavefunction under the molecular Hamiltonian $\hat{H}_0$ only. The following terms are perturbative terms that describe the effect of the interaction with the electrical field. These terms have an intuitive physical interpretation: The system propagates freely under the *system* Hamiltonian $\hat{H}_0$ until time τ_1, described by the time evolution operator $e^{-\frac{i}{\hbar}\hat{H}_0(\tau_1-t_0)}$. At time τ_1, it interacts with the perturbation $\hat{W}(\tau_1)$. Subsequently, it again propagates freely until time τ_2, and so on. This interpretation leads directly to the graphic representation of Feynman diagrams (Fig. 3.1), where the vertical arrow depicts the time axis, and the dotted arrows depict interaction with the perturbation $\hat{W}$ at the time points τ_1, τ_2, and so on. The perturbative expansion of a wavefunction is represented by a *single sided Feynman diagram*.

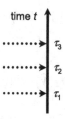

Figure 3.1 Single sided Feynman diagram.

3.2.3 Perturbative expansion of the density matrix

Along the same lines, we can develop a power expansion of the density matrix. To this end, we first define the density matrix in the interaction picture:

$$|\psi(t)\rangle\langle\psi(t)| = e^{-\frac{i}{\hbar}\hat{H}_0(t-t_0)}|\psi_I(t)\rangle\langle\psi_I(t)|e^{+\frac{i}{\hbar}\hat{H}_0(t-t_0)} \quad (3.47)$$

or

$$\rho(t) = e^{-\frac{i}{\hbar}\hat{H}_0(t-t_0)}\rho_I(t)e^{+\frac{i}{\hbar}\hat{H}_0(t-t_0)}. \quad (3.48)$$

Since this expression is linear in ρ, it also holds for a statistical average $\rho = \sum_s p_s|\psi_s\rangle\langle\psi_s|$. Since the time evolution of the wavefunction in the interaction picture $|\psi_I(t)\rangle$ is formally equivalent to the Schrödinger equation (see Eq. 3.40), the same is true for the density matrix in the interaction picture, for which we obtain an equation which is formally equivalent to the Liouville–von Neumann equation:

$$\frac{d}{dt}\rho_I(t) = -\frac{i}{\hbar}\left[\hat{W}_I(t), \rho_I(t)\right]. \quad (3.49)$$

Its perturbative expansion is obtained in an analogous way as for the wavefunction:

$$\rho_I(t) = \rho_I(t_0) + \sum_{n=1}^{\infty}\left(-\frac{i}{\hbar}\right)^n \int_{t_0}^{t} d\tau_n \int_{t_0}^{\tau_n} d\tau_{n-1} \cdots \int_{t_0}^{\tau_2} d\tau_1 \quad (3.50)$$

$$\left[\hat{W}_I(\tau_n), \left[\hat{W}_I(\tau_{n-1}), \cdots \left[\hat{W}_I(\tau_1), \rho_I(t_0)\right]\cdots\right]\right].$$

Going back to the Schrödinger picture for the density matrix yields:

$$\rho(t) = \rho^{(0)}(t) + \sum_{n=1}^{\infty}\left(-\frac{i}{\hbar}\right)^n \int_{t_0}^{t} d\tau_n \int_{t_0}^{\tau_n} d\tau_{n-1} \cdots \int_{t_0}^{\tau_2} d\tau_1$$

$$e^{-\frac{i}{\hbar}\hat{H}_0(t-t_0)}\left[\hat{W}_I(\tau_n), \left[\hat{W}_I(\tau_{n-1}), \cdots \left[\hat{W}_I(\tau_1), \rho(t_0)\right]\cdots\right]\right]e^{+\frac{i}{\hbar}\hat{H}_0(t-t_0)}$$

$$\equiv \rho^{(0)}(t) + \sum_{n=1}^{\infty}\rho^{(n)}(t). \quad (3.51)$$

Here, $\rho^{(0)}(t)$ is the zeroth-order density matrix (as the system would evolve without perturbation), and the succeeding terms, the nth-order density matrices $\rho^{(n)}(t)$, are ordered in powers of $\hat{W}_I$. The interaction Hamiltonian is still in the interaction picture and contains both the perturbation $\hat{W}(t)$ and time evolution operators, similar to Eq. 3.45. However, since the density matrix contains a *ket* and a *bra*, the interaction can be either from the left or the right. We have seen this in Sect. 2.6 when writing the commutators explicitly.

We now include the perturbation term from Eq. 3.2 and get:

$$\rho^{(n)}(t) = -\left(-\frac{i}{\hbar}\right)^n \int_{t_0}^{t} d\tau_n \int_{t_0}^{\tau_n} d\tau_{n-1} \cdots \int_{t_0}^{\tau_2} d\tau_1 E(\tau_n) E(\tau_{n-1}) \cdots E(\tau_1)$$

$$\cdot e^{-\frac{i}{\hbar}\hat{H}_0(t-t_0)} \left[\hat{\mu}_I(\tau_n), \left[\hat{\mu}_I(\tau_{n-1}), \cdots \left[\hat{\mu}_I(\tau_1), \rho(t_0)\right] \cdots \right]\right] e^{+\frac{i}{\hbar}\hat{H}_0(t-t_0)}. \tag{3.52}$$

Here, the dipole operator in the interaction picture is defined as:

$$\hat{\mu}_I(t) = e^{+\frac{i}{\hbar}\hat{H}_0(t-t_0)} \hat{\mu} e^{-\frac{i}{\hbar}\hat{H}_0(t-t_0)}. \tag{3.53}$$

In the Schrödinger picture, the dipole operator $\hat{\mu}$ is time independent. It is time dependent in the interaction picture since the system is evolving in time under the system Hamiltonian $\hat{H}_0$. The subscript I (which denotes the interaction picture) is commonly discarded, and the Schrödinger picture *versus* the interaction picture is specified implicitly by writing either $\hat{\mu}$ or $\hat{\mu}(t)$, respectively.

3.2.4 Nonlinear polarization

The macroscopic polarization is given by the expectation value of the dipole operator $\hat{\mu}$:

$$P(t) = \text{Tr}(\hat{\mu}\rho(t)) \equiv \langle\hat{\mu}\rho(t)\rangle. \tag{3.54}$$

When collecting the terms in powers of the electric field $E(t)$, we obtain for the nth-order polarization:

$$P^{(n)}(t) = \langle\hat{\mu}\rho^{(n)}(t)\rangle. \tag{3.55}$$

We insert Eq. 3.52 into Eq. 3.55 and obtain for the nth-order polarization

$$P^{(n)}(t) = -\left(-\frac{i}{\hbar}\right)^n \int_{t_0}^{t} d\tau_n \int_{t_0}^{\tau_n} d\tau_{n-1} \cdots \int_{t_0}^{\tau_2} d\tau_1 E(\tau_n) E(\tau_{n-1}) \cdots E(\tau_1)$$

$$\langle\hat{\mu}(t) \left[\hat{\mu}(\tau_n), \left[\hat{\mu}(\tau_{n-1}), \cdots \left[\hat{\mu}(\tau_1), \rho(t_0)\right] \cdots \right]\right]\rangle \tag{3.56}$$

where we translated the last dipole operator $\hat{\mu}(t)$ into the interaction picture as well, and made use of the invariance of the trace with respect to cyclic permutation, which makes the time evolution operators in Eq. 3.52 disappear. Finally, we assume that $\rho(t_0)$ is an equilibrium density matrix that does not evolve in time under the system Hamiltonian $\hat{H}_0$, and we can send $t_0 \to -\infty$.

$$P^{(n)}(t) = -\left(-\frac{i}{\hbar}\right)^n \int_{-\infty}^{t} d\tau_n \int_{-\infty}^{\tau_n} d\tau_{n-1} \cdots \int_{-\infty}^{\tau_2} d\tau_1 E(\tau_n) E(\tau_{n-1}) \cdots E(\tau_1)$$
$$\langle \hat{\mu}(t) [\hat{\mu}(\tau_n), [\hat{\mu}(\tau_{n-1}), \cdots [\hat{\mu}(\tau_1), \rho(-\infty)] \cdots]] \rangle. \quad (3.57)$$

Frequently, time intervals are used instead:

$$\begin{aligned} \tau_1 &= 0 \\ t_1 &= \tau_2 - \tau_1 \\ t_2 &= \tau_3 - \tau_2 \\ &\vdots \\ t_n &= t - \tau_n. \end{aligned} \quad (3.58)$$

We can choose $\tau_1 = 0$ since the zero time point is arbitrary. Throughout the book we will use τ's for absolute time points, while t's denote time intervals (Fig. 3.2). In a 2D IR experiment, it is typically these time intervals t_i which we control experimentally. Transforming Eq. 3.56 into this set of time variables gives:

$$P^{(n)}(t) = -\left(-\frac{i}{\hbar}\right)^n \int_0^\infty dt_n \int_0^\infty dt_{n-1} \cdots \int_0^\infty dt_1 \quad (3.59)$$
$$E(t - t_n) E(t - t_n - t_{n-1}) \cdots E(t - t_n - t_{n-1} - \cdots - t_1) \cdot$$
$$\langle \hat{\mu}(t_n + t_{n-1} + \cdots + t_1) [\hat{\mu}(t_{n-1} + \cdots + t_1), \cdots [\hat{\mu}(0), \rho(-\infty)] \cdots]\rangle.$$

Hence, very generally speaking, the nth-order nonlinear response can be written as a convolution of n electric fields

$$P^{(n)}(t) = \int_0^\infty dt_n \int_0^\infty dt_{n-1} \cdots \int_0^\infty dt_1 \quad (3.60)$$
$$E(t - t_n) E(t - t_n - t_{n-1}) \cdots E(t - t_n - \cdots - t_1) R^{(n)}(t_n, \cdots, t_1)$$

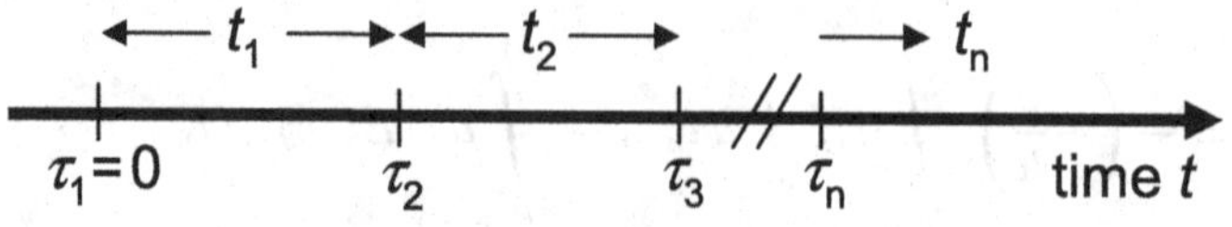

Figure 3.2 Time variables: the τ's refer to absolute times, and t's to time intervals.

with the nth-order nonlinear response function:

$$R^{(n)}(t_n, \cdots, t_1) = \qquad (3.61)$$
$$-\left(-\frac{i}{\hbar}\right)^n \langle \hat{\mu}(t_n + \cdots + t_1) [\hat{\mu}(t_{n-1} + \cdots + t_1), \cdots [\hat{\mu}(0), \rho(-\infty)] \cdots] \rangle.$$

Note the different role of the last interaction $\hat{\mu}(t_n + t_{n-1} + \cdots + t_1)$ compared to the previous interactions: The interactions at times $0, t_1, \cdots$ and $t_{n-1} + \cdots + t_1$ generate a nonequilibrium density matrix $\rho^{(n)}$, whose off-diagonal elements at time $t_n + t_{n-1} + \cdots + t_1$ emit a light field. Only the first n interactions are part of the commutators, while the last is not.

For most of the book, we will talk about the linear response, which reads:

$$P^{(1)}(t) = \int_0^\infty dt_1 E(t - t_1) R^{(1)}(t_1) \qquad (3.62)$$

with

$$R^{(1)}(t_1) \propto i \langle \hat{\mu}(t_1) [\hat{\mu}(0), \rho(-\infty)] \rangle \qquad (3.63)$$

or the third-order nonlinear response:

$$P^{(3)}(t) \propto \int_0^\infty dt_3 \int_0^\infty dt_2 \int_0^\infty dt_1 E_3(t - t_3) E_2(t - t_3 - t_2) \cdot$$
$$\cdot E_1(t - t_3 - t_2 - t_1) R^{(3)}(t_3, t_2, t_1) \qquad (3.64)$$

with

$$R^{(3)}(t_3, t_2, t_1) \propto -i \langle \hat{\mu}(t_3 + t_2 + t_1) [\hat{\mu}(t_2 + t_1), [\hat{\mu}(t_1), [\hat{\mu}(0), \rho(-\infty)]]] \rangle. \qquad (3.65)$$

These are the fundamental equations which we used in Section 2.4 to develop the Feynman diagrams. The response functions are single sided with $R^{(3)}(t_1, t_2, t_3) \neq 0$ only when $t_1 \geq 0$, $t_2 \geq 0$ and $t_3 \geq 0$. This fact reflects causality, i.e. the molecule emits a field only *after* interaction with the laser pulses.

An even-order response function, such as a second-order response function, disappears in an isotropic medium due to symmetry. If the symmetry is broken, e.g. at a surface or since a crystal is anisotropic, a second-order nonlinear response can indeed be measured and the formalism outlined above can be applied as well. We will not cover such experiments in this book.

Exercises

3.1 Show that one obtains $\rho = \rho^2$ for the density matrix of a pure state. Show that this is no longer true for a density matrix of a statistical average. Verify these results for the examples $\rho = \begin{pmatrix} 1/2 & 1/2 \\ 1/2 & 1/2 \end{pmatrix}$ and $\rho = \begin{pmatrix} 1/2 & 0 \\ 0 & 1/2 \end{pmatrix}$.

3.2 Verify that there is no wavefunction $|\psi\rangle$ whose density matrix is $\rho = \begin{pmatrix} 1/2 & 0 \\ 0 & 1/2 \end{pmatrix}$.

3.3 Show that exactly one eigenvalue of an $(n \times n)$-density matrix of a pure state is 1, while all others are zero. Hint: Start with $\rho = \rho^2$. Diagonalize with a matrix Q and compare the diagonal elements.

3.4 Starting from the definition of the trace of a matrix, $\text{Tr}(A) \equiv \sum_n A_{nn}$, show that $\text{Tr}(AB) = \text{Tr}(BA)$. Show furthermore that the trace is invariant with respect to cyclic permutation, $\text{Tr}(ABC) = \text{Tr}(CAB) = \text{Tr}(BCA)$.

3.5 Show that the trace of any density matrix is indeed $\text{Tr}(\rho) = 1$.

3.6 Show that the length of a Bloch vector corresponding to a pure state is unity.

3.7 Derive Eq. 3.33 starting from Eq. 3.32. Hint: Do a change of variables using $\rho_{nm}^{(1)}(\tau) = S_{nm}^{(1)}(\tau) e^{-\left(i\omega_{mn} + \frac{1}{T_2}\right)\tau}$. Then integrate from $\tau' = -\infty$ to τ. Finally, switch to relative time delays [16].

3.8 Derive Eq. 3.49.

4
Basics of 2D IR spectroscopy

In this chapter we apply the mathematical methodology that we have developed in the preceding chapters to predict what the 1D and 2D IR spectra will look like for some generic systems. It turns out that 2D IR lineshape and cross-peak patterns depend upon the experimental setup chosen to measure the 2D IR spectra, and some are better than others. Thus, this chapter is organized according to the common ways of collecting 2D IR spectra.

4.1 Linear spectroscopy

Before discussing 2D IR spectra, we illustrate the concepts of the preceding chapters by applying the methodology to linear infrared spectroscopy. For linear spectra measured using weak infrared light, and assuming that all the molecules are in their ground vibrational state before the laser pulse interacts with the sample, we only need to consider two vibrational levels and one Feynman diagram (Fig. 4.1a, b). Using this Feynman diagram, we develop the response function step by step:

- At negative times, the system is in the ground state, described by the density matrix $\rho = |0\rangle\langle 0|$.
- At time $t = 0$, we generate a ρ_{10} off-diagonal matrix element of the density matrix (we also generate a ρ_{01} element from the corresponding complex conjugate Feynman diagram, which is not necessary to consider because it is redundant). The probability that this happens is proportional to the transition dipole moment μ_{10}.

$$\rho_{10} \propto i\mu_{01}. \qquad (4.1)$$

- We have seen in Section 2.4 that this off-diagonal density matrix element (the coherence) oscillates at the frequency ω_{01} and decays with the homogeneous lifetime T_2:

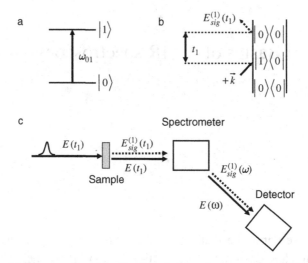

Figure 4.1 (a) The $v = 0$ and 1 vibrational states that we are considering, (b) the only important Feynman diagram and (c) a schematic of the experimental setup typically used in linear absorption spectroscopy.

$$\rho_{10} \propto i\mu_{01} e^{-i\omega_{01}t_1} e^{-t_1/T_2}. \tag{4.2}$$

- At time t_1, the first-order response of the molecular system, $R^{(1)}(t_1)$, to the laser pulse is then $\langle \mu \rangle = Tr[\mu_{01}\rho]$, which is:

$$R^{(1)}(t_1) \propto i\mu_{01}^2 e^{-i\omega_{01}t_1} e^{-t_1/T_2}. \tag{4.3}$$

$R^{(1)}$ is the quantity that we are interested in measuring because it contains information about the molecule. However, the molecular response is convoluted by the envelope of the laser pulse to create the actual macroscopic polarization $P^{(1)}(t)$ in the sample:

$$P^{(1)}(t) \propto \int_0^\infty dt_1 R^{(1)}(t_1) E(t - t_1) e^{-i\omega(t-t_1)+i\vec{k}_1 \cdot \vec{r}+i\phi}. \tag{4.4}$$

The macroscopic polarization gives rise to an emitted signal field with a 90° phase shift:

$$E_{sig}^{(1)}(t) \propto iP^{(1)}(t). \tag{4.5}$$

The homogeneous lifetimes of most vibrational modes are $T_2 = 1$–5 ps. Thus, one needs to use femtosecond pulses to make these measurements so that the emitted electric field (often called the *free induction decay*) reflects the molecular response $R^{(1)}(t_1)$ and not just the envelope of the laser pulse (Fig. 4.2a). Ideally, a δ-function laser pulse would be used so that the emitted field exactly measures the molecular

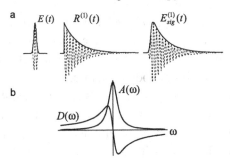

Figure 4.2 (a) Schematic of the convolution between a short laser pulse $E(t)$ (left), a molecular response $R^{(1)}(t)$ (middle) to give the polarization and the subsequent emitted field $E_{sig}^{(1)}$ (right). (b) Absorptive and dispersive parts of a complex-valued Lorentzian lineshape.

response. The need for femtosecond infrared pulses is the primary reason that 2D IR spectroscopy became practical only very recently.

Notice that the incident laser pulse imprints its wavevector $\vec{k}$ as a directionality into the density matrix in Eq. 4.4. Using this wavevector, we determine the direction that $E_{sig}^{(1)}$ is radiated using Rule 4 in Section 2.9, which states that the signal is a maximum in the direction that $\vec{k}_s = \sum \vec{k}_n$. In this case, where we have only one input pulse, the emitted field $E_{sig}^{(1)}(t)$ is radiated in the same direction as the incident field $\vec{k}_s = \vec{k}_1$ (Fig. 4.1c).

Once the field is emitted, it needs to be measured. Optical detectors are not fast enough to measure fs-ps electric fields, and so a spectrometer is usually used to convert the emitted field into the frequency domain. In this case, both the emitted field $E_{sig}^{(1)}(t)$ and the laser pulse E are reflected from a grating and then measured on an IR detector (Fig. 4.1c). Typical IR detectors are square-law detectors that can only measure intensities and not electric fields. They can also only measure real and not imaginary quantities. Thus, to calculate the signal, we mathematically mimic the sequence of physical operations that the spectrometer performs on the electric fields by first taking a Fourier transform and then the squared magnitude:

$$S(\omega) \propto \left| \int_0^\infty \{E(t) + E_{sig}^{(1)}(t)\} e^{i\omega t} dt \right|^2$$

$$\propto I_0(\omega) + 2\Re \left(\int_0^\infty E(t) e^{i\omega t} dt \cdot \int_0^\infty E_{sig}^{(1)}(t) e^{i\omega t} dt \right) + I_{sig}^{(1)}(\omega)$$

$$\approx I_0(\omega) + 2\Re \left(E(\omega) \cdot E_{sig}^{(1)}(\omega) \right). \tag{4.6}$$

Thus, the measured signal will be the sum of $I_0(\omega)$, which is just the spectrum of the laser pulse, an interference term between $E(t)$ and $E_{sig}^{(1)}(t)$, and $I_{sig}^{(1)}(\omega)$ which is the spectrum of the signal $E_{\text{sig}}^{(1)}$. $I_{\text{sig}}^{(1)}(\omega)$ is so much smaller than $I_0(\omega)$ that it is usually ignored. The interference term is a consequence of both the signal and the laser pulse hitting the detector. We say that the emitted field is *heterodyned* by E. Since the laser pulse both heterodynes and causes the signal, it is sometimes said that the signal is *self-heterodyned*. The two i's in Eqs. 4.3 and 4.5 cause the emitted light field to be opposite in sign relative to the incident light field E. Hence the $E_{\text{sig}}^{(1)}$ and E interfere destructively on the detector. As a result, the amount of light the detector sees is less than without the sample, which is consistent with the light being *absorbed* by the sample (see Section 2.5).

To isolate the interference term, which contains the information that we want, one usually calculates the absorbance to subtract off the spectrum of the laser pulse (to first order):

$$S'(\omega) \equiv -\log \frac{S(\omega)}{I_0(\omega)}$$
$$\approx 2\Re\left(E(\omega) \cdot E_{\text{sig}}^{(1)}(\omega)\right). \quad (4.7)$$

If one assumes for simplicity that the laser pulse is a δ-function pulse in time $E(t) \propto \delta(0)$ (i.e. white light),[1] then the interference term reduces to just the Fourier transform of $E_{\text{sig}}^{(1)}(t)$ (since $E(\omega) = \text{const}$). The convolution from Eq. 4.4 disappears, so that we can simply substitute the molecular response $R^{(1)}(t)$ from Eq. 4.3 into Eq. 4.7 to obtain

$$S'(\omega) = 2\Re \int_0^\infty iR^{(1)}(t)e^{i\omega t}dt$$
$$\propto \Re \int_0^\infty \mu_{01}^2 e^{i(\omega-\omega_{01})t_1} e^{-t_1/T_2} dt_1$$
$$\propto \Re \mu_{01}^2 \frac{1}{i(\omega-\omega_{01}) - 1/T_2}. \quad (4.8)$$

Due to heterodyning with the incident field, we see only the real part, which is the Lorentzian line:

$$A(\omega) \propto \mu_{01}^2 \frac{1/T_2}{(\omega-\omega_{01})^2 + 1/T_2^2}. \quad (4.9)$$

---

[1] One can show that the linear absorption spectrum is independent of the phase of the light field (see Problem 4.1), hence, for practical purposes, it does not make any difference whether we have spectrally broad light (i.e. "white" light) that is phase locked, i.e. a short pulse, or whether it is incoherent light. The mathematical treatment is much simpler when dealing with a short pulse. In nonlinear spectroscopy, the phase of the light field does play a role, and hence the nonlinear response on incoherent light or a phase-locked short pulse will be different.

We call this the *absorptive* part of the band (Fig. 4.2c). The full width half maximum (FWHM) bandwidth of a Lorentzian line is:

$$\Delta \nu = \frac{\Delta \omega}{2\pi} = \frac{1}{\pi T_2}. \quad (4.10)$$

For a typical dephasing time of $T_2 = 1$ ps, this reveals a FWHM bandwidth of 10 cm^{-1}.

If we could rotate the phase of the heterodyning electric field by $\phi = \pi/2$, so that $E(t) \propto \delta(0)e^{i\phi} = i\delta(0)$, then we could also measure the imaginary part, which gives the *dispersive* part of the lineshape.

$$\begin{aligned} D(\omega) &\propto \Im\, \mu_{01}^2 \frac{1}{i(\omega - \omega_{01}) - 1/T_2} \\ &= -\mu_{01}^2 \frac{\omega - \omega_{01}}{(\omega - \omega_{01})^2 + 1/T_2^2}. \end{aligned} \quad (4.11)$$

The dispersive part gives rise to the index of refraction. Far from the center frequency it decays as $1/(\omega - \omega_{01})$ whereas the absorptive lineshape decays as $1/(\omega - \omega_{01})^2$ and is thus much narrower (Fig. 4.2c).

4.2 Third-order response functions

For 2D IR spectroscopy, we need the emitted field generated from the third-order response functions. In this section, we calculate these response functions using the same methodology as for linear response. We start by drawing the possible Feynman diagrams. For a single oscillator system that is initially in the ground state, there are eight possible Feynman diagrams (six of which are shown in Fig. 4.3). To illustrate the methodology, we explicitly develop the emitted electric field for just one pathway, the non-rephasing diagram R_4:

- We start out in the ground state, $\rho = |0\rangle\langle 0|$ (not shown in the Feynman diagram).
- At time $t = 0$, we generate a ρ_{10} off-diagonal matrix element of the density matrix:

$$\rho_{10} \propto i\mu_{01}. \quad (4.12)$$

- The system dephases for time t_1:

$$\rho_{10} \propto i\mu_{01} e^{-i\omega_{01}t_1} e^{-t_1/T_2}. \quad (4.13)$$

- At time $t = t_1$, the system is switched into a population state by the second field interaction:

$$\rho_{11} \propto i\mu_{01}^2 e^{-i\omega_{01}t_1} e^{-t_1/T_2}. \quad (4.14)$$

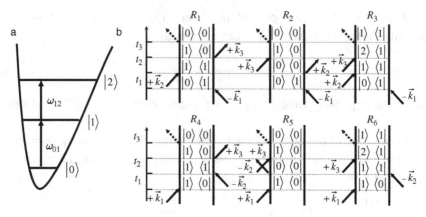

Figure 4.3 (a) The potential energy curve for an anharmonic oscillator. (b) Six possible Feynman diagrams for third-order nonlinear spectroscopy when the system starts in the ground state $\rho = |0\rangle\langle 0|$ (not shown). Top-row: rephasing diagrams; bottom row: non-rephasing diagrams.

- During the time period t_2, the system experiences population relaxation:

$$\rho_{11} \propto i\mu_{01}^2 e^{-i\omega_{01}t_1} e^{-t_1/T_2} e^{-t_2/T_1}. \qquad (4.15)$$

- At time $t = t_1 + t_2$, the system is switched back into a coherence state ρ_{10}, which then propagates and dephases during time period t_3:

$$\rho_{10} \propto i\mu_{01}^3 e^{-i\omega_{01}t_1} e^{-t_1/T_2} e^{-t_2/T_1} e^{-i\omega_{01}t_3} e^{-t_3/T_2}. \qquad (4.16)$$

- Finally, at time $t = t_1 + t_2 + t_3$, the third-order response that is responsible for the macroscopic polarization $P^{(3)}(t)$ and ultimately the emitted third-order signal field $E_{\text{sig}}^{(3)}(t)$ is given by $\text{Tr}[\mu_{01}\rho]$:

$$R_4(t_1, t_2, t_3) \propto i\mu_{01}^4 e^{-i\omega_{01}t_1} e^{-t_1/T_2} e^{-t_2/T_1} e^{-i\omega_{01}t_3} e^{-t_3/T_2}. \qquad (4.17)$$

Following this procedure, one can generate the molecular responses for the five other Feynman pathways. The only salient differences are the sign of the coherences for the rephasing diagrams and the anharmonic shift for the overtone coherences. Let us look at each diagram individually.

The rephasing diagram R_1 is different in one respect from R_4. It has a ρ_{01} coherence after the first field interaction rather than a ρ_{10} coherence. Hence, the matrix element oscillates as the complex conjugate during the time period t_1:

$$R_1(t_1, t_2, t_3) \propto i\mu_{01}^4 e^{+i\omega_{01}t_1} e^{-t_1/T_2} e^{-t_2/T_1} e^{-i\omega_{01}t_3} e^{-t_3/T_2}. \qquad (4.18)$$

R_2 and R_5 have the same sequences of coherences as R_1 and R_4, respectively, and so $R_2 = R_1$ and $R_5 = R_4$, although this approximation is not strictly true for more sophisticated theories of pure dephasing such as the Brownian oscillator model [141]. It might appear counterintuitive that the system undergoes T_1 relaxation in diagrams R_2 and R_5, since the system is in the ground state $\rho = |0\rangle\langle 0|$ during t_2. But the e^{-t_2/T_1} term in these pathways accounts for refilling of the ground state upon vibrational relaxation which decreases the signal strength, just like population relaxation out of the $\rho = |1\rangle\langle 1|$ excited state decreases the signal strength in R_1 and R_4.

The diagrams R_3 and R_6 account for pathways in which the laser pulses access $v = 2$. Their response functions are:

$$R_3(t_3, t_2, t_1) = -i\mu_{01}^2\mu_{12}^2 e^{+i\omega_{01}t_1 - t_1/T_2^{(01)}} e^{-t_2/T_1} e^{-i\omega_{12}t_3 - t_3/T_2^{(12)}}$$
$$R_6(t_3, t_2, t_1) = -i\mu_{01}^2\mu_{12}^2 e^{-i\omega_{01}t_1 - t_1/T_2^{(01)}} e^{-t_2/T_1} e^{-i\omega_{12}t_3 - t_3/T_2^{(12)}}. \quad (4.19)$$

At this point we need to recall that homogeneous dephasing T_2 includes pure dephasing T_2^* and the population relaxation contribution T_1 (see Section 2.3):

$$\frac{1}{T_2} = \frac{1}{2T_1} + \frac{1}{T_2^*}. \quad (4.20)$$

While it is often a good assumption that pure dephasing of the 0–1 level pair is the same as that of the 1–2 level pair (because both fluctuate in a correlated fashion, see Section 7.4.2), the same is not true for the T_1 contribution. By perturbation theory, one finds that the 1–2 population relaxation is twice as fast as the 0–1 population relaxation, $T_1^{(12)} = T_1^{(01)}/2$, and both contribute to 1–2 homogeneous dephasing:

$$\frac{1}{T_2^{(12)}} = \frac{1}{2T_1^{(01)}} + \frac{1}{2T_1^{(12)}} + \frac{1}{T_2^*} = \frac{3}{2T_1^{(01)}} + \frac{1}{T_2^*}. \quad (4.21)$$

Population relaxation of vibrational transitions is quite fast, often in the range of 1 ps, and hence contributes significantly to homogeneous dephasing (linewidth 5 cm^{-1} for $T_1 = 1$ ps). Hence, the population relaxation contribution to the 1–2 homogeneous dephasing is commonly responsible for the broader linewidth of the 1–2 transition in a 2D IR spectrum. Also note the minus signs in Eqs. 4.19, which originate from the fact that these diagrams have one rather than two interactions from the right (Rule 3 in Section 2.9).

Diagrams R_3 and R_6 are interesting because in these pathways the laser pulses climb the vibrational ladder to access higher quantum states. We did not consider states above $v = 1$ when discussing linear spectroscopy. Of course $v = 2$ (and higher states) can be accessed with linear spectroscopy, which will have the response function

$$R^{(1)}(t_1) \propto i\mu_{02}^2 e^{-i(\omega_{01}+\omega_{12})t_1} e^{-t_1/T_2'}. \quad (4.22)$$

However, direct overtone transitions are much weaker than fundamental transitions because the vibrational selection rules make μ_{02} very small (they are generally 10 times weaker for each skipped quantum level). In comparison, transition dipole strengths for resonant transitions increase when climbing up the vibrational ladder step by step, because $b^\dagger |n\rangle = \sqrt{n+1}|n+1\rangle$ where $b^\dagger$ is a ladder operator operating on a harmonic oscillator (see Appendix B). Thus, sequences of on-resonant pulses are a powerful way of accessing quantum states that are only weakly allowed in linear spectroscopy.

Rephasing diagrams R_1, R_2 and R_3 are all emitted in the $-\vec{k}_1 + \vec{k}_2 + \vec{k}_3$ direction and thus cannot be separated further by phase matching (although they might be distinguishable by other features such as their frequencies and phases); the same holds for the non-rephasing diagrams R_4, R_5 and R_6 which are emitted in the $+\vec{k}_1 - \vec{k}_2 + \vec{k}_3$ direction. Hence, one measures the sums:

$$R_{1,2,3}(t_3, t_2, t_1) = \sum_{n=1}^{3} R_n(t_3, t_2, t_1)$$

$$R_{4,5,6}(t_3, t_2, t_1) = \sum_{n=4}^{6} R_n(t_3, t_2, t_1). \quad (4.23)$$

One often assumes that the transition dipoles scale like a harmonic oscillator, $\mu_{12} = \sqrt{2}\mu_{01}$, and that the homogeneous dephasing time T_2 of the 1–2 transition is the same as that of the 0–1 transition (which is the case if population relaxation is very long and thus contributes only negligibly to homogeneous dephasing). In that case, the response functions can be lumped together:

$$R_{1,2,3} = 2i\mu_{01}^4 \left(e^{-i\omega_{01}(t_3-t_1)} - e^{-i((\omega_{01}-\Delta)t_3 - \omega_{01}t_1)}\right) e^{-(t_1+t_3)/T_2}$$
$$R_{4,5,6} = 2i\mu_{01}^4 \left(e^{-i\omega_{01}(t_3+t_1)} - e^{-i((\omega_{01}-\Delta)t_3 + \omega_{01}t_1)}\right) e^{-(t_1+t_3)/T_2}$$

$$(4.24)$$

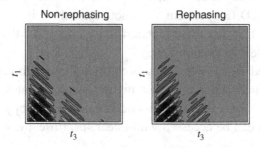

Figure 4.4 Rephasing and non-rephasing response functions of an anharmonic oscillator (Eq. 4.24, only the real part is plotted).

where $\Delta \equiv \omega_{01} - \omega_{12}$ is the anharmonic frequency shift. Figure 4.4 plots these response functions. They oscillate close to the fundamental frequency along both axes, but there is an additional beating along t_3 set by the difference frequency between the 0–1 and the 1–2 pathways. In fact, since these two pathways have opposite signs, the emitted field is zero when $t_3 = 0$. Also note the 45° tilt of the oscillatory features, which is opposite for rephasing and non-rephasing diagrams. This reflects the change of sign of the t_1 coherence.

4.3 Time domain 2D IR spectroscopy

The third-order response functions $R_n(t_3, t_2, t_1)$ developed in the last section contain all the physics of the molecular system, and thus are the quantities that we want to measure. There are many variants of third-order nonlinear spectroscopies (2D IR spectroscopy, pump–probe spectroscopy, transient grating spectroscopy, two-pulse and three-pulse photon echo spectroscopy) that aim to measure these response functions, but they differ in the way that the measurement is performed and the information that is extracted. That is, the response functions are convoluted with the electric fields of the laser pulses

$$P^{(3)}(t_3, t_2, t_1) \propto \int_0^\infty dt_3 \int_0^\infty dt_2 \int_0^\infty dt_1 \sum_n R_n(t_3, t_2, t_1) E_3(t - t_3)$$
$$\cdot E_2(t - t_3 - t_2) E_1(t - t_3 - t_2 - t_1) \quad (4.25)$$

and the convolution is different for each technique. In fact, most third-order spectroscopies incompletely measure the response functions by forcing some of the time variables t_3, t_2 or t_1 to zero, by integrating over them, and/or by measuring only the modulus of the complex-valued response function. 2D IR spectroscopy, in contrast, is capable of measuring the third-order response functions in their entirety by using a sequence of short pulses that minimize the convolutions in Eq. 4.25 and results in both amplitude and phase resolved spectra. As such, 2D IR spectroscopy extracts the maximum amount of information about the molecular system under study. In the following sections, we describe the most common third-order spectroscopies, starting with 2D IR spectroscopy.

In its most general form, a 2D IR experiment uses a beam geometry in which the input laser pulses all have different wavevectors. The generated field will then have a wavevector $\mp\vec{k}_1 \pm \vec{k}_2 + \vec{k}_3$ (where the signs depend on whether we consider rephasing or non-rephasing diagrams), that will be emitted noncollinearly with any of the input fields. This phase matching geometry is often referred to as the *box-CARS* geometry.[2] By placing a square-law detector in that direction, the intensity of the emitted field is measured, integrated over its duration:

$$S(t_1, t_2) = \int_0^\infty \left| E_{sig}^{(3)}(t_3; t_1, t_2) \right|^2 dt_3. \tag{4.26}$$

We refer to this as *homodyne detection*.[3] Because of the integration, the information on the time structure and the phase of the emitted light field is lost. This is what is typically done in so-called *integrated* three-pulse photon echo experiments (see Section 7.5).

However, the phase and time dependence of the emitted light field is required to perform the Fourier transform needed to obtain 2D IR spectra. Thus, we need a heterodyned signal like we had for linear spectroscopy (Eq. 4.6). In some phase matching geometries $E_{sig}^{(3)}$ is self-heterodyned. If it is not, then we can add a fourth laser pulse as the so-called *local oscillator* (Fig. 4.5). When heterodyned, both the local oscillator $E_{LO}(t - t_{LO})$ and the emitted signal field $E_{sig}^{(3)}(t)$ are incident on the square-law detector

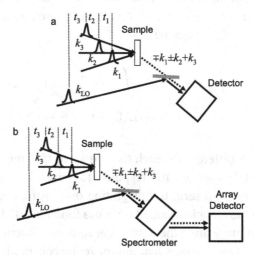

Figure 4.5 Principle of the experimental setup for 2D spectroscopy. (a) Direct detection of the third-order polarization and (b) after dispersing it in a spectrometer.

---

[2] The name of this phase matching geometry originates from its original usage in Coherent Anti-Stokes Raman Spectroscopy (CARS). It is also sometimes called the *boxcar* geometry.

[3] Not all scientific communities refer to the terms homodyne detection and heterodyne detection the same as we do. In continuous-wave spectroscopy, optical homodyne and heterodyne detection refers to overlaying a signal with a local oscillator that has either the same or a different frequency, respectively. A different frequency only helps if one has a detector with a fast enough response to resolve the time domain beats, which does not exist for ultrafast signals. In the nonlinear spectroscopy community, we use the terms "homo" and "hetero" to refer to "itself" or by "another" ultrashort laser pulse, respectively.

4.3 Time domain 2D IR spectroscopy

$$S(t_{LO}; t_1, t_2) \propto \int_0^\infty \left| E_{LO}(t_3 - t_{LO}) + E_{\text{sig}}^{(3)}(t_3; t_2, t_1) \right|^2 dt_3 \qquad (4.27)$$

$$\approx I_{LO} + 2\Re \int_0^\infty \left\{ E_{LO}(t_3 - t_{LO}) \cdot E_{\text{sig}}^{(3)}(t_3; t_1, t_2) \right\} dt_3.$$

As before (Eq. 4.6), we ignore $I_{\text{sig}}^{(3)}$ and can subtract off I_{LO} if desired. Regardless, we are interested in the cross-term, which allows us to measure the phase and time dependence of $E_{\text{sig}}^{(3)}(t)$ by scanning t_{LO}. When self-heterodyned, because the input field acts as both the pump and an intrinsic local oscillator, the phase and time dependence of the local oscillator cannot be easily altered.

Ideally, one uses laser pulses for the local oscillator as well as for the three incident pulses that are short compared to any time-scale of the system, but long compared to the oscillation period of the light field. In this case, we say that we are in the *semi-impulsive limit*, in which the envelopes of the pulses are approximated by δ-functions

$$E(t) \propto \delta(t) e^{\pm i\omega t \mp i\vec{k}\cdot\vec{r} \mp i\phi} \qquad (4.28)$$

but we retain the carrier frequency, phase and wavevector. In this limit, the convolution in Eq. 4.25 disappears, and the emitted signal field becomes proportional to the response functions directly[4]

$$E_{\text{sig}}^{(3)}(t_1, t_2, t_3) \propto e^{i(\mp \vec{k}_1 \pm \vec{k}_2 + \vec{k}_3)\cdot \vec{r}} e^{i(\mp \phi_1 \pm \phi_2 + \phi_3)} \sum_n i R_n(t_1, t_2, t_3) \qquad (4.29)$$

where the signs in the prefactor depend on the type of Feynman diagrams considered (e.g. rephasing versus non-rephasing diagrams), as does the summation index n. The convolutions in Eq. 4.27 disappear as well, $t_{LO} = t_3$, and we obtain after heterodyne detection with $\vec{k}_{LO} = \mp\vec{k}_1 \pm \vec{k}_2 + \vec{k}_3$:

$$S(t_1, t_2, t_3) \propto E_{LO} \cdot E_{\text{sig}}^{(3)} \propto e^{i(\mp \phi_1 \pm \phi_2 + \phi_3 - \phi_{LO})} \sum_n i R_n(t_1, t_2, t_3). \qquad (4.30)$$

Thus, in the semi-impulsive limit, we are directly measuring the third-order response of the system. Times t_1, t_2, and t_3 are the delay times between the four laser pulses that are directly controlled by the experiment. In addition, the phase of the emitted polarization is measured. Typically, in a properly phased spectrum, one tries to set the overall phase to zero $\mp\phi_1 \pm \phi_2 + \phi_3 - \phi_{LO} = 0$ (see Section 9.5.6), but varying these phases can also be used to enhance or suppress certain contributions (see Section 9.3).

---

[4] Note that when the pulses have finite pulse durations and overlap in time, the time ordering of the field interactions will change during the convolution in Eq. 4.25. When this occurs, one also has to switch sets of Feynman diagrams (see Problem 4.5).

2D IR spectroscopy is the ultimate nonlinear experiment, since it gathers the maximum amount of information within the framework of third-order nonlinear spectroscopy. That is, what we cannot learn with 2D IR spectroscopy, we will not be able to learn with any other type of third-order nonlinear spectroscopy (two-pulse or three-pulse photon echo, pump–probe, transient grating, etc.). In fact, one could deduce any other third-order experiment from a 2D IR spectrum.

In the time domain the third-order response can be quite difficult to visualize (see e.g. Fig. 4.4), but it is much easier to interpret in the frequency domain. Thus, we take a 2D Fourier transform of the measured emitted field, which is equivalent to Fourier transforming the responses themselves in the semi-impulsive limit (with the overall phase zero for simplicity):

$$S(\omega_3, t_2, \omega_1) = \int_0^\infty \int_0^\infty S(t_3, t_2, t_1) e^{i\omega_1 t_1} e^{i\omega_3 t_3} dt_1 dt_3$$
$$= \int_0^\infty \int_0^\infty \sum_n i R_n(t_3, t_2, t_1) e^{i\omega_1 t_1} e^{i\omega_3 t_3} dt_1 dt_3. \quad (4.31)$$

Most often, the Fourier transform is performed with respect to the t_1 and t_3 coherence times while time t_2, during which the system is in a population state, is not transformed. This leads to a sequence of 2D IR spectra for various waiting times t_2, which gives the full information about the third-order response function $R_n(\omega_3, t_2, \omega_1)$.

4.3.1 2D IR spectrum of diagonal peaks

So, what does the 2D IR spectrum look like? To answer this question, consider the two response functions R_1 and R_4 from Section 4.2 (Eqs. 4.17 and 4.18). Their 2D Fourier transforms are (setting $t_2 = 0$):

$$R_1(\omega_1, \omega_3) \propto \frac{1}{i(\omega_1 + \omega_{01}) - 1/T_2} \cdot \frac{1}{i(\omega_3 - \omega_{01}) - 1/T_2}$$
$$R_4(\omega_1, \omega_3) \propto \frac{1}{i(\omega_1 - \omega_{01}) - 1/T_2} \cdot \frac{1}{i(\omega_3 - \omega_{01}) - 1/T_2}. \quad (4.32)$$

Thus, each of these response functions produces a peak that lies on the diagonal of the spectrum with $|\omega_1| = |\omega_3|$. However, there are differences, which are that the spectra of these two response functions appear in different quadrants in frequency space and have different *phase twists*. We discuss the quadrants first. R_4 is a non-rephasing diagram, and so oscillates with the same positive frequency during t_1 and t_3, whereas R_1 is a rephasing pathway in which the coherence during t_1 is reversed. As a result, spectra of non-rephasing pathways appear in

4.3 Time domain 2D IR spectroscopy

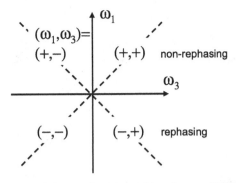

Figure 4.6 Schematic of the quadrants resulting from a 2D Fourier transform.

the $(\omega_1, \omega_3) = (+, +)$ quadrant whereas rephasing spectra appear in the $(-, +)$ quadrant. These quadrants are graphically shown in Fig. 4.6.

In an actual experiment, the measured data will be real valued. That is, rather than Eqs. 4.17 and 4.18, we will measure the equivalent of Eq. 4.27:

$$\Re(i R_1(t_1, t_2, t_3)) \propto \mu_{01}^4 \cos(+\omega_{01}t_1 - \omega_{01}t_3)e^{-t_1/T_2 - t_3/T_2}$$
$$\Re(i R_4(t_1, t_2, t_3)) \propto \mu_{01}^4 \cos(-\omega_{01}t_1 - \omega_{01}t_3)e^{-t_1/T_2 - t_3/T_2}. \quad (4.33)$$

However, we can write the oscillatory part as:

$$\cos(\pm \omega_{01}t_1 - \omega_{01}t_3) \propto e^{\pm i\omega_{01}t_1 - i\omega_{01}t_3} + e^{\mp i\omega_{01}t_1 + i\omega_{01}t_3} \quad (4.34)$$

the first term of which is exactly that of Eqs. 4.17 and 4.18, while the second has the signs of the coherences inverted during both time periods t_1 and t_3. As a result, a real-valued rephasing diagram will produce two peaks (see Appendix A) in the $(\omega_1, \omega_3) = (-, +)$ and $(+, -)$ quadrants of the 2D IR spectrum, while a non-rephasing diagram produces two peaks in the $(-, -)$ and $(+, +)$ quadrants. If we restrict ourselves only to the $\omega_3 \geq 0$ half, then this is effectively the same as measuring a complex-valued response function. This trick often comes in handy, such as for phase cycling (Section 9.3.2).

We now turn to the discussion of the phase twist of the two types of peaks in Eq. 4.32. Just like in 1D spectroscopy (Eq. 4.8), these spectra are complex valued. Each dimension (ω_1 and ω_3) contains an absorptive (real) and a dispersive (imaginary) contribution (in analogy to Fig. 4.2b). Unfortunately, we cannot single out a purely absorptive 2D spectrum by just taking the real or imaginary part of Eq. 4.32 like we can in linear spectroscopy. For example, if we take the real part of R_4:

$$\Re[(A(\omega_1) + iD(\omega_1))(A(\omega_3) + iD(\omega_3))] = A(\omega_1)A(\omega_3) - D(\omega_1)D(\omega_3) \quad (4.35)$$

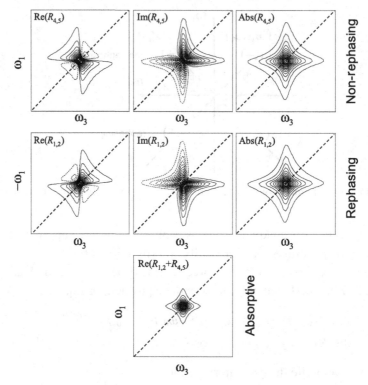

Figure 4.7 Real part, imaginary part, and absolute value 2D spectra of a two-level system with a 2D Lorentzian lineshape. Top-row: non-rephasing diagrams; middle row: rephasing diagrams; bottom: purely absorptive spectrum. Solid and dotted contour lines depict positive and negative contribution to the spectra, respectively. The dashed lines in both (a) and (b) are to highlight the diagonal of the spectra where $|\omega_1| = |\omega_3|$.

we see that it contains both absorptive and dispersive contributions. The dispersive contribution broadens the peaks and causes them to twist (called phase twist) into lineshapes with both positive and negative regions (Fig. 4.7, top and middle row) [105]. The phase twist is especially problematic because it complicates the interpretation of spectra when peaks overlap in systems with multiple vibrational modes. The spectra can be simplified by calculating their absolute value which removes the phase altogether (Fig. 4.7, right), but the spectra are still broad and complicated interferences appear. However, an interesting consequence of the rephasing and non-rephasing spectra being in different quadrants is that their phase twists cancel when the two spectra are added. As a result, it is most common nowadays to plot the real part of the sum of both rephasing and non-rephasing spectra, after inverting the sign of the ω_1-coordinate of the former:

$$R_{abs}(\omega_1, \omega_3) = \Re\left(R_1(-\omega_1, \omega_3) + R_4(\omega_1, \omega_3)\right) = 2A(\omega_1)A(\omega_3) \quad (4.36)$$

which gives the so-called *purely absorptive* 2D IR spectrum (Fig. 4.7, bottom) [105]. The purely absorptive spectrum yields the highest possible frequency resolution, since it decays as $1/(\omega - \omega_{01})^2$ far away from the resonance (in contrast to the dispersive contribution, which decays as $1/(\omega - \omega_{01})$). Furthermore, the purely absorptive 2D spectrum preserves the sign of the response function (in contrast to the absolute-valued spectrum). Finally, the contributions of overlapping peaks are just additive without interferences and so are much simpler to interpret.

Unfortunately, from the experimental point of view, the price one has to pay to obtain purely absorptive spectra is relatively high: One has to collect two sets of data (rephasing and non-rephasing diagrams) under exactly identical conditions. These measurements could be done by placing a detector in each of the $-\vec{k}_1 + \vec{k}_2 + \vec{k}_3$ and $\vec{k}_1 - \vec{k}_2 + \vec{k}_3$ phase matching directions. By convention, the subscripts of the k-vectors refer to the time ordering of the pulses. If all the pulses are identical, then interchanging the time ordering of the first two pulses switches the rephasing into a non-rephasing phase-matching condition and vice versa. Hence, rather than using two detectors to measure the two signals, it is more commonly done with one detector by interchanging the time ordering of the first two pulses, which automatically guarantees that both signals are equally strong. Either way, the absolute phases of the signals must be known for proper addition, which can be challenging. Alternatively, one can use a pump–probe beam geometry with either an etalon (also known as a Fabry–Perot filter), a Michelson interferometer, or a pulse shaper. In a pump–probe geometry the first two pulses are aligned collinearly ($\vec{k}_1 = \vec{k}_2$) so that the rephasing and non-rephasing diagrams have the same phase matching condition. As a result, the sum of both diagrams is measured automatically, producing a purely absorptive 2D IR spectrum directly. These methodologies are discussed below in Section 4.4 as well as Chapter 9.

So far, we have only considered four of the six Feynman diagrams for a single oscillator system. The other two diagrams, R_3 and R_6, create peaks that appear at the fundamental frequency $\pm \omega_{01}$ along ω_1, as do the peaks from R_1 and R_4, but along ω_3 they appear at the overtone frequency ω_{12}. Moreover, they have an inverted sign (Rule 3 from Section 2.9). We see that the second diagonal peak of 2D IR spectra comes from the R_3 and R_6 response functions. When the anharmonic shift is larger than the linewidths, the two peaks are well resolved, but many types of vibrational modes (such as carbonyls) have anharmonic shifts of $\Delta = 5$–$10 \, \text{cm}^{-1}$ and linewidths of 5–30 cm^{-1}, and so the two peaks overlap. Shown in Fig. 4.8 are simulated 2D IR spectra of partially overlapping diagonal peaks, and the apparent peak positions as a function of linewidth is shown in Fig. 9.21.

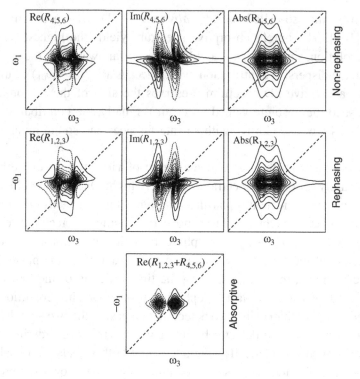

Figure 4.8 Real part, imaginary part, and absolute-value 2D spectra of an anharmonic oscillator with a 2D Lorentzian lineshape. Homogeneous dephasing was set to $T_2 = 1$ ps, and the anharmonic shift to $\Delta = 25$ cm^{-1} in this example. Top row: non-rephasing diagrams; middle row: rephasing diagrams; bottom: purely absorptive spectrum. Solid and dotted contour lines depict positive and negative contribution to the spectra, respectively. The dashed lines in both (a) and (b) are to highlight the diagonal of the spectra where $|\omega_1| = |\omega_3|$.

Before leaving this section we make one comment about data collection. To measure the emitted electric field, $E_{\text{sig}}^{(3)}(t_3)$, one can either collect data in the time domain by scanning the local oscillator explicitly (Fig. 4.5a) and performing the Fourier transform with a computer, or in the frequency domain by using a spectrometer to perform the Fourier transform along t_{LO} and collecting all the frequencies at once with an array detector (Fig. 4.5b). Most labs now use a spectrometer and an array detector, which reduces the data collection time significantly because one no longer needs to manually scan t_{LO}. In this case, one then transforms $E_{\text{sig}}^{(3)} + E_{LO}$ like we did $E_{\text{sig}}^{(1)} + E_{LO}$ for linear spectroscopy, in which we first take the Fourier transform for the grating and then the magnitude squared for the square-law array detector (Eq. 4.6). Data collection is explained more explicitly in Chapter 9.

4.3.2 2D IR spectra of coupled oscillators

Now that we have covered the basics of a 2D IR spectrum for a single transition dipole, let us consider a 2D IR spectrum for a set of coupled oscillators. As is apparent from the previous sections, there are two basic types of 2D IR spectra, which are the rephasing and non-rephasing spectra (there is also a so-called two-quantum 2D IR spectrum that we cover in Section 11.1). In this section, we discuss the cross-peak patterns for these two spectra, which are not the same for both. We also discuss how they can be added to get pseudo-absorptive spectra, or a narrow-band pump can be used to get true absorptive spectra. For all scenarios, we will use the eigenstate scheme that is shown in Fig. 4.9(a). On the left-hand side, the eigenstates are labeled by the nomenclature we used in Section 1.1. In this section we use a more compact notation (Fig. 4.9a, right-hand side) in which the fundamental states are i and j, their overtones are $2i$ and $2j$, respectively, and $i+j$ is the combination band. Furthermore, it is often convenient to lump the overtones and combination bands together with a common index, k.

Rephasing 2D IR spectra

We begin by considering the rephasing 2D IR spectrum as measured using impulsive IR pulses in a box-CARS geometry in the $-\vec{k}_1+\vec{k}_2+\vec{k}_3$ direction. Three types of Feynman diagrams need to be considered (Fig. 4.10). They correspond to the

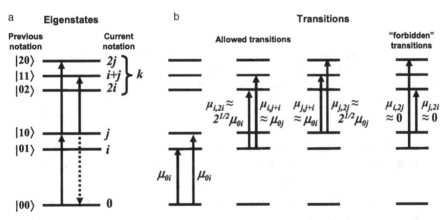

Figure 4.9 Eigenstates and transition dipole strengths for coupled oscillator systems. (a) Notation used in Section 1.1 for two coupled oscillators, which is modified in a more compact notation used in this chapter that works for any number of coupled oscillators. The index k refers to the overtones and combination bands. (b) An approximation for the transition dipole strengths of two oscillators in either the weak or strong coupling limit, in which the "*forbidden*" transitions are weak.

78   Basics of 2D IR spectroscopy

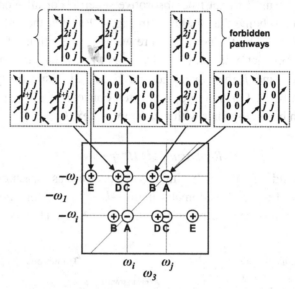

Figure 4.10 Feynman diagrams that can be used to generate the rephasing 2D IR spectra for any number of coupled eigenstates.

Figure 4.11 2D IR spectrum generated from the third-order rephasing diagrams. The pathways that have a forbidden vibrational transition are labeled as such. Peaks are labeled with their signs (by our convention, the signs of the peaks are opposite to the signs of the corresponding Feynman diagrams).

R_1, R_2 and R_3 rephasing diagrams discussed in Section 4.2, but now also account for pathways that involve multiple transition dipoles, which is why we label them with quotation marks. For a two-oscillator system, there are 20 diagrams in total. The transition dipoles responsible for the intensity of each pathway are also shown as are their respective signs.

Ten of these Feynman diagrams are drawn explicitly in Fig. 4.11, along with the peaks in the 2D IR spectrum that they create. The way that one determines the peak positions is by examining the coherence during times t_1 and t_3. For instance, one of the response functions that creates peak E (the left one) is:

$$R_E(t_1, t_2, t_3) \propto i\mu_{0j}^2 \mu_{j,2i}^2 e^{+i\omega_{0j}t_1} e^{-t_1/T_2} e^{-i\omega_{2i,j}t_3} e^{-t_3/T_2}. \tag{4.37}$$

It oscillates with $\rho = |0\rangle\langle j| \to e^{+i\omega_{0j}t_1}$ during t_1 and $\rho = |2i\rangle\langle j| \to e^{-i\omega_{2i,j}t_3}$ during t_3, so it is positioned at $(\omega_1, \omega_3) = (-\omega_{0j}, \omega_{2i,j})$. The signs of these coherences dictate which quadrant in 2D Fourier space that the spectra appear. For the rephasing spectra, since the evolution during t_1 and t_3 have opposite signs, the 2D IR spectra appear in the $(\omega_1, \omega_3) = (-, +)$ quadrant. Notice that two Feynman pathways contribute to each peak in the spectrum.

Not all pathways produce equally strong signals. Pathways that go through forbidden transitions are much weaker. Forbidden transitions are ones in which more than one vibrational quantum must change simultaneously, such as $j \to 2i$, which would require losing one quantum in the j mode and gaining two quanta in the i mode. When the oscillators are very weakly coupled, or when the local mode anharmonicity is small, the vibrational states are not mixed enough to make such transitions appreciable, although they have been observed. In some coupling situations, however, they can become appreciable (see Section 6.4).

It is also interesting to note that half of the Feynman pathways that create the cross-peaks have an *interstate coherence* during t_2. That is, during t_2, the density matrix is not in a population state, but is instead in an interstate coherence $\rho = |i\rangle\langle j|$. For instance, the response function corresponding to the other (right) Feynman pathways that create peak E is:

$$R_E(t_1, t_2, t_3) \propto i\mu_{0j}\mu_{0i}\mu_{i,2i}\mu_{j,2i}e^{+i\omega_{0j}t_1}e^{-t_1/T_2}e^{-i\omega_{i,j}t_2}e^{-i\omega_{2i,j}t_3}e^{-t_3/T_2}. \quad (4.38)$$

Thus, the cross-peaks will oscillate at the difference frequency between the two modes during t_2 (see Fig. 4.14 below). This fact can be used to further distinguish between the two sets of pathways, or might also be used to generate a frequency axis in a 2D or 3D IR spectrum. Interstate coherences have been observed in 2D IR spectroscopy [106] and used in 2D electronic spectroscopy to measure the interstate dephasing time of coupled electronic chromophores [52].

Non-rephasing 2D IR spectra

The non-rephasing spectrum can be measured by detecting the signal in the $\vec{k}_s = \vec{k}_1 - \vec{k}_2 + \vec{k}_3$ phase matching direction, and the Feynman diagrams are those shown in Fig. 4.12. The spectrum generated by these non-rephasing pathways lies in the $(\omega_1, \omega_3) = (+, +)$ quadrant, and so has the opposite phase twist of the rephasing spectra. Regarding the peak pattern, there are 20 individual pathways for a two-oscillator system, just like for the rephasing diagrams. However, the non-rephasing 2D IR spectra contain more peaks because fewer pathways are degenerate.

Shown in Fig. 4.13 are 10 of the Feynman diagrams, only two of which create peaks at the same position in the 2D spectrum. Typically, we imagine that cross

Figure 4.12 Feynman diagrams that can be used to generate the non-rephasing 2D IR spectra for any number of coupled eigenstates.

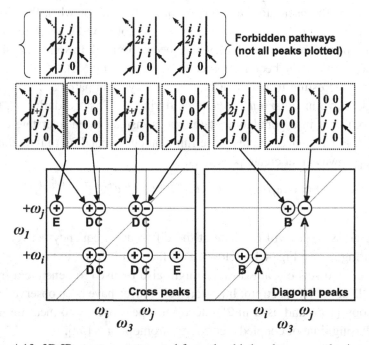

Figure 4.13 2D IR spectrum generated from the third-order non-rephasing spectrum. For clarity, the cross-peaks are plotted separately from the diagonal peaks. Not all of the weak pathways are plotted. Note that some of the cross-peaks actually lie on the diagonal of the 2D IR spectrum and thus overlap with the diagonal cross-peaks.

peaks appear off the diagonal, which is true for the rephasing diagrams, but the non-rephasing diagrams actually create cross peaks that are both on and off the diagonal, as we explain now. In Fig. 4.13 we have artificially separated the spectrum into "cross" peaks and "diagonal" peaks. We define a "diagonal peak" as one that includes either oscillator i or j, and a "cross" peak pathway as one that includes both i and j. We have separated them for the sake of clarity, but both sets of peaks are superimposed in a typical experiment. The non-rephasing spectrum contains more peaks than the rephasing 2D IR spectrum because the transition pathways all have unique frequencies along ω_3, but more importantly, because the four Feynman pathways that create the "cross" peaks C and D are split into two pairs, one of which lies off-diagonal and the other is on-diagonal. The positive peak C lies at the same position as the diagonal peak A, and so they interfere with each other. In the rephasing spectra, both of these two sets of pathways created peaks that appeared on the off-diagonal. Thus, in the non-rephasing spectra, the off-diagonal cross-peaks are only half as intense as in the rephasing spectra and the on-diagonal cross-peaks overlap with the diagonal features (the splitting between the positive and negative peaks will be different for the on-diagonal cross-peaks and the diagonal peaks).

From the perspective that the frequencies of the eigenstates are all that is needed to determine the molecular structure, the non-rephasing peak positions contain the same information as the rephasing spectrum. However, in practice, it can be much more difficult to determine the peak positions in non-rephasing spectra of an inhomogeneous system, because the peaks are not line-narrowed (see Section 7.4.1 and Fig. 7.6 for a discussion of line narrowing). As a result, it is usually much more difficult to accurately deconvolute the lineshapes for precise frequency determination [61].

Quasi-absorptive 2D IR spectra

While the rephasing spectra have narrower antidiagonal linewidths than the non-rephasing spectra, both types of spectra suffer from phase twists. Thus, it is often advantageous to add them together to obtain absorptive spectra with the best possible frequency resolution, as discussed in Section 4.3.1 for the diagonal peaks. However, in coupled oscillator systems, addition does not give perfectly absorptive spectra like for single oscillator systems, at least when the spectra are measured by impulsive pulses.

One can understand the limitation by comparing Figs. 4.11 and 4.13. If these spectra are added, the phase twist will not be entirely removed because the peak intensities and positions are not perfectly equal. The intensity mismatch is largest for the cross-peaks, which are only half as intense in the non-rephasing as in the rephasing spectrum, and thus the phase twist is only halfway compensated. This fact is demonstrated in Fig. 4.14 for a simulated purely absorptive 2D IR spectrum of a metal dicarbonyl. The diagonal peak intensities are also not equal, because of the on-diagonal cross-peaks and the forbidden pathways. The two spectra are nevertheless added because the phase twist is mostly removed from the diagonal peaks and it is the diagonal peaks that usually dominate 2D IR spectra. Notice that the phase twist rotates with the interstate coherence frequency $\omega_{01} - \omega_{02}$ because of the interference between the unequally weighted pathways (Fig. 4.14).

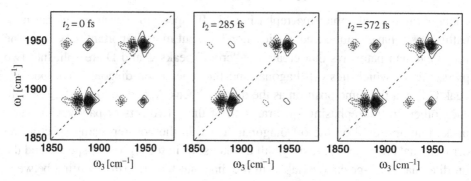

Figure 4.14 Simulated purely absorptive 2D IR spectrum of a metal dicarbonyl for a series of population times $t_2 = 0$ fs, $t_2 = 285$ fs and $t_2 = 572$ fs, corresponding to 0, half cycle and full cycle of the interstate coherence $|1\rangle\langle 1'|$. See Section 10.2 for details of the simulation. Notice that the cross-peaks are phase twisted even in this absorptive spectrum.

4.4 Frequency domain 2D IR spectroscopy

An alternative method to measure purely absorptive 2D IR spectra is to collect the spectra using a narrowband pump–probe methodology. This is the method that was used to collect the first 2D IR spectrum. It does not require multiple pulses nor a sophisticated phase matching geometry, but just uses a picosecond pump pulse in a typical pump–probe geometry [83]. The experiment is shown schematically in Fig. 4.15(a) and was described qualitatively in Section 1.1. Now, with our mathematical formalism in place, we can examine this method of collecting 2D IR spectra more exactly.

To collect 2D IR spectra in this manner, one generates a spectrally narrow, long-in-time (typically 1 ps) pump pulse either using an etalon [83] or a pulse shaper [166] so that its center frequency can be scanned across the vibrational modes of interest. Either way, the narrowband pump pulse serves as both of the first two excitation pulses so that $\vec{k}_1 = \vec{k}_2$. With these conditions, we can write down the Feynman diagrams. At first glance one might think that all of the Feynman diagrams from both the rephasing and non-rephasing pathways should be considered, and for the diagonal peaks that is true (R_1 through R_6 in Fig. 4.3), but not all the Feynman diagrams for the cross-peaks will contribute. Those that contain interstate coherences during t_2 will not, at least not when the vibrational modes are well resolved and the pump bandwidth is narrow enough that it only spectrally overlaps with one vibrational mode at a time. As a consequence, only a subset of the Feynman diagrams in Figs. 4.10 and 4.12 will contribute since $i = j$, which are shown in Fig. 4.15(b). The peak pattern is the same as that for the rephasing 2D IR spectrum (compare Fig. 4.16 with Fig. 4.13), but an interesting consequence is that

4.4 Frequency domain 2D IR spectroscopy

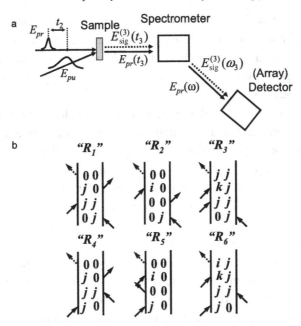

Figure 4.15 (a) Principle of the experimental setup for pump–probe 2D IR spectroscopy. (b) Feynman diagrams used to generate the narrowband pump–probe version of 2D IR spectra for any number of coupled eigenstates.

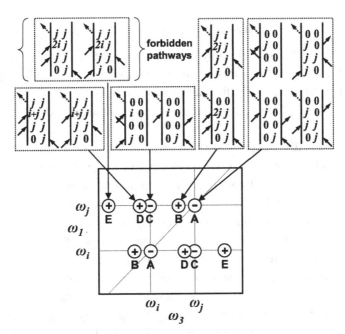

Figure 4.16 2D IR spectrum generated when data collection is done in the pump–probe geometry using a narrowband pump pulse that can selectively pump the fundamental transitions.

each peak in the narrowband 2D IR spectrum is created from an equal weighting of rephasing and non-rephasing pathways. As a result, phase twist is completely removed, to generate a spectrum that is fully absorptive for both the cross-peaks and the diagonal peaks. Of course, phase twist occurs if the peaks overlap and thus cannot be frequency resolved.

This style of collecting 2D IR spectra is often considered a frequency domain method since both axes are measured in the frequency domain (the frequency of the pump is scanned and a spectrometer is usually used to collect the emitted electric field). But one must remember that like all the third-order techniques, the macroscopic polarization is a convolution of the molecular response with the electric fields of the laser pulses. Thus, using a narrowband pump pulse degrades the time, and thus, the frequency resolution. The frequency resolution is altered the most along the ω_1-axis, because the lineshapes become convoluted with the pump pulse spectrum (in the simplest sense). The best compromise between time and frequency resolution is obtained when setting the pump pulse duration to the homogeneous linewidth of the transition. Along the ω_3-axis the linewidths resemble the intrinsic molecular lineshapes because a femtosecond pulse is still being used to measure t_3. The relationship between time domain and frequency domain 2D IR spectroscopy has been worked out in detail [24, 166].

4.5 Transient pump–probe spectroscopy

Transient pump–probe spectroscopy is the predecessor of 2D IR spectroscopy. It contains just a subset of the information in a full 2D IR data set. The experimental arrangement is the same as for the narrowband pump–probe method of collecting 2D IR spectra (Fig. 4.15a), except that a broadband femtosecond pulse is now used as the pump.

Like in the preceding sections, we start by considering which Feynman diagrams should be included in the signal. Because it is a pump–probe experiment, $\vec{k}_1 = \vec{k}_2$ so that the signals from both the rephasing and non-rephasing Feynman diagrams emerge in the direction $\vec{k}_s = \vec{k}_3$. Thus, for a system composed of a single vibrational mode, one gets absorptive peaks. And by incrementing the t_2 delay, which is the only experimentally controllable delay, one measures the population time of that mode, because the response functions depend upon e^{-t_2/T_1}.

However, the situation is not so simple for a multi-oscillator system. If the broadband pump spans multiple vibrational modes, all of the Feynman diagrams in Figs. 4.10 and 4.12 contribute, including the pathways that create the cross-peaks in 2D IR spectra. Since the pump pulse is short, the time separation between the first two field interactions is small ($t_1 = 0$ for a semi-impulsive pulse), and the spectrum is not resolved with respect to the ω_1-axis. In fact, the broadband

4.5 Transient pump–probe spectroscopy

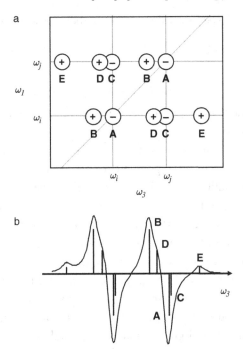

Figure 4.17 (a) Purely absorptive 2D IR spectrum of a set of two coupled oscillators, taken from Fig. 4.16. (b) The broadband pump–probe spectrum is revealed after projection of the 2D IR spectrum onto the ω_3-axis.

pump–probe response is the purely absorptive 2D IR spectrum projected onto the ω_3-axis (Fig. 4.17). This can be seen from:

$$\int_{-\infty}^{\infty} d\omega_1 R(\omega_3, t_2, \omega_1) = \int_{-\infty}^{\infty} d\omega_1 \left(\int_0^{\infty} \int_0^{\infty} dt_1 dt_3 R(t_3, t_2, t_1) e^{i\omega_1 t_1} e^{i\omega_3 t_3} \right)$$

$$= \int_0^{\infty} \int_0^{\infty} dt_1 dt_3 R(t_3, t_2, t_1) e^{i\omega_3 t_3} \delta(t_1)$$

$$= \int_0^{\infty} dt_3 R(t_3, t_2, t_1 = 0) e^{i\omega_3 t_3} \qquad (4.39)$$

which is the so-called *projection slice theorem* [62].

In particular, the cross-peak pathways are not resolved from the diagonal peaks in a transient pump–probe spectrum. Since the diagonal peaks are usually much more intense than the cross-peaks, it may be appropriate to neglect the cross-peak pathways and assign the lifetime to the diagonal peaks, but one should remember that the cross-peak lifetimes and anisotropies contribute as well (Fig. 4.17b). In fact, in principle, one can Fourier transform a transient pump–probe data set along

t_2 to generate a 2D IR spectrum that resolves the cross-peaks by their interstate coherences [86].

Exercises

4.1 Show that a linear absorption spectrum is independent of the phase of the incident light field.
4.2 For an isolated vibrator, discuss the t_2 dependence of the intensity of the 2D IR peaks if the system undergoes a chemical reaction from the first excited state so that population relaxation does not refill the ground state.
4.3 Plot the rephasing and non-rephasing spectra of a set of two coupled oscillators for $t_2 = 0$ and t_2 equal to half the interstate coherence times. How could these spectra be used to simplify the absorption spectra?
4.4 Should pump–probe spectra have signals at negative time delays? Hint: Consider a frequency resolved pump–probe experiment (Fig. 4.15a) with interchanged time ordering of pump and probe pulses (i.e. negative delay times). Assume semi-impulsive pulses. Collect the Feynman diagrams that describe this experiment for a slightly anharmonic oscillator, develop the response function and the signal as a function of the pump–probe delay time. You will have to take into account that the third-order polarization starts to emit only after the last field interaction, which is the pump pulse, and not the probe pulse. As a consequence, probe pulse and third-order polarization have a time lag when they interfere. Show that this leads to characteristic beats at negative delay times, as shown in Fig. 4.18. This effect is called a *perturbed free induction decay*.

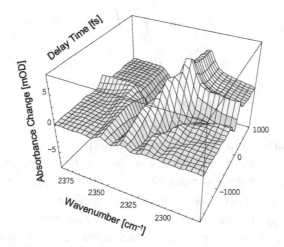

Figure 4.18 Perturbed free induction decay, measured for CO_2 in water.

4.5 Now consider a pump–probe experiment of a vibrational transition with short but finite pump and probe pulses and the pump–probe delay time set to zero. Draw the Feynman diagrams of all possible time orderings that occur during pulse overlap. The additional Feynman diagrams lead to an effect that is sometimes called a *coherence spike* or *coherence artifact*.

5
Polarization control

Polarization plays a central role in the measurement and interpretation of 2D IR spectra. In standard pump–probe spectroscopies, polarization has been used for many years to measure the rotational times of molecules or eliminate rotational motion from dynamics measurements. The polarization dependence of the diagonal peaks provide the same capabilities, but polarization can do much more in 2D and 3D spectroscopies. Recall Fig. 4.11 from Chapter 4, which is a schematic of a rephasing 2D IR spectrum with each peak labeled by its respective Feynman pathway. The objective of using polarization in 2D IR spectroscopy is to enhance or suppress particular Feynman pathways based on the relative angles of the transition dipoles. Selection is possible because each Feynman pathway has a different ordering of quantum states (e.g. $j \to j \to i \to i$ versus $j \to i \to j \to i$). Thus, the ordering of polarized pulses in a pulse sequence will scale one pathway differently from another, thereby altering the intensities and phases of the diagonal and cross-peaks. By measuring these effects, the relative angles between transition dipoles can be measured [69, 190, 202]. Angles are an extremely insightful tool for monitoring the structures of molecules, perhaps more so than actual couplings. In fact, properly polarized pulses can actually eliminate the diagonal peaks from the 2D IR spectra [201], thereby better resolving the cross-peaks, which is illustrated in Fig. 5.1. Suppressing the diagonal peaks is a particularly important capability since the diagonal peaks often obscure the much weaker cross-peaks. Polarization will play an even more important role in 3D and higher-order spectroscopies [43]. In what follows, we take a stepwise approach to polarization control by introducing key concepts one at a time.

5.1 Using polarization to manipulate the molecular response

To begin, we expand some of the equations we used in earlier chapters. We now need to explicitly take into account that the electric fields of the laser pulses are vectorial (Eq. 2.1):

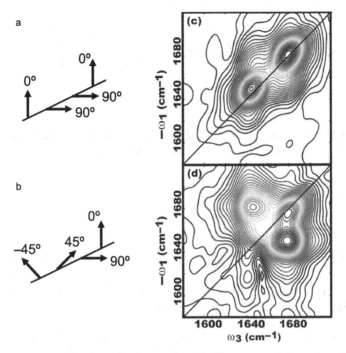

Figure 5.1 (a,c) Rephasing, absolute-value 2D IR spectra of the carbonyl stretch region of a guanosine-cytidine strand of DNA using the two sets of polarized pulses shown on the left. (b,d) This polarization condition removes the diagonal peaks from the 2D IR spectra, which better resolves the cross-peaks. The remaining intensity along the diagonal are from cross-peaks that were obscured by the diagonal peaks in the upper polarization condition. Adapted from Ref. [113] with permission.

$$\vec{E}_p = \vec{E}'_p(t) \cos(\vec{k} \cdot \vec{r} - \omega t + \phi) \tag{5.1}$$

where p indexes its polarization. The interaction energy in Eq. 2.3 is thus a dot-product rather than the product of two scalars:

$$W(t) = -\vec{\mu}_\alpha \cdot \vec{E}_a(t) \tag{5.2}$$

where we use the subscripts to signify that each pulse has a polarization a and each pulse interacts with a potentially unique transition dipole, α, which is illustrated in Fig. 5.2. Thus, we need to incorporate these dot-products into the equations from the earlier chapters. The critical equations are the perturbative expansions in Chapter 3, Eqs. 3.64 and 3.65, which were (written together and using τ's for absolute times)

$$P^{(3)}(t) \propto -i \int_{-\infty}^{t} d\tau_3 \int_{-\infty}^{\tau_3} d\tau_2 \int_{-\infty}^{\tau_2} d\tau_1 E(\tau_3) E(\tau_2) E(\tau_1) \cdot$$
$$\cdot \langle \mu(t) [\mu(\tau_3), [\mu(\tau_2), [\mu(\tau_1), \rho(-\infty)]]] \rangle \tag{5.3}$$

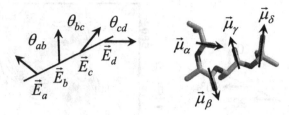

Figure 5.2 A schematic diagram of the arbitrarily polarized pulse sequence and arbitrarily oriented transition dipoles that we wish to quantify.

which becomes

$$\vec{P}^{(3)}(t) \propto -i \int_{-\infty}^{t} d\tau_3 \int_{-\infty}^{\tau_3} d\tau_2 \int_{-\infty}^{\tau_2} d\tau_1 \cdot \tag{5.4}$$
$$\cdot \left\langle \vec{\mu}_\delta(t) \left[\vec{\mu}_\gamma(\tau_3) \cdot \vec{E}_c(\tau_3), \left[\vec{\mu}_\beta(\tau_2) \cdot \vec{E}_b(\tau_2), \left[\vec{\mu}_\alpha(\tau_1) \cdot \vec{E}_a(\tau_1), \rho(-\infty) \right] \right] \right] \right\rangle.$$

In the semi-impulsive limit and for one particular term of the commutator (i.e. one particular Feynman pathway), this reduces to:

$$\vec{P}^{(3)}(t_3) \propto -i \left\langle \vec{\mu}_\delta(t_3) \left(\vec{\mu}_\gamma(t_2) \cdot \vec{E}_c(t_2) \right) \left(\vec{\mu}_\beta(t_1) \cdot \vec{E}_b(t_1) \right) \left(\vec{\mu}_\alpha(0) \cdot \vec{E}_a(0) \right) \right\rangle \tag{5.5}$$

where we have neglected to write $\rho(-\infty)$. Notice that $\vec{P}^{(3)}$ is now a vector as well because there are only three dot-products and $\vec{\mu}_\delta$ remains a vector. When a polarizer is placed in the signal beam in a homodyne measurement or the third-order polarization is overlapped with a local oscillator $\vec{E}_d$ (see e.g. Eq. 4.27) we must take another projection

$$S \propto \vec{E}_d \cdot \vec{E}_{\text{sig}} \propto i \vec{E}_d \cdot \vec{P}^{(3)} \tag{5.6}$$

so that only the component of the signal field $\vec{E}_{\text{sig}}$ onto the polarization of the local oscillator (or polarizer) results in an interference term. Another dot-product appears

$$S \propto \left\langle \left(\vec{\mu}_\delta \cdot \vec{E}_d \right) \left(\vec{\mu}_\gamma \cdot \vec{E}_c \right) \left(\vec{\mu}_\beta \cdot \vec{E}_b \right) \left(\vec{\mu}_\alpha \cdot \vec{E}_a \right) \right\rangle \tag{5.7}$$

so that we get a *four-point correlation function*. In contrast to the previous chapter, where only the eigenstate energies of the molecule influenced the signal, now the orientations of the transition dipoles and the rotational dynamics contributes as well.

How do we proceed? The transition dipoles $\vec{\mu}_n$ are a function of both vibrations and orientation in Eq. 5.7. So, we write these two properties separately as

$$W = -\vec{\mu} \cdot \vec{E} = -\left(\hat{\mu} \cdot \hat{E} \right) \mu E \tag{5.8}$$

where μ and $\hat{\mu}$ are the magnitude and direction of the transition dipole operator, respectively, and the same for the electric field (in this chapter we use hats to represent unit vectors, not operators). Then, if the rotational motions of the molecules are uncorrelated from the vibrational motions (which is usually true, but not always) [119], then the four-point correlation function separates into rotational and vibrational terms

$$S = \langle (\hat{\mu}_\alpha \cdot \hat{E}_a)(\hat{\mu}_\beta \cdot \hat{E}_b)(\hat{\mu}_\gamma \cdot \hat{E}_c)(\hat{\mu}_\delta \cdot \hat{E}_d) \rangle \langle \mu_\alpha \mu_\beta \mu_\gamma \mu_\delta \rangle E_a E_b E_c E_d. \quad (5.9)$$

The first term describes the orientational component of the signal and the second term contains everything else. In the previous chapters we ignored the influence of rotational motions and transition dipole angles on the signal strength, so that we only considered the second term in Eq. 5.9. In this chapter, we look at how the first term can be used to enhance 2D IR spectroscopy.

To use this new response, we examine the orientational four-point correlation function for each of the Feynman diagrams in our 2D IR spectrum. For example, consider the pulse sequence that gives rise to peak A in Fig. 4.11. The response is

$$S_{\text{diag}} = \langle (\hat{\mu}_\alpha \cdot \hat{E}_a)(\hat{\mu}_\alpha \cdot \hat{E}_b)(\hat{\mu}_\alpha \cdot \hat{E}_c)(\hat{\mu}_\alpha \cdot \hat{E}_d) \rangle \quad (5.10)$$

since the four laser fields all interact with the same transition dipole $\hat{\mu}_\alpha$ for a diagonal peak. For the cross-peaks in the rephasing diagrams (peak C in Fig. 4.11, for example), there are two pathways and so the polarization dependence is the sum of two four-point correlation functions

$$\begin{aligned} S_{\text{cross}} &= \langle (\hat{\mu}_\alpha \cdot \hat{E}_a)(\hat{\mu}_\alpha \cdot \hat{E}_b)(\hat{\mu}_\beta \cdot \hat{E}_c)(\hat{\mu}_\beta \cdot \hat{E}_d) \rangle \\ &+ \langle (\hat{\mu}_\alpha \cdot \hat{E}_a)(\hat{\mu}_\beta \cdot \hat{E}_b)(\hat{\mu}_\alpha \cdot \hat{E}_c)(\hat{\mu}_\beta \cdot \hat{E}_d) \rangle. \end{aligned} \quad (5.11)$$

Learning to manipulate the laser polarizations to obtain a different (and informative) polarization response from the cross-peaks than from the diagonal peaks is the objective of this chapter.

These particular Feynman pathways are only a few of the many that could be accessed for a particular pulse sequence and two coupled oscillators. The more general problem to solve is for four arbitrarily polarized pulses and four arbitrary transition dipoles. Such an approach will provide a mathematical formalism that can be applied for any 2D IR pulse sequence on any set of molecular transitions as well as be generalizable to higher-dimensional experiments. Thus, we aim to solve the situation depicted in Fig. 4.11 and the general four-point correlation function

$$S_{\text{sig}} = \langle (\hat{\mu}_\alpha \cdot \hat{E}_a)(\hat{\mu}_\beta \cdot \hat{E}_b)(\hat{\mu}_\gamma \cdot \hat{E}_c)(\hat{\mu}_\delta \cdot \hat{E}_d) \rangle. \quad (5.12)$$

This equation is also necessary to precisely calculate the polarization response of Feynman pathways that contain transitions to the overtone and combination bands,

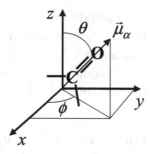

Figure 5.3 A molecule in which we consider a single transition dipole $\hat{\mu}_\alpha$ that sits in the laboratory frame.

because the transition dipole directions of these bands are not the same as the fundamental transitions due to the mixing of the eigenstates.

5.2 Diagonal peak, no rotations

Let us start with the simplest case, which is the polarization dependence of the diagonal peak. Consider the molecule in the laboratory frame shown in Fig. 5.3. Let the polarization of the four pulses be along the z-axis, $\hat{E}_a = \hat{Z}$, and we use $\hat{\alpha}$ as a short nomenclature for $\hat{\mu}_\alpha$. Since our samples are isotropically oriented, we must average over all possible molecular orientations so that the orientational component of the four-point correlation function becomes

$$\langle (\hat{Z}\cdot\hat{\alpha})(\hat{Z}\cdot\hat{\alpha})(\hat{Z}\cdot\hat{\alpha})(\hat{Z}\cdot\hat{\alpha})\rangle = \int_0^{2\pi} d\phi \int_0^{\pi} \sin\theta\, d\theta\, (\hat{Z}\cdot\hat{\alpha})^4 p_0, \quad (5.13)$$

where $p_0 = 1/4\pi$ is used to normalize the distribution since

$$\int_0^{2\pi} d\phi \int_0^{\pi} \sin\theta\, d\theta\, \frac{1}{4\pi} = 1. \quad (5.14)$$

Thus,

$$S_{\text{diag}} = \frac{1}{4\pi} \int_0^{2\pi} d\phi \int_0^{\pi} \sin\theta d\theta \cos^4\theta = \frac{1}{5}. \quad (5.15)$$

That is, in an isotropic sample the signal is only 1/5 as intense as in a sample in which all the molecules are oriented parallel to the laser pulse.

Another polarization condition that is often used when only the rephasing spectrum is being measured is $\langle (\hat{Z}\cdot\hat{\alpha})(\hat{X}\cdot\hat{\alpha})(\hat{X}\cdot\hat{\alpha})(\hat{Z}\cdot\hat{\alpha})\rangle$, for reasons which will become apparent in the next section. To solve this equation, we must project $\hat{\alpha}$ onto the x-axis, which we accomplish using the following equations:

$$(\hat{Z} \cdot \hat{\alpha}) = \cos\theta$$
$$(\hat{X} \cdot \hat{\alpha}) = \sin\theta \cos\phi$$
$$(\hat{Y} \cdot \hat{\alpha}) = \sin\theta \sin\phi \qquad (5.16)$$

which gives

$$\langle(\hat{Z}\cdot\hat{\alpha})(\hat{X}\cdot\hat{\alpha})(\hat{X}\cdot\hat{\alpha})(\hat{Z}\cdot\hat{\alpha})\rangle$$
$$= \frac{1}{4\pi} \int_0^{2\pi} d\phi \int_0^{\pi} \sin\theta d\theta \cos^2\theta \sin^2\theta \cos^2\phi = \frac{1}{15}. \qquad (5.17)$$

And since we currently are not considering molecular rotations and only one transition dipole, which two pulses are perpendicular does not matter so that

$$\langle(\hat{Z}\cdot\hat{\alpha})(\hat{Z}\cdot\hat{\alpha})(\hat{X}\cdot\hat{\alpha})(\hat{X}\cdot\hat{\alpha})\rangle = \langle(\hat{Z}\cdot\hat{\alpha})(\hat{X}\cdot\hat{\alpha})(\hat{X}\cdot\hat{\alpha})(\hat{Z}\cdot\hat{\alpha})\rangle$$
$$= \langle(\hat{Z}\cdot\hat{\alpha})(\hat{X}\cdot\hat{\alpha})(\hat{Z}\cdot\hat{\alpha})(\hat{X}\cdot\hat{\alpha})\rangle \qquad (5.18)$$

although this will not be the case when we examine the cross-peaks. Thus, the diagonal peaks decrease in intensity by a factor of 3 between parallel and perpendicularly polarized pulses. We will use this finding in the next section to help measure the relative orientations of coupled transition dipoles.

Notice that in our equations the polarizations of the pulses have to come in pairs because of spherical symmetry (e.g. $\langle ZZXX \rangle$ is allowed but not $\langle ZZZX \rangle$). That is, if there is an odd number of cosine terms then the signal integrates to zero. This is true in the dipole approximation, in which the expansion of the transition charges of the molecule are truncated at the dipole term and multipole terms are neglected. Moreover, we are ignoring signals that come from the magnetic field. In general, these approximations are well justified because such signals are usually 10^2–10^4 times smaller than the dipole terms. However, such fields contain interesting information, such as vibrational circular dichroism, although we do not address such measurements in this book [91, 156].

5.3 Cross-peaks and orientations of coupled transition dipoles

Having looked at the basic polarization dependence of the diagonal peaks, let us continue with the cross-peaks, with the aim of using them to measure the relative angles between two coupled transition dipoles. Once again, let us neglect rotational motion (or other processes such as energy transfer which also "rotates" the transition dipole moment). Consider a cross-peak that appears between two transition dipoles α and β of a rigid molecule. The two vectors are illustrated in Fig. 5.4. As above, the calculation involves projecting the transition dipoles onto the respective laboratory axes, so that, for instance:

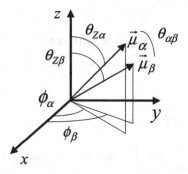

Figure 5.4 A molecule in which we consider two coupled transition dipoles, $\hat{\mu}_\alpha$ and $\hat{\mu}_\beta$ that sit in the laboratory frame.

$$\langle(\hat{Z}\cdot\hat{\alpha})(\hat{Z}\cdot\hat{\alpha})(\hat{Z}\cdot\hat{\beta})(\hat{Z}\cdot\hat{\beta})\rangle = \frac{1}{4\pi}\int d\Omega \cos^2\theta_{Z\alpha} \cos^2\theta_{Z\beta} \quad (5.19)$$

where $\int d\Omega$ is once again the integral of the molecule over all possible orientations. However, we cannot integrate over $\theta_{Z\alpha}$ and $\theta_{Z\beta}$ separately, because rotation of one transition dipole alters the other since they are connected by the structure of the molecule. That is, they need to be rotated together. We present two ways to solve this problem.

The first method is to place the two transition dipoles into the molecular frame and then to rotationally average the molecular frame with regard to the laboratory frame. *Euler angles* can be used to rotate one frame against another. As long as we use the same rotation matrices for both vectors, then they will retain their proper relative orientations. The procedure is as follows.

$\hat{\alpha}$ and $\hat{\beta}$ have a fixed relationship to each other, but have an arbitrary orientation in 3D space. Let us start by having the molecular and laboratory frames aligned and arbitrarily place $\hat{\alpha}$ along the z-axis and $\hat{\beta}$ in the (x,z)-plane (in the molecular frame it does not matter how the molecule sits since everything is going to be rotated), which gives

$$\hat{\alpha} = (0, 0, 1)$$
$$\hat{\beta} = (\sin\theta_{\alpha\beta}, 0, \cos\theta_{\alpha\beta}) \quad (5.20)$$

when $\hat{\beta}$ is expressed using the relative angle between the transition dipoles and the coordinates are (x,y,z). To rotate one frame against the other, one can use Euler angles (ϕ, θ, ψ) and rotation matrices [129]. That is, one performs three counterclockwise rotations, first through an angle ϕ around z, then the angle θ around x, and finally ψ around z. This is done with three rotation matrices, which together give

5.3 Cross-peaks and orientations

$$R(\phi, \theta, \psi) = \qquad (5.21)$$

$$\begin{pmatrix} \cos\psi & \sin\psi & 0 \\ -\sin\psi & \cos\psi & 0 \\ 0 & 0 & 1 \end{pmatrix} \begin{pmatrix} 1 & 0 & 0 \\ 0 & \cos\theta & \sin\theta \\ 0 & -\sin\theta & \cos\theta \end{pmatrix} \begin{pmatrix} \cos\phi & \sin\phi & 0 \\ -\sin\phi & \cos\phi & 0 \\ 0 & 0 & 1 \end{pmatrix}.$$

So, once rotated, $\hat{\alpha}$ and $\hat{\beta}$ become

$$\hat{\alpha}' = R(\phi, \theta, \psi)\hat{\alpha} = (\sin\psi \sin\theta, \cos\psi \sin\theta, \cos\theta)$$

$$\hat{\beta}' = R(\phi, \theta, \psi)\hat{\beta}$$
$$= (\sin\theta_{\alpha\beta}[\cos\psi \cos\phi - \cos\theta \sin\phi \sin\psi] + \cos\theta_{\alpha\beta} \sin\psi \sin\theta,$$
$$\sin\theta_{\alpha\beta}[-\sin\psi \cos\phi - \cos\theta \sin\phi \cos\psi] + \cos\theta_{\alpha\beta} \cos\psi \sin\theta,$$
$$\sin\theta_{\alpha\beta} \sin\theta \sin\phi + \cos\theta_{\alpha\beta} \cos\theta). \qquad (5.22)$$

We can now average the molecular frame relative to the laboratory frame by integrating over ϕ, θ, ψ. Thus, Eq. 5.19 becomes

$$\langle (\hat{Z}\cdot\hat{\alpha})(\hat{Z}\cdot\hat{\alpha})(\hat{Z}\cdot\hat{\beta})(\hat{Z}\cdot\hat{\beta})\rangle = p_0' \int d\Omega' \cos^2\theta_{Z\alpha} \cos^2\theta_{Z\beta} \qquad (5.23)$$

$$= p_0' \int d\Omega' \cos^2\theta (\sin\theta_{\alpha\beta} \sin\theta \sin\phi + \cos\theta_{\alpha\beta} \cos\theta)^2$$

where $p_0' = 1/8\pi^2$ is the normalization constant for the Euler integrals which are $\int d\Omega' = \int_0^{2\pi} d\psi \int_0^{\pi} \sin\theta d\theta \int_0^{2\pi} d\phi$. In the second line of Eq. 5.23, we have substituted the corresponding elements from the rotated vectors in Eq. 5.22. This concept of rotating one frame against another is useful in understanding how one arrives at the general polarization response of arbitrarily oriented transition dipoles and arbitrarily polarized laser pulses, which we present below.

The second and more elegant method is to use spherical harmonics, which simplifies the algebra and will become useful later in this chapter when we consider rotational dynamics. Since the transition dipole geometries are ultimately dictated by the structure of the molecule that we know or want to measure, if we express the relative angle between the transition dipoles, $\theta_{\alpha\beta}$, as a function of $\theta_{Z\alpha}$ and $\theta_{Z\beta}$, then we can replace $\theta_{Z\beta}$ in Eq. 5.19 so that we only need to integrate over $\theta_{Z\alpha}$. A convenient method for relating these three angles to one another is to use spherical harmonics $Y_{\ell,m}(\theta, \phi)$ (see Appendix D). The addition theorem for spherical harmonics relates the relative angle θ between two vectors with coordinates θ_1, ϕ_1 and θ_2, ϕ_2

$$P_\ell(\cos\theta) = \frac{4\pi}{2\ell+1} \sum_{m=-\ell}^{+\ell} (-1)^m Y_{\ell,m}(\theta_1, \phi_1) Y_{\ell,-m}(\theta_2, \phi_2) \qquad (5.24)$$

where P_ℓ are the Legendre polynomials. Since $P_1(\cos\theta) = \cos\theta$ (Appendix D), by choosing $\ell = 1$ we get

$$\cos\theta_{Z\beta} = \frac{4\pi}{3}[-Y_{1,-1}(\theta_{Z\alpha},\phi_{Z\alpha})Y_{1,1}(\theta_{\alpha\beta},\phi_{\alpha\beta}) + Y_{1,0}(\theta_{Z\alpha},\phi_{Z\alpha})Y_{1,0}(\theta_{\alpha\beta},\phi_{\alpha\beta})$$
$$-Y_{1,1}(\theta_{Z\alpha},\phi_{Z\alpha})Y_{1,-1}(\theta_{\alpha\beta},\phi_{\alpha\beta})] \quad (5.25)$$

which can be substituted into Eq. 5.19 so that the entire integral depends on $\theta_{Z\alpha}$ and the relative (molecular frame) geometry of $\theta_{\alpha\beta}$. The integral can be calculated directly on the spherical harmonics (see Appendix D) or by converting them into spherical sine and cosine terms (Problem 5.1).

Whether or not one uses spherical harmonics or Euler angles to solve Eq. 5.19, one gets

$$\langle(\hat{Z}\cdot\hat{\alpha})(\hat{Z}\cdot\hat{\alpha})(\hat{Z}\cdot\hat{\beta})(\hat{Z}\cdot\hat{\beta})\rangle = (4P_2+5)/45$$
$$= \frac{1}{15}(2\cos^2\theta_{\alpha\beta}+1). \quad (5.26)$$

But this equation is not enough. When measuring a rephasing 2D IR spectrum, there are two Feynman pathways that contribute to the cross-peaks (see Eq. 5.11 and Fig. 4.11), and so we also need to calculate $\langle(\hat{Z}\cdot\hat{\alpha})(\hat{Z}\cdot\hat{\beta})(\hat{Z}\cdot\hat{\alpha})(\hat{Z}\cdot\hat{\beta})\rangle$. Since all of our pulses have the same polarization, no extra work is needed because we can rearrange the pulse ordering in Eq. 5.19 without changing the integral (which will not necessarily be possible when we consider rotational dynamics below). Thus, we can write

$$\langle(\hat{Z}\cdot\hat{\alpha})(\hat{Z}\cdot\hat{\alpha})(\hat{Z}\cdot\hat{\beta})(\hat{Z}\cdot\hat{\beta})\rangle = \langle(\hat{Z}\cdot\hat{\alpha})(\hat{Z}\cdot\hat{\beta})(\hat{Z}\cdot\hat{\alpha})(\hat{Z}\cdot\hat{\beta})\rangle$$
$$= \langle(\hat{Z}\cdot\hat{\alpha})(\hat{Z}\cdot\hat{\beta})(\hat{Z}\cdot\hat{\beta})(\hat{Z}\cdot\hat{\alpha})\rangle. \quad (5.27)$$

Equation 5.26 is a mathematical formula that scales the intensity of the cross-peak for the $\langle ZZZZ\rangle$ polarization condition, which depends upon $\theta_{\alpha\beta}$. In order to actually measure the angle $\theta_{\alpha\beta}$, one needs in addition an orthogonally polarized pulse sequence like $\langle ZXXZ\rangle$ for the cross-peaks. These responses can be derived using the procedures outlined above, but once again the algebra is quite tedious. An alternative method is to use a tensor approach to derive a general equation, which Hochstrasser published a number of years ago [95], and is

$$\langle(\hat{\alpha}\cdot\hat{E}_a)(\hat{\beta}\cdot\hat{E}_b)(\hat{\gamma}\cdot\hat{E}_c)(\hat{\delta}\cdot\hat{E}_d)\rangle \quad (5.28)$$
$$= \frac{1}{30}\{\cos\theta_{\alpha\beta}\cos\theta_{\gamma\delta}(4\cos\theta_{ab}\cos\theta_{cd} - \cos\theta_{ac}\cos\theta_{bd} - \cos\theta_{ad}\cos\theta_{bc})$$
$$+ \cos\theta_{\alpha\gamma}\cos\theta_{\beta\delta}(4\cos\theta_{ac}\cos\theta_{bd} - \cos\theta_{ab}\cos\theta_{cd} - \cos\theta_{ad}\cos\theta_{bc})$$
$$+ \cos\theta_{\alpha\delta}\cos\theta_{\beta\gamma}(4\cos\theta_{ad}\cos\theta_{bc} - \cos\theta_{ab}\cos\theta_{cd} - \cos\theta_{ac}\cos\theta_{bd})\}$$

5.3 Cross-peaks and orientations

Table 5.1 *Polarization response for a given pathway (e.g. jjjj) and polarization condition (e.g. ZZZZ) for two transition dipoles separated by $\theta_{\alpha\beta}$ where $P_2 = \frac{1}{2}(3\cos^2\theta_{\alpha\beta} - 1)$. Modified from Ref. [95].*

Pathway	ZZZZ	ZZXX	ZXZX	ZXXZ
jjjj	1/5	1/15	1/15	1/15
jiji	$(4P_2+5)/45$	$P_2/15$	$(5-2P_2)/45$	$P_2/15$
jjii	$(4P_2+5)/45$	$(5-2P_2)/45$	$P_2/15$	$P_2/15$
jiij	$(4P_2+5)/45$	$P_2/15$	$P_2/15$	$(5-2P_2)/45$

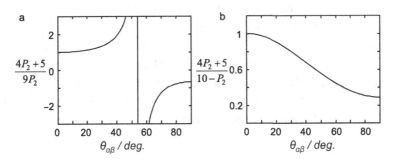

Figure 5.5 The signal strength of the ratio of the cross-peaks measured with (a) $\langle ZZZZ\rangle/3\langle ZXXZ\rangle$ polarizations for a rephasing spectrum and (b) $\langle ZZZZ\rangle/3\langle ZZXX\rangle$ for a collinear absorptive spectrum.

where the polarizations of the laser pulses are now written relative to one another just as the transition dipoles are. In Section 5.6 we rewrite this equation in a more compact form and extend it to higher-order pulse sequences. Table 5.1 contains the signal strengths for two coupled transition dipoles i and j for the irreducible polarization tensors, which includes the two other equations that we need:

$$\langle(\hat{Z}\cdot\hat{\alpha})(\hat{X}\cdot\hat{\alpha})(\hat{X}\cdot\hat{\beta})(\hat{Z}\cdot\hat{\beta})\rangle = \langle(\hat{Z}\cdot\hat{\alpha})(\hat{X}\cdot\hat{\beta})(\hat{X}\cdot\hat{\alpha})(\hat{Z}\cdot\hat{\beta})\rangle$$
$$= P_2/15$$
$$= \frac{1}{30}(3\cos^2\theta_{\alpha\beta} - 1). \tag{5.29}$$

We now have all the equations we need to describe the signal strength of the cross-peaks and diagonal peaks when the rephasing spectra are measured with $\langle ZZZZ\rangle$ and $\langle ZXXZ\rangle$ polarizations. Using these equations, we can determine $\theta_{\alpha\beta}$ from the polarization dependence of the cross-peaks relative to the diagonal peaks. For instance, if we multiply the spectrum collected with $\langle ZXXZ\rangle$ by a factor of 3, then the diagonal peaks of $\langle ZXXZ\rangle$ and $\langle ZZZZ\rangle$ are equal. However, the ratio of the cross-peaks depends on $\theta_{\alpha\beta}$, which produces the function shown in Fig. 5.5(a):

$$\frac{S_{ZZZZ}}{3S_{ZXXZ}} = \frac{4P_2 + 5}{9P_2}. \tag{5.30}$$

Notice that for angles larger than the *magic angle* of $\theta_{\alpha\beta} = 54.7°$, the function becomes negative, indicating that the cross-peaks have flipped sign in the 2D IR spectra of $\langle ZXXZ \rangle$. Thus, one can collect $\langle ZZZZ \rangle$ and $\langle ZXXZ \rangle$ rephasing 2D IR spectra for a given molecule, scale the one to the other using the diagonal peaks (the scaling factor should be about 3), and then use the ratio of the cross-peak intensities to extract the relative angle between the transition dipoles using Fig. 5.5(a). Methods like this are also finding use in 2D electronic spectroscopy for measuring the angles between coupled electronic transition dipoles [153].

In the example above we have focused on 2D IR spectra collected using a rephasing pulse sequence. Similar ratios can be derived for the other types of 2D IR spectra. For absorptive 2D IR spectra, one must also include the non-rephasing diagrams (Fig. 4.13), which have the same polarization dependence as in Fig. 5.5(a). If one instead decides to use $\langle ZZZZ \rangle$ and $\langle ZZXX \rangle$ polarizations to measure $\theta_{\alpha\beta}$ with an absorptive 2D IR spectrum, then one gets the ratio

$$\frac{S_{ZZZZ}}{3S_{ZZXX}} = \frac{4P_2 + 5}{10 - P_2} \tag{5.31}$$

which is plotted in Fig. 5.5(b). This ratio is not as sensitive to $\theta_{\alpha\beta}$ because the pathways for the cross-peaks destructively interfere (which is why we chose $\langle ZXXZ \rangle$ to begin with). One can also derive a ratio for 2D IR spectra measured in a narrowband pump experiment, which we save for Problem 5.2, although the result is very similar to Eq. 5.31.

It is important to make three comments on using these relations to extract accurate angles. First, we note that the cross-peaks need to be well separated from the diagonal peaks for accurate angle measurements, lest the polarization dependence of the diagonal peaks perturb the measurement. The diagonal peaks are usually more intense than the cross-peaks, and so it is often best to simulate the 2D IR spectra, including both the diagonal and cross-peaks, to better extract the angular dependence. Second, we have not (so far) considered rotational motion. If the molecule rotates during the laser pulse sequence, then this must be considered as well in the equations. Finally, if population or coherence transfer occurs to another mode (see Chapter 8), then the combination band of the new transition dipole probably absorbs in the same frequency range. If it does, then the measured transition angle will actually be to the new mode, not the mode of interest. One way of taking into account the last two points is to measure the polarization dependence as a function of the t_2 time, and then extrapolate the results back to $t_2 = 0$.

5.4 Combining pulse polarizations: Eliminating diagonal peaks

Another use for polarizations is to eliminate or enhance features in the 2D IR spectra by adding and/or subtracting various combinations of 2D IR spectra. For example, in the sections above we learned that the diagonal peaks decrease by a factor of 3 between $\langle ZZZZ \rangle$ and $\langle ZZXX \rangle$, but the cross-peaks will decrease by less if $\theta_{\alpha\beta} \neq 0$. Thus, if one subtracts the two spectra according to:

$$S = \langle ZZZZ \rangle - 3\langle ZZXX \rangle \tag{5.32}$$

then the diagonal peaks will disappear and leave only those cross-peaks that have nonzero θ_{ab} [190]. This is an important ability, because it enables one to eliminate the diagonal peaks to reveal the cross-peaks, which are oftentimes obscured by the more intense diagonal features.

Instead of measuring two independent spectra and manually subtracting them, one can instead choose pulse polarizations that cause the emitted electric fields to interfere so that the measured signal is already a linear combination of orientational responses. For instance, to subtract the diagonal peaks, one could also measure [201]

$$S(-45°, +45°, 0°, 90°) \equiv \frac{1}{2} \langle (Z-X)(Z+X)ZX \rangle$$
$$= \frac{1}{2} \langle ZXZX \rangle - \langle XZZX \rangle \tag{5.33}$$

where the angles are the polarizations of the pulses in the laboratory frame. This method works because the diagonal peaks have the same intensity in both $\langle ZXZX \rangle$ and $\langle XZZX \rangle$, while the cross-peaks do not, unless they happen to be exactly parallel ($\theta_{\alpha\beta}=0$). Thus, the measured 2D IR spectrum contains only cross-peaks from nonparallel transition dipoles, which is illustrated in Fig. 5.1.

Are there other linear combinations that might be useful for enhancing or reducing particular features in the 2D IR spectra? Probably so. Each pathway has a different combination of transition dipoles and so it should be possible to selectively enhance or suppress individual pathways. Table 5.2 lists several possibly useful pulse combinations. However, it may be difficult to analytically recognize the best polarizations. Mukamel has predicted through calculations that complicated polarization shaped pulses might be very useful for improving the spectral resolution of 2D IR spectroscopy [187], which initial experiments have validated [133]. Regardless of whether the best polarization is analytically derived or not, due to the isotropic symmetry of bulk samples, there are only three independent polarization conditions that are all related by the equation:

$$\langle ZZZZ \rangle = \langle ZZXX \rangle + \langle ZXZX \rangle + \langle ZXXZ \rangle. \tag{5.34}$$

Table 5.2 *Potentially useful pulse sequence polarizations. "Circ" is for circular polarization.*

$\hat{E}_a$	$\hat{E}_b$	$\hat{E}_c$	$\hat{E}_d$	Polarization combination
0	$\pi/4$	$-\pi/4$	0	$\frac{1}{2}(\langle ZZXX\rangle + \langle ZXZX\rangle)$
$\pi/3$	$-\pi/3$	0	0	$\frac{1}{4}(\langle ZZZZ\rangle - 3\langle ZZXX\rangle)$
θ	$-\theta$	$\pi/2$	0	$\frac{1}{2}\sin 2\theta(\langle ZXZX\rangle - \langle ZXXZ\rangle)$
circ	circ	0	0	$\frac{1}{2}(\langle ZZZZ\rangle + \langle ZZXX\rangle)$

Thus, if one measured three of these polarizations, one would know everything there is about the orientational response of the system and so one could build linear combinations to mathematically enhance or suppress particular peaks.

However, when considering which polarized spectra to combine, one should also account for dynamics (such as rotations) that occur during the laser pulse sequences. For example, to eliminate the diagonal peaks, ideally one would subtract spectra that have equivalent dynamics during the time delays [201]. We will find in the next section that the rotational dynamics are not the same for $\langle ZZZZ\rangle$ and $\langle ZZXX\rangle$, and thus subtraction using the method in Eq. 5.32 cannot perfectly remove the diagonal peaks. However, the rotational dynamics of $\langle ZXZX\rangle$ and $\langle XZZX\rangle$ are identical, and so the method in Eq. 5.33 improves their elimination. Moreover, since both electric fields are being emitted from the sample simultaneously, laser noise in the two spectra is well correlated. Thus, the overall subtraction is typically of higher quality when done *in situ*.

5.5 Including (or excluding) rotational motions

The above sections apply to molecules in which the rotational motions of the molecule are much slower than the longest time delays needed to measure the spectra, in which case they can be neglected. However, at least for very small molecules, rotational motion significantly contributes to homogeneous dephasing. In fact, polarization is often used to measure the rotational diffusion times of molecules. In this section, we include a basic description of rotational motion that allows one to incorporate rotational diffusion times into the molecular response functions above.

The essential problem is that we need to allow the molecules to rotate in between the laser pulses, which means that we need to pay attention to the time ordering in the four-point correlation function:

5.5 Including (or excluding) rotational motions

$$\langle (\hat{\mu}_\alpha \cdot \hat{E}_a(0)) \, (\hat{\mu}_\beta \cdot \hat{E}_b(t_1)) \, (\hat{\mu}_\gamma \cdot \hat{E}_c(t_2)) \, (\hat{\mu}_\delta \cdot \hat{E}_d(t_3)) \rangle. \tag{5.35}$$

Let us consider how this correlation function evolves. Before a laser pulse interacts with the sample, the ensemble is randomly oriented and thus we use the same normalization constant as in Eq. 5.14

$$t = -\infty \quad \Longrightarrow \quad p_0(\Omega_0) = 1/4\pi \tag{5.36}$$

where Ω_0 represents the angles of the distribution of molecules p_0. At time $t = 0$, the first pulse then operates on this distribution so that

$$t = 0 \quad \Longrightarrow \quad \int d\Omega_0 (\hat{E}_a(0) \cdot \hat{\alpha}) p_0(\Omega_0) \tag{5.37}$$

where $\int d\Omega = \int_0^{2\pi} d\phi \int_0^\pi \sin\theta d\theta$. This integral weights the contribution of each molecule to the signal strength according to the direction it points with respect to the electric field. After the laser pulse the new distribution will evolve in time because the molecules will rotate (neglecting rotation during the laser pulse). For now, let us abstractly describe the rotation by a time propagator $G(\Omega_n t_n | \Omega_m)$ that allows the angle of a molecule to evolve from Ω_m to Ω_n during the time interval t_n, which is when the next pulse arrives. Thus, right after the second pulse at $t = t_1$ we have

$$t = t_1 \quad \Longrightarrow \quad \int d\Omega_1 \int d\Omega_0 (\hat{E}_b(t_1) \cdot \hat{\beta}) G(\Omega_1 t_1 | \Omega_0) (\hat{E}_a(0) \cdot \hat{\alpha}) p_0(\Omega_0) \tag{5.38}$$

a new distribution Ω_1 that $\hat{E}_b(t_1)$ operates on (which could be the same transition dipole or a different one). This process continues for each relative time in the pulse sequence until the sample emits, which means that we write the full correlation function as [69]

$$\langle (\hat{\alpha} \cdot \hat{E}_a(0)) \, (\hat{\beta} \cdot \hat{E}_b(t_1)) \, (\hat{\gamma} \cdot \hat{E}_c(t_2)) \, (\hat{\delta} \cdot \hat{E}_d(t_3)) \rangle$$
$$= \int d\Omega_3 \int d\Omega_2 \int d\Omega_1 \int d\Omega_0 (\hat{E}_d(t_3) \cdot \hat{\delta}) G(\Omega_3 t_3 | \Omega_2) \tag{5.39}$$
$$(\hat{E}_c(t_2) \cdot \hat{\gamma}) G(\Omega_2 t_2 | \Omega_1) (\hat{E}_b(t_1) \cdot \hat{\beta}) G(\Omega_1 t_1 | \Omega_0) (\hat{E}_a(0) \cdot \hat{\alpha}) p_0(\Omega_0).$$

To describe the rotational motion, we need to create a model for $G(\Omega_n t_n | \Omega_m)$ and then plug it into Eq. 5.39.

Anyone who has studied molecular quantum mechanics knows that rotational dynamics of molecules can be simple or extraordinarily complicated depending on the molecular symmetry. Thin linear molecules and spherically symmetric molecules can be described by a single moment of inertia, and so need just one rotational constant; oblate and prolate tops require two moments of inertia; and nonsymmetric molecules require three, which causes very complicated rotational motion. Moreover, if we are probing more than one transition dipole on a single molecule, then the rotational motions of the dipoles are probably correlated. These aspects are beyond the scope of this book. Indeed, in condensed-phase spectroscopy in which the linewidths are usually dominated by vibrational and not by rotational decoherence, rotational motion is usually approximated as that of a single moment of inertia, which is also the approach we take here. Thus, the description that follows is only rigorously valid for linear and spherical molecules.

This section is based on the excellent book by Berne and Pecora [12]. Consider that our molecule is a rigid rod which has only one moment of inertia, like a diatomic molecule. In a condensed-phase system, it will experience collisions due to its surrounding environment, which will change the direction in which the rod points. If the collisions occur frequently enough that only a small change in θ and ϕ occurs between collisions, then the angular motion will perform a random walk in spherical coordinates. Thus, after a laser pulse interacts with the sample, the ensemble of transition dipoles that are excited by the laser pulse will all point in the same direction, so that all molecules have the same θ and ϕ in spherical coordinates. But as time progresses, each molecule will undergo a different random walk. This will lead to a diffusion of pointing directions so that ultimately, the ensemble of molecules will all be pointing in different directions, which will create a uniform distribution in spherical coordinates. Thus, in order to describe the rotational motion, we will use statistical mechanics, and treat the randomization of the transition dipole directions as diffusion in spherical coordinates. In that case, the orientation of a transition dipole can be modeled with a rotational diffusion equation

$$\frac{\partial G(\Omega, t)}{\partial t} = -D\hat{J}^2 G(\Omega, t). \tag{5.40}$$

In this equation, which is known as the Debye equation, D is the rotational diffusion constant, $\hat{J}$ is the angular momentum operator, and Ω is the direction in which the transition dipole points in spherical coordinates. This equation is similar to the Schrödinger equation for a rigid rotor, so it will have the same set of eigenfunctions.[1] Regardless, the eigenfunctions of the Debye equation are the spherical harmonics $Y_{\ell m}(\theta\phi) \equiv Y_{\ell m}(\Omega)$

$$\hat{J}^2 Y_{\ell m}(\Omega) = \ell(\ell+1) Y_{\ell m}(\Omega) \tag{5.41}$$

that form an orthonormal basis set

$$\int d\Omega \, Y_{\ell'm'}(\Omega) Y^*_{\ell m}(\Omega) = \delta_{\ell'\ell}\delta_{m'm} \tag{5.42}$$

---

[1] But unlike the Schrödinger equation, the Debye equation is real, not complex. Thus, it has exponential decays rather than oscillatory eigenfunctions.

and so have the closure property

$$\delta(\Omega - \Omega_0) = \sum_{\ell=0}^{\infty} \sum_{m=-\ell}^{+\ell} Y_{\ell m}(\Omega_0) Y_{\ell m}^*(\Omega)$$
$$\equiv \sum_{\ell m} Y_{\ell m}(\Omega_0) Y_{\ell m}^*(\Omega). \tag{5.43}$$

The solution to the Debye equation is

$$G(\Omega, t) = e^{-tD\hat{J}^2} G(\Omega, 0) \tag{5.44}$$

which becomes

$$G(\Omega, t) = e^{-tD\hat{J}^2} \sum_{\ell m} Y_{\ell m}(\Omega_0) Y_{\ell m}^*(\Omega) \tag{5.45}$$

when we use the initial condition $G(\Omega, 0) = \delta(\Omega - \Omega_0)$ from Eq. 5.43. We can now operate $\hat{J}^2$ onto the eigenstates $Y_{\ell m}$ using Eq. 5.41, to get the final equation

$$G(\Omega, t) = \sum_{\ell m} e^{-\ell(\ell+1)Dt} Y_{\ell m}(\Omega_0) Y_{\ell m}^*(\Omega). \tag{5.46}$$

$G(\Omega, t)$ is the distribution we would get after time t if we had started out with perfectly oriented molecules in direction Ω_0. In other words, $G(\Omega, t)$ is the time-propagator that we want for Eq. 5.39

$$G(\Omega, t) \equiv G(\Omega, t|\Omega_0). \tag{5.47}$$

If the initial distribution is not perfectly oriented, then we must integrate over all the starting orientations, which is why we have an integral for each laser pulse.

Let us use these results to calculate the orientation contribution to the macroscopic polarization of a linear spectrum of a single transition dipole. In other words, let us calculate

$$\langle (\hat{\alpha} \cdot \hat{E}_a(0))(\hat{\alpha} \cdot \hat{E}_a(t_1)) \rangle$$
$$= \int d\Omega_1 \int d\Omega_0 (\hat{E}_a(t_1) \cdot \hat{\alpha}) G(\Omega_1 t_1 | \Omega_0)(\hat{E}_a(0) \cdot \hat{\alpha}) p_0(\Omega_0). \tag{5.48}$$

Appendix D contains a few helpful spherical harmonics, which allows us to write the projections in Eq. 5.16 in spherical harmonics

$$(\hat{Z} \cdot \hat{\alpha}) = \cos\theta$$
$$= \left(\frac{4\pi}{3}\right)^{1/2} Y_{10}(\Omega)$$

$$(\hat{X} \cdot \hat{\alpha}) = \sin\theta \cos\phi$$
$$= -\frac{1}{2}\left(\frac{8\pi}{3}\right)^{1/2}(Y_{11}(\Omega) - Y_{1,-1}(\Omega))$$
$$(\hat{Y} \cdot \hat{\alpha}) = \sin\theta \sin\phi$$
$$= \frac{i}{2}\left(\frac{8\pi}{3}\right)^{1/2}(Y_{11}(\Omega) + Y_{1,-1}(\Omega)). \tag{5.49}$$

Since the two pulses are identical, we can choose any of these projections, although $(\hat{Z} \cdot \hat{\alpha})$ is easiest to solve analytically. Substituting into Eq. 5.48 and solving each integral one at a time, we get

$$\langle(\hat{\alpha} \cdot \hat{E}_a(0))(\hat{\alpha} \cdot \hat{E}_a(t_1))\rangle \tag{5.50}$$
$$= p_0 \int d\Omega_1 \int d\Omega_0 (\hat{E}(t_1) \cdot \hat{\alpha}) \sum_{\ell m} e^{-\ell(\ell+1)Dt_1} Y_{\ell m}^*(\Omega_1) Y_{\ell m}(\Omega_0) \left(\frac{4\pi}{3}\right)^{1/2} Y_{10}(\Omega).$$

We make use of the fact that the spherical harmonics are orthonormal

$$\int d\Omega_0 Y_{\ell m}(\Omega_0) Y_{10}(\Omega_0) = \delta_{\ell 1}\delta_{m0} \tag{5.51}$$

which makes the integrals disappear and we get the result after substituting in the second $(\hat{Z} \cdot \hat{\alpha})$ to get

$$\frac{1}{4\pi}\left(\frac{4\pi}{3}\right)^{1/2} e^{-2Dt_1} \int d\Omega_1 \left(\frac{4\pi}{3}\right)^{1/2} Y_{10}(\Omega_1) Y_{10}^*(\Omega_1)$$
$$= \frac{1}{3}e^{-2Dt_1}. \tag{5.52}$$

Thus, the rotational contribution to homogeneous dephasing of the density matrix after the first pulse is e^{-2Dt_1}. In fact, for the R_1 to R_6 response functions, rotation motion causes the signal to decay as e^{-2Dt} whenever the system is in a coherence state. For the typical rephasing and non-rephasing third-order pulse sequences in 2D IR spectroscopy there are two coherence times and one population time (see Chapter 4). The rotational contribution to the decay of the diagonal terms for a linear or spherical molecule is e^{-6Dt}, which can be calculated by including another pulse in the sequence to generate a population state $(\hat{E}_a(t_2) \cdot \hat{\alpha})G(\Omega_2 t_2|\Omega_1)$. And by including a fourth pulse, the full orientational response for a 2D IR pulse sequence with two coherence and one population decay can be computed with the formalism described above (Problem 5.3), and gives

$$\langle(\hat{\alpha} \cdot \hat{E}_a(0))(\hat{\alpha} \cdot \hat{E}_a(t_1))(\hat{\alpha} \cdot \hat{E}_a(t_2))(\hat{\alpha} \cdot \hat{E}_a(t_3))\rangle$$
$$= \frac{1}{9}e^{-2Dt_1}e^{-2Dt_3}\left[1 + \frac{4}{5}e^{-6Dt_2}\right]. \tag{5.53}$$

Table 5.3 *Table of polarization factors that include rotational decays for Feynman pathways of a two-oscillator system with transition dipoles i and j. $P_2 = \frac{1}{2}(3\cos^2\theta - 1)$. Modified from Ref. [95].*

quantity × $e^{-2D(t_1+t_3)}$	jjjj	jiji	jjii	jiij
$\langle ZZZZ \rangle$	$\frac{1}{45}[4e^{-6Dt_2}+5]$	$\frac{1}{27}[2P_2+1+\frac{2}{5}e^{-6Dt_2}[P_2+5]]$	$\frac{1}{45}[4P_2e^{-6Dt_2}+5]$	$\langle Z_jZ_iZ_jZ_i \rangle$
$\langle ZXZX \rangle$	$\frac{1}{15}e^{-6Dt_2}$	$\frac{1}{90}e^{-6Dt_2}[P_2+5]+\frac{1}{18}[1-P_2]e^{-2Dt_2}$	$\frac{1}{15}e^{-6Dt_2}P_2$	$\langle Z_jZ_iZ_jZ_i \rangle$
$\langle ZZXX \rangle$	$\frac{1}{45}[5-2e^{-6Dt_2}]$	$\frac{1}{27}[2P_2+1-\frac{1}{5}e^{-6Dt_2}[P_2+5]]$	$\frac{1}{45}[5-2P_2e^{-6Dt_2}]$	$\langle Z_jZ_iZ_jZ_i \rangle$
$\langle ZXXZ \rangle$	$\frac{1}{15}e^{-6Dt_2}$	$\frac{1}{90}e^{-6Dt_2}[P_2+5]-\frac{1}{18}[1-P_2]e^{-2Dt_2}$	$\frac{1}{15}e^{-6Dt_2}P_2$	$\langle Z_jZ_iZ_jZ_i \rangle$

When $t_1 = t_3 = 0$, as it does in pump–probe experiments, Eq. 5.53 reduces to

$$\langle (\hat{\alpha} \cdot \hat{E}_a(0))\,(\hat{\alpha} \cdot \hat{E}_a(0))\,(\hat{\alpha} \cdot \hat{E}_a(t_2))\,(\hat{\alpha} \cdot \hat{E}_a(t_2)) \rangle$$
$$= \frac{1}{9} + \frac{4}{45}e^{-6Dt_2}. \tag{5.54}$$

Working out the integrals for more general polarizations is quite tedious. Table 5.3 gives the results for common polarizations and Feynman pathways, which were originally written by Hochstrasser [95].

We end this section by using these equations to demonstrate how one measures population relaxation without the complications of rotational motions. The traditional method to do this is to perform a transient pump–probe measurement with the pump and probe pulses oriented at the magic angle ($\theta = 54.7°$) to one another. If we only consider the diagonal peaks, then we need to calculate

$$\langle (\hat{\alpha} \cdot \hat{Z}(0))\,(\hat{\alpha} \cdot \hat{Z}(0))\,(\hat{\alpha} \cdot \hat{M}(t_2))\,(\hat{\alpha} \cdot \hat{M}(t_2)) \rangle$$
$$= \langle (\hat{\alpha} \cdot \hat{Z}(0))\,(\hat{\alpha} \cdot \hat{Z}(0))\,(\hat{\alpha} \cdot \hat{Z}(t_2))\,(\hat{\alpha} \cdot \hat{Z}(t_2)) \rangle \cos^2\theta$$
$$+ \langle (\hat{\alpha} \cdot \hat{Z}(0))\,(\hat{\alpha} \cdot \hat{Z}(0))\,(\hat{\alpha} \cdot \hat{X}(t_2))\,(\hat{\alpha} \cdot \hat{X}(t_2)) \rangle \sin^2\theta \tag{5.55}$$

since

$$\hat{M}(t_2) = \hat{Z}(t_2)\cos\theta + \hat{X}(t_2)\sin\theta. \tag{5.56}$$

Using the quantities from Table 5.3 (with $t_1 = t_2 = 0$), we get

$$\langle (\hat{\alpha} \cdot \hat{Z}(0))\,(\hat{\alpha} \cdot \hat{Z}(0))\,(\hat{\alpha} \cdot \hat{M}(t_2))\,(\hat{\alpha} \cdot \hat{M}(t_2)) \rangle$$
$$= \frac{1}{9} + \frac{4}{45}e^{-6Dt_2}P_2(\cos\theta). \tag{5.57}$$

Thus, θ scales the contribution of the rotational motion to the signal. At 54.7°, $P_2(\cos\theta) = 0$, and so one will only measure the population relaxation during t_2

because the rotational contribution has been removed. If one wants to measure the rotational motion instead, then one calculates the anisotropy by combining the signals

$$\alpha(t_2) \equiv \frac{\langle Z(0)Z(0)Z(t_2)Z(t_2)\rangle - \langle Z(0)Z(0)X(t_2)X(t_2)\rangle}{\langle Z(0)Z(0)Z(t_2)Z(t_2)\rangle + 2\langle Z(0)Z(0)X(t_2)X(t_2)\rangle}$$
$$\equiv \frac{I_\| - I_\perp}{I_\| + 2I_\perp} = \frac{2}{5}e^{-6Dt_2} \tag{5.58}$$

(for a diagonal peak).

5.6 Polarization conditions for higher-order pulse sequences

In Chapter 11 we discuss fifth-order pulse sequences and 3D IR spectra. These higher-order experiments obey the six-point orientational correlation function

$$E_{\text{diag}}^{(5)} = \langle (\hat{\mu}_\alpha \cdot \hat{E}_a)(\hat{\mu}_\beta \cdot \hat{E}_b)(\hat{\mu}_\gamma \cdot \hat{E}_c)(\hat{\mu}_\delta \cdot \hat{E}_d)(\hat{\mu}_\epsilon \cdot \hat{E}_e)(\hat{\mu}_\xi \cdot \hat{E}_f)\rangle$$
$$\equiv \langle a_\alpha b_\beta c_\gamma d_\delta e_\epsilon f_\xi \rangle. \tag{5.59}$$

As outlined in this chapter for 2D IR spectroscopy, polarization control will be useful in 3D IR spectroscopy to eliminate diagonal peaks and enhance cross-peaks [18, 43]. In fact, polarization selectivity for 3D IR spectroscopy should be able to discriminate even better between pathways, since there are more transition dipoles, but this has not yet been experimentally explored.

Before presenting the fifth-order results, let us rewrite the general solution to the third-order polarization from Eq. 5.28 in a more compact form:

$$\langle (\hat{\alpha} \cdot \hat{E}_a)(\hat{\beta} \cdot \hat{E}_b)(\hat{\gamma} \cdot \hat{E}_c)(\hat{\delta} \cdot \hat{E}_d)\rangle$$
$$= \frac{1}{30}\begin{pmatrix}\cos\theta_{ab}\cos\theta_{cd}\\ \cos\theta_{ac}\cos\theta_{bd}\\ \cos\theta_{ad}\cos\theta_{bc}\end{pmatrix}^{\mathrm{T}}\begin{pmatrix}4 & -1 & -1\\ -1 & 4 & -1\\ -1 & -1 & 4\end{pmatrix}\begin{pmatrix}\cos\theta_{\alpha\beta}\cos\theta_{\gamma\delta}\\ \cos\theta_{\alpha\gamma}\cos\theta_{\beta\delta}\\ \cos\theta_{\alpha\delta}\cos\theta_{\beta\gamma}\end{pmatrix}$$
$$\equiv \mathbf{P}^{\mathrm{T}}\mathbf{M}\mathbf{D}. \tag{5.60}$$

Written in matrix form, the calculation of the polarization response is computationally straightforward. Moreover, it is easily generalizable to higher dimensions if one knows the matrix $\mathbf{M}$. This matrix has been derived for fifth-order and seventh-order responses [1, 34, 44]. One way to arrive at this equation is to orientationally average the molecular frame and a polarization frame relative to the laboratory frame. In Section 5.3 we used Euler angles to rotate just the molecular frame. A more efficient way is to use tensors. The matrix $\mathbf{M}$ is the result of summing together a series of products between tensors for the laboratory and polarization frame [95].

5.6 Higher-order pulse sequences

The fifth-order matrix **M** is given by

$$\mathbf{M} = \frac{1}{210}$$

$$\times \begin{pmatrix}
16 & -5 & -5 & -5 & 2 & 2 & -5 & 2 & 2 & 2 & 2 & -5 & 2 & 2 & -5 \\
-5 & 16 & -5 & 2 & -5 & 2 & 2 & 2 & -5 & -5 & 2 & 2 & 2 & -5 & 2 \\
-5 & -5 & 16 & 2 & 2 & -5 & 2 & -5 & 2 & 2 & -5 & 2 & -5 & 2 & 2 \\
-5 & 2 & 2 & 16 & -5 & -5 & -5 & 2 & 2 & 2 & -5 & 2 & 2 & -5 & 2 \\
2 & -5 & 2 & -5 & 16 & -5 & 2 & -5 & 2 & -5 & 2 & 2 & 2 & 2 & -5 \\
2 & 2 & -5 & -5 & -5 & 16 & 2 & 2 & -5 & 2 & 2 & -5 & -5 & 2 & 2 \\
-5 & 2 & 2 & -5 & 2 & 2 & 16 & -5 & -5 & -5 & 2 & 2 & -5 & 2 & 2 \\
2 & 2 & -5 & 2 & -5 & 2 & -5 & 16 & -5 & 2 & -5 & 2 & 2 & 2 & -5 \\
2 & -5 & 2 & 2 & 2 & -5 & -5 & -5 & 16 & 2 & 2 & -5 & 2 & -5 & 2 \\
2 & -5 & 2 & 2 & -5 & 2 & -5 & 2 & 2 & 16 & -5 & -5 & -5 & 2 & 2 \\
2 & 2 & -5 & -5 & 2 & 2 & 2 & -5 & 2 & -5 & 16 & -5 & 2 & -5 & 2 \\
-5 & 2 & 2 & 2 & 2 & -5 & 2 & 2 & -5 & -5 & -5 & 16 & 2 & 2 & -5 \\
2 & 2 & -5 & 2 & 2 & -5 & -5 & 2 & 2 & -5 & 2 & 2 & 16 & -5 & -5 \\
2 & -5 & 2 & -5 & 2 & 2 & 2 & -5 & 2 & -5 & 2 & -5 & 16 & -5 \\
-5 & 2 & 2 & 2 & -5 & 2 & 2 & -5 & 2 & 2 & 2 & -5 & -5 & -5 & 16
\end{pmatrix}$$

(5.61)

with

$$\mathbf{D}^T = \tag{5.62}$$

$(\cos\theta_{\alpha\beta}\cos\theta_{\gamma\delta}\cos\theta_{\varepsilon\xi},\ \cos\theta_{\alpha\beta}\cos\theta_{\gamma\varepsilon}\cos\theta_{\delta\xi},\ \cos\theta_{\alpha\beta}\cos\theta_{\gamma\xi}\cos\theta_{\delta\varepsilon}$

$\cos\theta_{\alpha\gamma}\cos\theta_{\beta\delta}\cos\theta_{\varepsilon\xi},\ \cos\theta_{\alpha\gamma}\cos\theta_{\beta\varepsilon}\cos\theta_{\delta\xi},\ \cos\theta_{\alpha\gamma}\cos\theta_{\beta\xi}\cos\theta_{\varepsilon\delta},$

$\cos\theta_{\alpha\delta}\cos\theta_{\beta\gamma}\cos\theta_{\varepsilon\xi},\ \cos\theta_{\alpha\delta}\cos\theta_{\beta\varepsilon}\cos\theta_{\gamma\xi},\ \cos\theta_{\alpha\delta}\cos\theta_{\beta\xi}\cos\theta_{\gamma\varepsilon},$

$\cos\theta_{\alpha\varepsilon}\cos\theta_{\beta\gamma}\cos\theta_{\delta\xi},\ \cos\theta_{\alpha\varepsilon}\cos\theta_{\beta\delta}\cos\theta_{\gamma\xi},\ \cos\theta_{\alpha\varepsilon}\cos\theta_{\beta\xi}\cos\theta_{\gamma\delta},$

$\cos\theta_{\alpha\xi}\cos\theta_{\beta\gamma}\cos\theta_{\delta\varepsilon},\ \cos\theta_{\alpha\xi}\cos\theta_{\beta\delta}\cos\theta_{\gamma\varepsilon},\ \cos\theta_{\alpha\xi}\cos\theta_{\beta\varepsilon}\cos\theta_{\gamma\delta})$

and an analogous vector for **P**. When the polarization is the same for all six pulses, this equation reduces to

$$\langle Z_\alpha Z_\beta Z_\gamma Z_\delta Z_\varepsilon Z_\xi \rangle = \frac{1}{105} \tag{5.63}$$

$\times \{\cos\theta_{\alpha\beta}\cos\theta_{\gamma\delta}\cos\theta_{\varepsilon\xi} + \cos\theta_{\alpha\beta}\cos\theta_{\gamma\varepsilon}\cos\theta_{\delta\xi} + \cos\theta_{\alpha\beta}\cos\theta_{\gamma\xi}\cos\theta_{\delta\varepsilon} +$

$\cos\theta_{\alpha\gamma}\cos\theta_{\beta\delta}\cos\theta_{\varepsilon\xi} + \cos\theta_{\alpha\gamma}\cos\theta_{\beta\varepsilon}\cos\theta_{\delta\xi} + \cos\theta_{\alpha\gamma}\cos\theta_{\beta\xi}\cos\theta_{\varepsilon\delta} +$

$\cos\theta_{\alpha\delta}\cos\theta_{\beta\gamma}\cos\theta_{\varepsilon\xi} + \cos\theta_{\alpha\delta}\cos\theta_{\beta\varepsilon}\cos\theta_{\gamma\xi} + \cos\theta_{\alpha\delta}\cos\theta_{\beta\xi}\cos\theta_{\gamma\varepsilon} +$

$\cos\theta_{\alpha\varepsilon}\cos\theta_{\beta\gamma}\cos\theta_{\delta\xi} + \cos\theta_{\alpha\varepsilon}\cos\theta_{\beta\delta}\cos\theta_{\gamma\xi} + \cos\theta_{\alpha\varepsilon}\cos\theta_{\beta\xi}\cos\theta_{\gamma\delta} +$

$\cos\theta_{\alpha\xi}\cos\theta_{\beta\gamma}\cos\theta_{\delta\varepsilon} + \cos\theta_{\alpha\xi}\cos\theta_{\beta\delta}\cos\theta_{\gamma\varepsilon} + \cos\theta_{\alpha\xi}\cos\theta_{\beta\varepsilon}\cos\theta_{\gamma\delta}\}.$

Another special case is when two of the pulses are polarized perpendicular to the other two, which is

$$\langle X_\alpha X_\beta Z_\gamma Z_\delta Z_\varepsilon Z_\xi \rangle = \frac{1}{2}\langle Z_\gamma Z_\delta Z_\varepsilon Z_\xi \rangle \cos\theta_{\alpha\beta} - \frac{1}{2}\langle Z_\alpha Z_\beta Z_\gamma Z_\delta Z_\varepsilon Z_\xi \rangle \tag{5.64}$$

where $\langle Z_\gamma Z_\delta Z_\varepsilon Z_\xi \rangle$ is the fourth-rank tensor given in Eq. 5.28. With fifth-order spectroscopy, one can also have three orthogonal polarizations such as $\langle XXYYZZ \rangle$. For a more general discussion of these equations and their discussion with regards to a 3D IR spectrum of a two-oscillator system, see Refs. [43] and [44].

Exercises

5.1 Derive Eq. 5.26.

5.2 Derive the ratio analogous to Eq. 5.31 but for a 2D IR spectrum measured using a narrowband pump method like an etalon (Section 4.4).

5.3 Using spherical harmonics, derive Eq. 5.53.

5.4 Derive Eq. 5.57 for a diagonal peak. At magic angle ($\theta = 54.7°$), one only measures population relaxation during t_2. Is that also true for the cross-peaks?

5.5 When switching between $\langle ZZZZ \rangle$ and $\langle ZXXZ \rangle$ with a rephasing 2D IR pulse sequence the ratio of the peaks on the diagonal should be 3. Explain why this is not the case for (a) non-rephasing and (b) absorptive 2D IR spectra collected with impulsive pulses in a collinear beam geometry.

5.6 Show that the angles between two transition dipoles can be measured using the ratio of $\langle 45°, -45°, 0°, 0° \rangle$ and $\langle 75°, -75°, 0°, 0° \rangle$ in a non-rephasing spectrum. Is this ratio preferable to $\langle 90°, 90°, 0°, 0° \rangle$ and $\langle 0°, 0°, 0°, 0° \rangle$ used in pump–probe style 2D IR methods? [153]

5.7 Explain how one experimentally measures the fifth-order orientational response $\langle XXYYZZ \rangle$.

6
Molecular couplings

The most common picture of vibrational spectroscopy is that of *normal modes*. However, the normal mode picture is not sufficient to describe 2D IR spectroscopy without modification, because normal modes are harmonic. Anharmonicity is necessary to create 2D IR spectra. This fact can be seen from Fig. 1.5, where the negative and positive peaks of the peak pairs would overlap and cancel if the anharmonic shifts Δ_{ij} were all zero. However, describing anharmonicity with normal modes is somewhat complicated.

The purpose of this chapter is to introduce a *local mode* description of molecular vibrations. It is a useful description for simulating 2D IR spectra and providing a conceptual framework for visualizing the vibrations of molecules, especially of molecules built from repeating units, like proteins. In the local mode description, we treat each repeat unit as a local coordinate. In a 3D structure, these local modes will be coupled, so that they vibrate in unison, forming delocalized states which are called *vibrational excitons*. The coupling between local modes depends on their relative distances and orientations, and thus are probes of the 3D structure.

6.1 Vibrational excitons

The term *vibrational exciton* is borrowed from *molecular excitons*, which come from studies of closely packed aggregates of optical chromophores that create delocalized electronic excitations [36]. Vibrational excitons, which are sometimes also called *vibrons*, deal with vibrational rather than electronic excitations. In fact, their Hamiltonians (Eq. 6.2 below) look formally the same.[1]

---

[1] In the language of semiconductor physics, a molecular exciton would be a strongly bound Frenkel exciton for electronic excitation, or an internal optical phonon for a vibrational excitation in molecular crystals.

The exciton model starts out from a system of coupled local modes (for clarity we consider only two coordinates):

$$H = \hbar\omega_1\left(b_1^\dagger b_1 + \frac{1}{2}\right) + \hbar\omega_2\left(b_2^\dagger b_2 + \frac{1}{2}\right) + \beta_{12}(b_1^\dagger b_2 + b_2^\dagger b_1). \quad (6.1)$$

Here, $b_n^\dagger$ and b_n are the creation and annihilation operators of the local oscillators, respectively (a brief summary of the ladder operator formalism is given in Appendix B). We ignore zero-point energies from here onwards so that the Hamiltonian reads:

$$H = \hbar\omega_1 b_1^\dagger b_1 + \hbar\omega_2 b_2^\dagger b_2 + \beta_{12}(b_1^\dagger b_2 + b_2^\dagger b_1). \quad (6.2)$$

Writing the Hamiltonian using creation and annihilation operators leads to a very intuitive explanation of coupling, in which the coupling terms describe a hopping of the excitation from one site to the other. For example, $b_1^\dagger b_2|01\rangle = |10\rangle$, where $|ij\rangle$ is in a local mode basis.

If there was no coupling, then the two local oscillators would not influence one another because the excitation would not hop from one site to the other. The coupling term $b_1^\dagger b_2 + b_2^\dagger b_1$ originates from the *bilinear term* $q_1 q_2$ of a Taylor expansion of the potential energy surface:

$$V(q_1, q_2) = \frac{1}{2}V_{11}q_1^2 + \frac{1}{2}V_{22}q_2^2 + 2\beta_{12}q_1 q_2 \quad (6.3)$$

where q_1 and q_2 are local mode coordinates. By substituting $q_n = 1/\sqrt{(2)}(b_n^\dagger + b_n)$ (see Appendix B) into the mixed term of the potential energy surface, one gets:

$$q_1 q_2 = 1/2(b_1 + b_1^\dagger)(b_2 + b_2^\dagger) = 1/2(b_1^\dagger b_2 + b_1 b_2^\dagger + b_1 b_2 + b_1^\dagger b_2^\dagger). \quad (6.4)$$

In the exciton model, only the *quantum conserving terms* $b_1^\dagger b_2$ and $b_1 b_2^\dagger$ are retained since they very efficiently couple two closely resonant states, as we did in Eq. 6.2. Terms of the sort $b_1^\dagger b_2^\dagger$ couple the ground state to a state with one quantum each in mode 1 and 2. The latter is much higher in energy than the ground state, hence, unless the coupling β_{12} is exceptionally strong, these terms will have only a minor effect, and so are commonly neglected.

We expand the Hamiltonian 6.2 in a site basis $\{|ij\rangle\}$ where the two digits refer to the number of quanta in the two modes. For third-order nonlinear spectroscopy, it turns out that it is sufficient to consider basis states only up to double excitations $(\{|ij\rangle\} = \{|00\rangle, |10\rangle, |01\rangle, |20\rangle, |02\rangle, |11\rangle\})$, because the pulses in third-order nonlinear spectroscopy do not probe higher eigenstates. In this basis, the Hamiltonian matrix is (see Problem 6.1):

6.1 Vibrational excitons

$$H = \begin{pmatrix} 0 & & & & & \\ & \hbar\omega_1 & \beta_{12} & & & \\ & \beta_{12} & \hbar\omega_2 & & & \\ \hline & & & 2\hbar\omega_1 & 0 & \sqrt{2}\beta_{12} \\ & & & 0 & 2\hbar\omega_2 & \sqrt{2}\beta_{12} \\ & & & \sqrt{2}\beta_{12} & \sqrt{2}\beta_{12} & \hbar\omega_1 + \hbar\omega_2 \end{pmatrix}. \quad (6.5)$$

Because the Hamiltonian (Eq. 6.2) only includes quantum conserving terms, it separates into blocks: the ground state, the one-exciton Hamiltonian H_1, and the two-exciton Hamiltonian H_2. The zero-, one-, and two-exciton manifolds have been separated by lines in Eq. 6.5. The $\sqrt{2}$ factors in the two-exciton Hamiltonian originate from the ladder operators operating on the double excited states with $n = 2$, e.g. $b|n\rangle = \sqrt{n}|n-1\rangle$ (see Appendix B).

Even though this Hamiltonian includes coupling between the local modes, it is still harmonic, and thus we will not get a 2D IR spectrum. In order to see why this is so, we diagonalize the one-exciton Hamiltonian:

$$U^T H_1 U = U^T \begin{pmatrix} \hbar\omega_1 & \beta_{12} \\ \beta_{12} & \hbar\omega_2 \end{pmatrix} U = \begin{pmatrix} \hbar\omega_1' & 0 \\ 0 & \hbar\omega_2' \end{pmatrix} \quad (6.6)$$

which gives a new basis with new ladder operators and new one-exciton matrix with no off-diagonal elements. The new creation and annihilation operators $B_{1,2}^\dagger$ and $B_{1,2}$ can be written in the local mode basis using:

$$B_i = \sum_j U_{ij} b_j$$

$$B_i^\dagger = \sum_j U_{ij} b_j^\dagger. \quad (6.7)$$

In this basis, the Hamiltonian is written as:

$$H' = \hbar\omega_1' B_1^\dagger B_1 + \hbar\omega_2' B_2^\dagger B_2 \quad (6.8)$$

which does not have the *bilinear coupling* term $B_1^\dagger B_2 + B_2^\dagger B_1$. Since there is no bilinear coupling, this Hamiltonian is diagonal not only for the one-exciton states, but also for all higher-exciton states, including the two-exciton states. Thus, we do not need to explicitly diagonalize the two-exciton matrix in Eq. 6.5, because our coordinate transformation already did it for us. Thus, the eigenstate energies can be written:

$$E = \sum_i \hbar\omega_i' n_i. \quad (6.9)$$

This new basis is the normal mode basis. Notice that there are no anharmonic shifts. That is, the combination band is equal to the sum of two fundamentals, and the overtones are equal to twice their respective fundamentals. Hence, there is no 2D IR signal, because we started out with a potential energy surface (Eq. 6.3) that only includes terms up to second order, and so is perfectly harmonic.

What characteristic must a potential energy surface have to create a 2D IR spectrum? We need to include terms in the Hamiltonian that are higher than second order. While there are many of them, we use the ones that allow us to describe the local modes as anharmonic potentials (because the potential of a bond stretch is better described by a Morse oscillator than a harmonic oscillator) and are quantum-conserving (to retain the block-diagonal form of the Hamilton matrix). With these terms, the Hamiltonian is now:

$$H = \hbar\omega_1 b_1^\dagger b_1 + \hbar\omega_2 b_2^\dagger b_2 + \beta_{12}(b_1^\dagger b_2 + b_2^\dagger b_1)$$
$$- \frac{\Delta}{2} b_1^\dagger b_1^\dagger b_1 b_1 - \frac{\Delta}{2} b_2^\dagger b_2^\dagger b_2 b_2. \quad (6.10)$$

The terms $-b_n^\dagger b_n^\dagger b_n b_n$ lower the site energies of the doubly excited local states by an energy Δ. Δ is the *local mode anharmonic shift*. It is proportional to the quartic expansion coefficients of the potential energy surface (see Problem 6.3). Expanding the Hamiltonian in the same basis as before $\{|ij\rangle\} = \{|00\rangle, |10\rangle, |01\rangle, |20\rangle, |02\rangle, |11\rangle\}$, the Hamiltonian now reads [83] (see Problem 6.2):

$$H = \begin{pmatrix} 0 & & & & & \\ & \hbar\omega_1 & \beta_{12} & & & \\ & \beta_{12} & \hbar\omega_2 & & & \\ & & & 2\hbar\omega_1 - \Delta & 0 & \sqrt{2}\beta_{12} \\ & & & 0 & 2\hbar\omega_2 - \Delta & \sqrt{2}\beta_{12} \\ & & & \sqrt{2}\beta_{12} & \sqrt{2}\beta_{12} & \hbar\omega_1 + \hbar\omega_2 \end{pmatrix}. \quad (6.11)$$

The one-exciton matrix of this Hamiltonian is the same as in the harmonic case (Eq. 6.6), but the two-exciton matrix is not because of the local mode anharmonicity. Therefore, we must diagonalize the two-exciton matrix separately from the one-exciton block. In the limiting case when $\omega_1 = \omega_2 \equiv \omega$, the Hamiltonian becomes

$$H = \qquad\qquad\qquad\qquad\qquad\qquad\qquad\qquad\qquad (6.12)$$

$$\begin{pmatrix} 0 & & & & & \\ & \hbar\omega - \beta & & & & \\ & & \hbar\omega + \beta & & & \\ & & & 2\hbar\omega - \frac{1}{2}\Delta - \frac{1}{2}\sqrt{\Delta^2 + 16\beta^2} & & \\ & & & & 2\hbar\omega - \frac{1}{2}\Delta + \frac{1}{2}\sqrt{\Delta^2 + 16\beta^2} & \\ & & & & & 2\hbar\omega - \Delta \end{pmatrix}$$

after diagonalization. Thus, the local mode anharmonicity mixes into all of the double-excited states, creating diagonal and off-diagonal anharmonic shifts, which results in a 2D IR spectrum.

6.1 Vibrational excitons

This exciton Hamiltonian is a good approximation when we deal with close-to-resonant vibrational states, such as amide I vibrations of peptides and proteins [83], as well as the –C≡O manifold of states in metal-carbonyl complexes [44, 71]. Most 2D IR experiments are performed on nearly resonant vibrational states because they are "one-color" experiments. That is, all the laser pulses are identical and so only cover vibrational states within the bandwidth of a femtosecond laser pulse, which is typically 100–200 cm^{-1}. The exciton model fails to describe the coupling between modes with much different frequencies, such as the coupling between –CD and –C=O vibrations [117]. In this case, the exciton model would predict negligible coupling (see Eq. 6.24 below), which is usually not true. One would have to include many higher-order terms in the Taylor expansion of the potential energy surface, including non-quantum-conserving terms, to calculate the anharmonic shifts Δ_{ij} perturbatively [35] (see Section 6.7).

6.1.1 Transforming the transition dipole matrix

The second ingredient we need to calculate a 2D IR spectrum is the transition dipoles. These describe the transition strength between the eigenstates, and so give the intensities of the peaks in a 2D IR spectrum. We start by defining the *transition dipole operator* in the local mode basis, $\vec{\mu} \equiv \vec{\mu}_1 + \vec{\mu}_2$, where $\vec{\mu}_1$ and $\vec{\mu}_2$ act only on mode 1 or 2, respectively

$$\langle i_1 i_2 | \vec{\mu}_1 | j_1 j_2 \rangle \equiv \langle i_1 | \vec{\mu}_1 | j_1 \rangle \langle i_2 | j_2 \rangle = \langle i_1 | \vec{\mu}_1 | j_1 \rangle \delta_{i_2, j_2}$$
$$\langle i_1 i_2 | \vec{\mu}_2 | j_1 j_2 \rangle \equiv \langle i_2 | \vec{\mu}_2 | j_2 \rangle \langle i_1 | j_1 \rangle = \langle i_2 | \vec{\mu}_2 | j_2 \rangle \delta_{i_1, j_1}. \qquad (6.13)$$

Cross-excitations are not possible within the dipole approximation, since the dipole operator only acts on one coordinate (e.g. $\langle 10 | \vec{\mu}_1 + \vec{\mu}_2 | 02 \rangle = 0$). Furthermore, we use the harmonic approximation for climbing up the vibrational ladder (see Problem 6.5), so that

$$\langle 10 | \vec{\mu}_1 | 20 \rangle = \sqrt{2} \langle 00 | \vec{\mu}_1 | 10 \rangle. \qquad (6.14)$$

By expanding the dipole operator in the same basis as the Hamiltonian, we obtain the *transition dipole matrix* in the local mode basis:

$$\vec{\mu} = \begin{pmatrix} & \vec{\mu}_1 & \vec{\mu}_2 & & & \\ \vec{\mu}_1 & & & \sqrt{2}\vec{\mu}_1 & 0 & \vec{\mu}_2 \\ \vec{\mu}_2 & & & 0 & \sqrt{2}\vec{\mu}_2 & \vec{\mu}_1 \\ \sqrt{2}\vec{\mu}_1 & 0 & & & & \\ 0 & \sqrt{2}\vec{\mu}_2 & & & & \\ \vec{\mu}_2 & \vec{\mu}_1 & & & & \end{pmatrix}. \qquad (6.15)$$

This transition dipole matrix is transformed into the eigenstate basis by the same unitary transformation that diagonalizes the Hamiltonian matrix (Eq. 6.11). In this way, both the transition dipoles for the 0→1 exciton transitions ($\vec{\mu}_{0i}$ for the transition from the ground state to exciton $|i\rangle$) and for the 1→2 exciton transitions ($\vec{\mu}_{ik}$ for the transition from the one-exciton state $|i\rangle$ to the two-exciton state $|k\rangle$) are obtained. Since the transition dipoles are vectors, they are transformed as such during the unitary transformation (see Section 6.2.1).

6.2 Spectroscopy of a coupled dimer

With the exciton Hamiltonian introduced above, we now turn to spectroscopy. The type of molecule we have in mind is something like a protein where we use the coupled carbonyl stretch vibrations (amide I) of the protein backbone to study its structure. However, the following approach is general and can be applied in an analogous way to the carbonyl vibrations of the base pairs in DNA [114, 122], the electronic states of chlorophyll molecules in antenna complexes [100], J-aggregates [42], and many other systems.

In either case, it turns out that there is a very characteristic connection between the 3D structure of the molecular system (i.e. the relative distance and orientation of individual units) and the appearance of both linear and 2D IR spectra. To see that, we need one more ingredient, which is a model that relates the coupling constant β_{ij} to the structure. Models of this sort are a major focus of this chapter. To illustrate the concept, we introduce transition dipole coupling, which is the simplest possible coupling model and is what we used in Chapter 1. In this model, we calculate the coupling between each pair of units using

$$\beta_{ij} = \frac{1}{4\pi\epsilon_0} \left[\frac{\vec{\mu}_i \cdot \vec{\mu}_j}{r_{ij}^3} - 3\frac{(\vec{r}_{ij} \cdot \vec{\mu}_i)(\vec{r}_{ij} \cdot \vec{\mu}_j)}{r_{ij}^5} \right] \tag{6.16}$$

where $\vec{\mu}_i$ are the transition dipoles of the local modes and $\vec{r}_{ij}$ are the vectors connecting the sites i and j (Fig. 1.2). For polypeptides and proteins, this model was first put forward by Krimm and Bandekar [112], and later modified by Torii and Tasumi [178]. Note that it is the transition dipole, and not the static dipole, that enters for the calculation of the coupling strength. The strength of the transition dipole can be deduced from the absorption cross-section (see Problem 6.9).

6.2.1 Linear absorption spectrum of a coupled dimer

We start by discussing how the linear absorption spectrum for a coupled dimer depends on its geometry. For linear spectroscopy, we only have to consider the one-exciton Hamiltonian:

$$H_1 = \begin{pmatrix} \hbar\omega_1 & \beta_{12} \\ \beta_{12} & \hbar\omega_2 \end{pmatrix} \tag{6.17}$$

whose exciton (ex) eigenvalues are:

$$E_{1,2}^{(ex)} = \frac{\hbar\omega_1 + \hbar\omega_2 \mp \sqrt{4\beta_{12}^2 + (\hbar\omega_2 - \hbar\omega_1)^2}}{2}. \quad (6.18)$$

The eigenstates of $E_1^{(ex)}$ and $E_2^{(ex)}$ are:

$$|\Phi_1\rangle = +\cos\alpha|10\rangle - \sin\alpha|01\rangle$$
$$|\Phi_2\rangle = +\sin\alpha|10\rangle + \cos\alpha|01\rangle \quad (6.19)$$

with the mixing angle:

$$\tan 2\alpha = 2\frac{\beta_{12}}{\hbar\omega_2 - \hbar\omega_1}. \quad (6.20)$$

The transition dipole moments transform the same way

$$\vec{\mu}_1^{(ex)} = +\cos\alpha \cdot \vec{\mu}_1 - \sin\alpha \cdot \vec{\mu}_2$$
$$\vec{\mu}_2^{(ex)} = +\sin\alpha \cdot \vec{\mu}_1 + \cos\alpha \cdot \vec{\mu}_2 \quad (6.21)$$

so that the intensities of the two observed transitions, given by $|\vec{\mu}_1^{(ex)}|^2$ and $|\vec{\mu}_2^{(ex)}|^2$, depend on the relative orientation of the two coupled transition dipoles as well as their coupling strength β_{12} and frequency separation $\hbar\omega_2 - \hbar\omega_1$.

We discuss two limiting cases. When the coupling is small compared to the frequency splitting, $|\beta_{12}| \ll |\hbar\omega_2 - \hbar\omega_1|$, their mixing angle α is small and the exciton states will be localized mainly on the individual sites. By expanding Eq. 6.18, the energies of these states become

$$E_{1,2}^{(ex)} \approx \hbar\omega_{1,2} \mp \frac{2\beta_{12}^2}{\hbar\omega_2 - \hbar\omega_1}. \quad (6.22)$$

In the strong coupling regime, $|\beta_{12}| \gg |\hbar\omega_2 - \hbar\omega_1|$, we simply ignore $\hbar\omega_2 - \hbar\omega_1$ in Eq. 6.18 to find that the one-exciton eigenstates are perfectly delocalized with $\alpha = \pi/4$, revealing $1/\sqrt{2}(|10\rangle \mp |01\rangle)$. In this case, the energies of the excitonic states will split to give

$$E_{1,2}^{(ex)} \approx \frac{\hbar\omega_1 + \hbar\omega_2}{2} \mp \beta_{12}. \quad (6.23)$$

Depending on the sign of β_{12}, which depends on the geometry of the dimer (Fig. 6.1), either the symmetric or the antisymmetric combination will be the lower-energy solution. For example, when the two chromophores are parallel like in Fig. 6.1(a), the coupling β_{12} is positive (Eq. 6.16) and the higher-frequency solution will be the symmetric eigenstate with a transition dipole $\vec{\mu}_1 + \vec{\mu}_2$. It carries most of the oscillator strength since $\vec{\mu}_1 || \vec{\mu}_2$. In contrast, the antisymmetric,

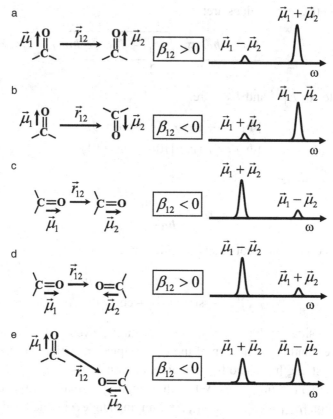

Figure 6.1 Typical geometries and the resulting linear absorption spectra for a coupled dimer of C=O vibrations. The sign of the coupling predicted by transition dipole coupling is given in the boxes.

lower-frequency solution with transition dipole $\vec{\mu}_1 - \vec{\mu}_2$ will be weak. If $\vec{\mu}_1$ and $\vec{\mu}_2$ are antiparallel (Fig. 6.1b), the coupling β_{12} is negative (Eq. 6.16) so that the antisymmetric eigenstate is the higher-energy solution. Yet the spectrum is identical to the parallel geometry since now $\vec{\mu}_1 - \vec{\mu}_2$ adds up constructively. Thus, we cannot distinguish parallel and antiparallel configurations.

If the two chromophores are situated as depicted in Fig. 6.1(c,d) (either head-to-tail or head-to-head), the lower-frequency transition will be the one that carries the most oscillator strength. It is possible to get equally intense transitions if the two transition dipoles are perpendicular but still coupled (Fig. 6.1e). Of course, most geometries are in between the limits discussed here.

Note that the coupling shifts intensities between transitions, but that the sum of intensities stays constant in a linear absorption spectrum, which is not the case for 2D IR spectra, as we discuss below.

6.2 Spectroscopy of a coupled dimer

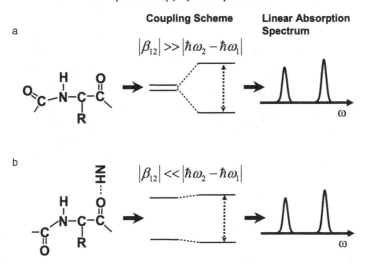

Figure 6.2 Hypothetical linear absorption spectra of (a) a homopolymer with strongly coupled carbonyl oscillators and (b) a heteropolymer with weak coupling because one of the carbonyls is hydrogen bonded. These two situations can produce spectra that are hard to distinguish with linear infrared spectroscopy.

6.2.2 2D IR spectrum of a coupled dimer

This coupling model allows us to illustrate why 2D IR spectroscopy is a better probe of structure than linear spectroscopy. Consider the carbonyl stretches of the two peptides shown in Fig. 6.2. In the upper case, the two local modes are identical, and strongly coupled. In the lower case, we have the same peptide in a different geometry and with a hydrogen bond to one of the local modes, which will shift its frequency. Depending on the angles of the transition dipole moments (i.e. the situation shown in Fig. 6.1e), the linear absorption spectra of these two peptides could be indistinguishable. Thus, with linear absorption spectroscopy, one cannot definitively distinguish between structures.

2D IR spectroscopy can make exactly this distinction since the local mode anharmonicity Δ mixes into the two-exciton states depending on the amount of delocalization (Fig. 6.3a,b). Figure 6.3 shows the resulting 2D IR spectra in two coupling regimes. In the strong coupling regime with $|\beta_{12}| \gg |\hbar\omega_2 - \hbar\omega_1|$, the cross-peaks in the 2D IR spectrum have a larger separation than the diagonal peaks (Fig. 6.3c) because the excitons are delocalized over the two local modes. In the weak coupling regime $|\beta_{12}| \ll |\hbar\omega_2 - \hbar\omega_1|$, the excitations are largely localized so that the two-excitonic states closely resemble the uncoupled local modes. In this case, the diagonal anharmonicities are approximately $\Delta_{11} = \Delta_{22} = -\Delta$ and the off-diagonal anharmonic constants can be calculated perturbatively [84]:

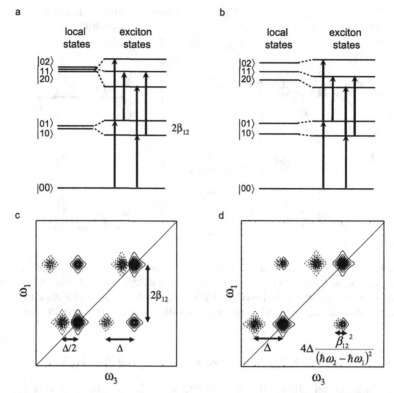

Figure 6.3 Level scheme of two coupled oscillators in (a) the strong $|\beta_{12}| \gg |\hbar\omega_2 - \hbar\omega_1|$ and (b) the weak $|\beta_{12}| \ll |\hbar\omega_2 - \hbar\omega_1|$ coupling regimes. (c,d) Resulting purely absorptive 2D IR spectra. The solid and dashed contour lines depict the negative and positive signs, respectively, of the corresponding bands. The parameters for these spectra are (in cm^{-1}): (c) $\omega_1 = \omega_2 = 1650$, $\beta_{12} = 10$, and $\Delta = 10$. (d) $\omega_1 = 1640$, $\omega_2 = 1660$, $\beta_{12} = 3$ and $\Delta = 10$. The spectra were simulated with perpendicularly oriented transition dipoles with a magic angle condition for the laser pulses, so that the mixing of the states does not redistribute the oscillator strength.

$$\Delta_{12} = -4\Delta \frac{\beta_{12}^2}{(\hbar\omega_2 - \hbar\omega_1)^2}. \quad (6.24)$$

Thus, in the weak coupling limit the splitting of the cross-peaks is proportional to the strength of the coupling squared. Because the splitting is usually smaller than the linewidths (Figure 6.3d), the intensity of the cross-peaks is diminished because they will partially overlap and cancel. Therefore, 2D IR spectroscopy is able to distinguish between the two peptides of Fig. 6.2 by the intensities and frequencies of the cross-peaks.

Besides the two coupling regimes discussed above, we also need to consider the two additional limiting cases $\Delta \ll |\beta_{12}|$ and $\Delta \gg |\beta_{12}|$ when the strong coupling

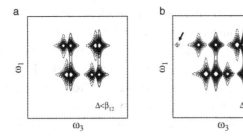

Figure 6.4 Purely absorptive 2D IR spectra for small and large anharmonicities $\Delta < \beta_{12}$ and $\Delta > \beta_{12}$, respectively, when $|\beta_{12}| \gg |\hbar\omega_2 - \hbar\omega_1|$. The arrows mark the "forbidden" transitions. The parameters for these spectra were (in cm^{-1}): (a) $\Delta = 5$, $\beta_{12} = 10$, and $\omega_1 = \omega_2 = 1650$; (b) $\Delta = 12$, $\beta_{12} = 10$, and $\omega_1 = \omega_2 = 1650$.

regime of $|\beta_{12}| \gg |\hbar\omega_2 - \hbar\omega_1|$ applies. In the first case, the anharmonicity is only a weak perturbation, and the two-exciton states can still be identified as overtones or combination modes of the one-exciton states. Consequently, only transitions that change one vibrational quantum are observed, such as $|10\rangle \rightarrow |20\rangle$ and $|10\rangle \rightarrow |11\rangle$, while $|10\rangle \rightarrow |02\rangle$ is dipole-forbidden (Fig. 6.4a). In the opposite limit $\Delta > |\beta_{12}|$, the "forbidden" transitions become weakly allowed (see the arrows in Fig. 6.4b). In the limit of large anharmonicity, a local mode point of view, as we use here, rather than a normal mode basis, is closer to the true vibrational eigenstates (see Section 6.7).

2D IR peak intensities

In order to emphasize the splittings and not the intensities of the peaks in Figs. 6.3 and 6.4, we simulated the spectra using perpendicularly oriented transition dipoles so that the two sets of diagonal peaks have the same intensities. But the peak intensities will depend on the coupling strengths, frequency mismatch, and relative orientations of the local modes, just like the linear spectra do (Fig. 6.1). In general, the diagonal peaks in 2D IR spectra scale as $|\vec{\mu}|^4$ versus $|\vec{\mu}|^2$ in linear absorption spectroscopy. Hence, strong bands will become much more prominent in 2D IR spectra. Cross-peaks scale as $|\vec{\mu}_1|^2|\vec{\mu}_2|^2$ so if a diagonal peak disappears, the cross-peak will as well. Also note that, unlike in a linear absorption spectrum, the sum of the intensities of the diagonal peaks is not conserved when the coupling strength changes.

2D IR spectroscopy to determine molecular structures

How well can we determine the molecular structure from the 2D IR spectrum? The answer depends on two factors. First, we must be able to accurately extract the potential energy surface from the spectrum. The 2D IR spectrum provides the peak positions and their intensities that can be fit to determine the local mode parameters and the coupling. In the most generic case, the frequencies are determined from five unknowns, ω_1, ω_2, Δ_1, Δ_2, and β_{12}. The 2D IR spectrum provides five observables, $E_1^{(ex)}$, $E_2^{(ex)}$, Δ_{11}, Δ_{22} and Δ_{12}. Thus, even with 2D IR spectroscopy, it can be difficult to precisely determine the parameters. However, one can usually measure Δ_1 and Δ_2 with model compounds, leaving just three unknowns. Additional observables can be gained from fifth-order spectra [44] and information about the relative angle θ can be obtained from the peak intensities and their polarization dependence (Chapter 5). Congestion is probably the limiting factor to the accuracy with which the potential can be extracted from the spectra. Individual vibrational modes are usually not resolved in infrared spectra, and so it can be difficult to extract precise coupling parameters. Isotope labeling helps, which is discussed below. Second, these parameters must be used to generate a structure, the accuracy of which depends on the coupling model. Transition dipole coupling is the crudest model. Later in the chapter we introduce more sophisticated coupling models, although testing these models and improving their accuracy is still an active area of research. Nonetheless, 2D IR spectroscopy is proving to be a highly valuable structural tool, especially for systems that other techniques have difficulty addressing, such as membrane proteins and kinetically evolving systems.

6.3 Extended excitons in regular structures

Infrared spectroscopy has been used for decades to probe the secondary structures of proteins and DNA. Its sensitivity to structure is a result of strong coupling that causes the vibrational modes to delocalize over large spatial regions, such as along α-helices and across β-sheets. While there are as many excitonic eigenstates in the diagonalized Hamiltonian as there are local modes to start with, due to the translational symmetry of these secondary structures, only one or a few infrared transitions are observed. The observed modes can be quite intense, because their oscillator strength is the sum of many individual local modes, in analogy to the two modes of a parallel or antiparallel dimer (Fig. 6.1a–c). The intensity effect is more pronounced in 2D IR than linear infrared spectroscopy, because 2D IR intensities scale as $|\vec{\mu}|^4$ rather than $|\vec{\mu}|^2$. Moreover, delocalized states can be identified in 2D IR spectra by their anharmonic shift, because the anharmonic shift decreases with the degree of delocalization. Thus, it is the combination of frequency shifts caused by couplings and the selection rules caused by symmetry that make 1D and 2D infrared spectroscopies sensitive to secondary structures.

The discussion that follows is applicable to any excitonically coupled system, but the vibrational modes that we have in mind are the amide I vibrations of proteins. The amide I mode is predominantly caused by the carbonyl stretch of the

6.3 Extended excitons in regular structures

protein backbone, with some contribution from the C–N stretch and N–H bend. In water, it has a local mode frequency of about 1645 cm^{-1}. The amide bond is the covalent linkage between residues, so the amide I mode actually spans two amino acids, but it is usually labeled according to the amino acid that contains the carbonyl group. Since the C–N stretch contributes, the transition dipole points towards the nitrogen atom with an angle of 20° with respect to the C=O bond, and its origin is located in between the carbon and oxygen atom with a distance of about 0.868 Å from the carbon (Fig. 6.5). It has a transition dipole strength of 0.374 D, obtained from model compounds (see Problem 6.9). These numbers have been fine-tuned to best simulate the IR spectra of regular polypetide structures and a set of proteins [112, 178] using transition dipole coupling, although they are far from definitive.

6.3.1 Linear chain of coupled oscillators

Before presenting α-helices and β-sheets, we illustrate the delocalization of vibrational states across a linear chain of coupled oscillators, such as a line of carbonyl groups as shown in Fig. 6.6. We will only consider linear chains of identical molecules with equal spacings, so that all nearest neighbors are coupled by $\beta_{i,i+1} \equiv \beta_1$, next-to-nearest neighbors by $\beta_{i,i+2} \equiv \beta_2$, etc. When only the nearest neighbor coupling terms are considered, then we have the so-called *tight binding model*, which has the one-exciton Hamiltonian:

$$\hat{H} = \sum_i \hbar\omega_0 b_i^\dagger b_i + \sum_i \beta_1 (b_i^\dagger b_{i+1} + b_i^\dagger b_{i-1}) \tag{6.25}$$

with the corresponding Hamilton matrix

$$H = \begin{pmatrix} \hbar\omega_0 & \beta_1 & & & \\ \beta_1 & \hbar\omega_0 & \beta_1 & & \\ & \beta_1 & \hbar\omega_0 & \beta_1 & \\ & & \ddots & \ddots & \ddots \end{pmatrix}. \tag{6.26}$$

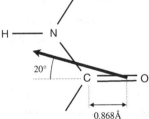

Figure 6.5 The direction and location of the transition dipole of the amide I vibration [112, 178].

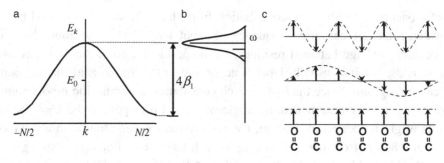

Figure 6.6 Linear chain of coupled oscillators with only nearest neighbor coupling. (a) Dispersion curve, (b) predicted infrared spectrum, and (c) three representative eigenvectors where the vectors represent the coefficients of the local mode basis. In the spectrum, the strongest IR allowed transition is the $k = 0$ transition, but other weaker bands will appear for finite-length chains or chains with disorder.

The eigenstates of this Hamiltonian must obey translation symmetry, due to the periodic nature of the linear chain, which leads to eigenstates that are composed of plane waves. This result is known as the *Bloch theorem* [4], which gives the eigenstates

$$\Psi_k = \frac{1}{\sqrt{N}} \sum_j e^{i\frac{2\pi jk}{N}} |j\rangle \qquad (6.27)$$

where N is the number of chromophores, $-N/2 + 1 \leq k \leq N/2$, and $|j\rangle$ is a state with a local excitation at site j. The $1/\sqrt{N}$ factor normalizes the wavefunction. Plugging the Bloch ansatz into the Hamiltonian 6.25, we obtain for the eigenenergies of state Ψ_k:

$$E_k = \hbar\omega_0 + 2\beta_1 \cos\left(\frac{2\pi k}{N}\right). \qquad (6.28)$$

This is the *dispersion relation* of the exciton, which relates the energy to its quantum number (Fig. 6.6a)[2] [4, 150].

Shown in Fig. 6.6(b,c) is the illustrative infrared spectrum, and three representative eigenstates. The vectors in Fig. 6.6 are the $e^{i\frac{2\pi jk}{N}}$ coefficients of each local mode. When $k = 0$, all of the local modes contribute equally and are in phase. When $k = N - 1$, all of the local modes contribute equally, but have opposite phases. At intermediate k, the contributions and phases are periodically varying. When β is negative, then the $k = 0$ state is the lowest energy state. For $\beta > 0$, it is at the high end of the dispersion curve. The $k = 0$ state holds significant importance in optical spectroscopy, because for infinitely long chains, it is the only

---

[2] We also now see why a vibrational exciton is the same as an optical phonon.

infrared allowed transition [150]. That is, when one sums its transition dipoles, one obtains

$$\mu_{k=0}^{(ex)} = \frac{1}{\sqrt{N}} \sum_j \mu_j = \sqrt{N}\mu. \tag{6.29}$$

All other states are optically dark since the $e^{i\frac{2\pi jk}{N}}$ terms eventually sum to zero in the equations analogous to Eq. 6.29. For finite chains, other exciton states will gain transition strength if the sums are not extensive enough for out-of-phase coefficients to cancel, which are illustrated by the weak lines in Fig. 6.6(b). Structural and environmental disorder can also cause forbidden transitions to become allowed, which is a topic that we touch on below.

If we include couplings other than nearest neighbor in the Hamiltonian matrix:

$$H = \begin{pmatrix} \hbar\omega_0 & \beta_1 & \beta_2 & \beta_3 & \cdots \\ \beta_1 & \hbar\omega_0 & \beta_1 & \beta_2 & \\ \beta_2 & \beta_1 & \hbar\omega_0 & \beta_1 & \\ \beta_3 & \beta_2 & \beta_1 & \hbar\omega_0 & \\ \vdots & & & & \ddots \end{pmatrix} \tag{6.30}$$

then the dispersion relation becomes

$$E_k = \hbar\omega_0 + 2 \sum_j \beta_j \cos\left(\frac{2\pi jk}{N}\right). \tag{6.31}$$

Thus, if we have multiple couplings, then they can increase the frequency span of the dispersion curve or compete to narrow or reverse the eigenstate energies, which is what we will see occurs in α-helices.

The two-exciton Hamiltonian is more difficult to diagonalize, because one must account for the local mode anharmonicity and coupling to the combination bands. Nonetheless, the strongest transition between the one- and two-exciton states will be to the two-exciton $k = 0$ state and the lower edge of the two-exciton band, just like the $k = 0$ state contained the oscillator state for the one-exciton states. Interestingly, the diagonal anharmonic shift decreases in proportion to the extent that the exciton is delocalized. As one example, consider two strongly coupled oscillators, like in Fig. 6.3(c). In that case, the diagonal anharmonic shift is $\Delta/2$, which is because the combination band mixes with the symmetric but not the antisymmetric stretch (see Problem 6.6). Simulations and simple models predict that for a system of N coupled oscillators, the diagonal anharmonic shift becomes Δ/N, at least under some conditions [50, 188]. What is clear is that if a diagonal anharmonic shift is observed that is smaller than the anharmonic shift of the local modes, then there is delocalized mode. Such an observation is a useful way of identifying secondary structures. For example, α-helices often have similar frequencies to random coil peptides, but since α-helices have large excitons, they can be distinguished by their smaller diagonal anharmonic shift [127].

6.3.2 Diagonal and off-diagonal disorder

The above discussion only rigorously applies to perfectly ordered systems. In condensed-phase systems, solvent molecules will interact with the oscillators through electrostatic forces or hydrogen bonding, for example, that will change the local mode frequencies. Structural changes, such as bends or kinks in the periodicity, can also change the local mode frequencies as well as the couplings. To model these effects, we add disorder terms to the diagonal and off-diagonal elements of the Hamiltonian

$$H = \begin{pmatrix} \hbar(\omega_0 - \delta\omega) & \beta_1 - \delta\beta & & \\ \beta_1 - \delta\beta & \hbar(\omega_0 - \delta\omega) & \beta_1 - \delta\beta & \\ & \beta_1 - \delta\beta & \hbar(\omega_0 - \delta\omega) & \beta_1 - \delta\beta \\ & & \ddots & \ddots & \ddots \end{pmatrix} \quad (6.32)$$

where $\delta\omega$ and $\delta\beta$ may be different for each matrix element. Consider *diagonal disorder*. $\delta\omega$ splits the energy levels, so that the coupling terms β do not mix the modes as effectively. If $\delta\omega$ becomes larger than β, then the frequency fluctuations of the local modes will dominate the spectra more so than the effects of coupling. Thus, $\delta\omega$ diminishes the importance of the coupling, causing the excitons to localize onto the local modes (as if they were not coupled). *Off-diagonal disorder* does not decrease the effectiveness of mixing, but randomizes it. The result is often the appearance of vibrational modes that are forbidden in rigorously periodic structures, such as the weak lines in Fig. 6.6(b).

To realistically simulate linear and 2D IR spectra of flexible molecules like peptides and proteins, one must often include diagonal and off-diagonal disorder. To include diagonal disorder, one can add a randomly generated $\delta\omega$ to each diagonal matrix element in the Hamiltonian and then sum the spectra calculated from many Hamiltonians. The magnitude of $\delta\omega$ can be estimated from the linewidths. For example, a random coil peptide has a center absorption of about 1645 cm^{-1} and a width of typically 35 cm^{-1}, which is caused largely by differences in hydrogen bonding and solvation of the peptide backbone (i.e. diagonal disorder). Off-diagonal disorder can also be randomly generated, but since it is so closely tied to molecular structure, it may be better to generate a realistic distribution of structures, produce a Hamiltonian for each, and then sum the resulting spectra. In Chapters 7 and 8, we discuss the dynamical aspects of diagonal and off-diagonal disorder.

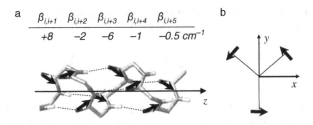

Figure 6.7 (a) Idealized α-helix with the local mode transition dipoles indicated. The helix is considered to lie along the $\hat{z}$-axis. Also shown are approximate coupling terms between the transition dipoles. (b) Transition dipoles looking down the helix axis for a helix that has three oscillators per turn (an α-helix has 3.6).

6.3.3 α-Helices

α-Helices differ from linear chains in two regards. First, they are 3D structures, and so have more than one symmetry-allowed mode. Second, since an α-helix has 3.6 residues per turn (Fig. 6.7), nearest neighbor couplings are not the only prominent coupling terms. Since the Hamilton matrix (Eq. 6.30) is still the same as for a linear chain, it will have the same eigenstates and eigenenergies (Eqs. 6.27 and 6.31). Thus, the dispersion curve will be the same as a linear chain that contains multiple coupling terms. However, when calculating the transition dipoles of an excitonic state, we now need to keep in mind that the local mode transition dipoles are vectors arranged in a 3D array (Fig. 6.7):

$$\vec{\mu}_k^{(ex)} = \frac{1}{\sqrt{N}} \sum_j e^{i\frac{2\pi jk}{N}} \vec{\mu}_j. \tag{6.33}$$

As a result, we now obtain three IR active excitonic states [136]. Since the z-component along the helical axis is the same for all peptide units ($\langle 0|\mu_z|j\rangle$ = const.) (Fig. 6.7), the total z-component averages to zero except when $k = 0$. This is the A mode with a large transition dipole parallel to the helical axis. It has the eigenstate energy

$$E_A = \hbar\omega_0 + 2\sum_j \beta_j. \tag{6.34}$$

The other two IR active excitonic states are along the x- and y-axes. The x-component of peptide unit j is $\langle 0|\mu_x|j\rangle \propto \cos(j\gamma) \propto e^{ij\gamma} + e^{-ij\gamma}$, where $\gamma = 2\pi/l$ and l is the number of oscillators per turn of the helix (see Fig. 6.7b). Plugging μ_x into Eq. 6.33, one finds that the total x-component averages to zero unless $2\pi k/N = \pm\gamma$, which gives $k = N/l$ (and the same holds for the y-component). Thus, the x- and y-components give rise to doubly degenerate E

modes whose transition dipoles lie in the plane perpendicular to the helical axis and have energies

$$E_E = \hbar\omega_0 + 2\sum_j \beta_j \cos\left(\frac{2\pi j}{l}\right). \tag{6.35}$$

The separation between the A and E modes depends on l, the number of oscillators per turn of the helix. Thus, the separation is different for 3_{10} than α-helices, for example. As for their intensities, the strengths of the two bands are determined by the angle of the transition dipole moment to the helix axis [137]. In an α-helix, the E modes are much weaker than the A mode, because the local mode transition dipoles lie nearly parallel to the helix axis (Fig. 6.7a). In fact, the E modes are not easily observed in infrared spectra. In DNA, where the base-pair carbonyls are nearly perpendicular to the helix axis, the E mode is more intense.

So, what are the local mode frequencies and couplings strengths for an α-helix? The local mode frequency should be the same as that for a random coil peptide, which is about 1645 cm^{-1}. The coupling strengths have been estimated from a number of experimental and theoretical studies, and generally agree to within a couple of wavenumbers. The approximate values for an α-helix are given in Fig. 6.7(a). Whether or not these numbers are precisely correct, it is certainly true that the nearest neighbor coupling $\beta_{i,i+1}$ is large and positive whereas the other couplings are negative. Since the frequencies of the E_A mode of an α-helix is given by the sum of the coupling strengths (Eq. 6.34), there is competition between nearest neighbor couplings, which shift the amide I band in one direction, and longer-range couplings which shift it in the other direction. Thus, the frequency at which an α-helix absorbs depends on the length of the helix and the long-range disorder of the structure. The shorter the helix and the more disordered, the lower the contribution of the negative couplings terms so that the nearest neighbor couplings dominate, which shifts the absorption band to higher frequencies. One can also observe the competition between positive and negative couplings in isotope labeled peptides, which we discuss below [39].

6.3.4 β-Sheets

In contrast to linear chains and α-helices, which are nominally 1D systems, β-sheets are created from peptide strands that form 2D arrays of coupled oscillators. The strands are aligned in either an antiparallel or parallel fashion, as shown in Fig. 6.8(a,b). β-Sheet Hamiltonians are more difficult to solve analytically, but like all exciton systems, the eigenstates form a dispersion curve of energies and the symmetries dictate which of these modes are IR active. Shown in Fig. 6.8(c) is a simulated IR spectrum of an antiparallel sheet composed of five strands with

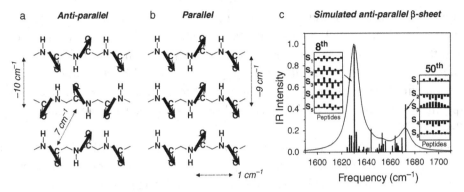

Figure 6.8 Structure and transition dipoles for (a) antiparallel and (b) parallel β-sheets. Also shown are some of the largest couplings. (c) Simulated IR spectrum of five-strand, 10-residue per strand, antiparallel sheet. Diagrams for the eighth and fiftieth eigenstates (the strongest) are shown that give the relative contributions of the local modes. Taken from Ref. [77] with permissions.

10 residues (local modes) in each strand [77]. The simulations are quite sophisticated in that they include diagonal and off-diagonal disorder, which is why many eigenstates contribute (the thin lines in Fig. 6.8c). Nonetheless, the spectrum is dominated by just two eigenstates, which are the eighth and the fiftieth modes. In simulations of perfectly ordered and large antiparallel β-sheets, these are the only two modes that are observed. The eighth mode, $A_\perp$, is bright because it is caused by the vector sum of transition dipoles that lie in lines across the strands, which can be seen in the inset of Fig. 6.8(c) that shows the relative amplitudes of the local modes. Residues that are in register from strand to strand have the same phase, in analogy to the $k=0$ eigenvector of a linear chain. These linear chains create a transition dipole that lies perpendicular to the strands. Since $k=0$ eigenvectors for each linear chain are out of phase with each other, the transition dipole of the eighth mode has no component along the strands. The strong intersheet coupling is about 10 cm^{-1}, which is why the observed mode appears at a lower frequency than the local modes just like a linear chain would be with negative couplings. In contrast, for the fiftieth eigenstate, $A_\parallel$, the local modes between strands are out of phase so that the interstrand linear chains are disrupted. As a result, residues along the strands are in phase, causing the transition dipole of $A_\parallel$ to lie parallel to the strands. It is well established by empirical observations that the characteristic infrared absorption of an antiparallel β-sheet is a very intense mode around 1620 cm^{-1} and a weaker but prominent mode at about 1670 cm^{-1}.

Parallel β-sheets have parallel and perpendicularly oriented transition dipoles similar to those in antiparallel β-sheets, but the parallel transition does not appear at the highest frequency, but is near the tenth eigenstate (for a 50-residue sheet).

Thus, it lies at much lower frequency. In principle, one can distinguish between parallel and antiparallel β-sheets by the prominence of $A_{||}$, but in practice it is difficult because disorder causes the symmetry forbidden transitions to gain intensity (i.e. all the weak modes in Fig. 6.8c) to make the comparison less obvious in actual proteins. However, it has been proposed that polarized 2D IR spectroscopy is a more reliable way of distinguishing between anti and parallel β-sheets [77].

Finally, we note that the frequency of the transition dipole that runs perpendicular to the strands (the low-frequency mode) is quite sensitive to the number of residues over which the exciton is delocalized. In aggregation studies of peptides that form amyloid fibers of parallel β-sheets, this mode has been observed to red-shift as the size of the β-sheets grows until it reached about 1620 cm^{-1} which is the typical absorption frequency for large and well-ordered β-sheets [172]. Moreover, in the 2D IR spectra, the diagonal anharmonic shift decreases with the size of sheet, as was discussed above on the section about linear chains. We return to these observations and discuss how they can be exploited for structure determination in the next section on isotope labeling.

6.4 Isotope labeling

Isotope labeling is an extremely powerful tool in infrared spectroscopy. It can be used to identify specific bonds as well as probe exciton delocalization by disrupting couplings. Nitriles, ^{15}N, C–D and other groups have been used in conjunction with 2D IR spectroscopy as well, but in this section we limit ourselves to isotope labeling with regard to the amide I mode of proteins. Infrared spectroscopy of the amide I mode has been used for decades to probe protein secondary structure via the characteristic exciton states that we discussed above. That is, excitonic states are good global structural indicators. However, delocalized states are not useful if one wants residue- or bond-specific structural resolution. Nowadays, with advances in structural biology, bond-specific resolution is usually necessary to obtain structural information of sufficient detail to be of interest to protein scientists.

The amide I band of proteins is about 85% carbonyl stretch, 10% C–N stretch, and some N–H bend. Thus, isotope-labeling either the carbon or oxygen of the carbonyl produces a sizeable frequency shift. Shown in Fig. 6.9 is a spectrum of a 27-residue transmembrane peptide without isotope labels, with a single ^{13}C=O labeled residue, and a single ^{13}C=^{18}O labeled residue. The ^{13}C=O band is moderately well resolved, but still lies on the tail of the unlabeled amide I band. Moreover, this label competes with the natural abundance of ^{13}C=O, which is about 1%. Thus, even a

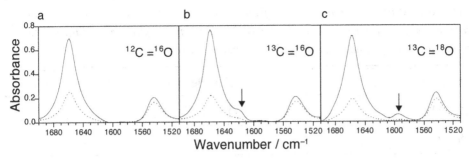

Figure 6.9 Linear (FTIR) spectrum of the 27-residue CD3ζ transmembrane peptide in a lipid bilayer with (a) no labeled residues, (b) a $^{13}C=O$ and (c) a $^{13}C=^{18}O$ labeled residue in D_2O. Adapted from Ref. [180] with permissions.

short 40-residue peptide will have a natural abundance $^{13}C=O$ peak that is 40% the intensity of a $^{13}C=O$ label. A more advantageous label is $^{13}C=^{18}O$, which has about a 60 cm^{-1} shift that places it in between the unlabeled amide I and II bands [180]. ^{18}O labeling is performed by either acid catalyzed exchange on an unprotected ^{13}C labeled amino acid [134] or using a multiple turn-over method on either protected or unprotected amino acids [164].

One must be aware that some side-chains absorb in the same region as the isotope labels (side-chains also absorb in the unlabeled region as well) [10]. The most problematic side-chain absorbances for $^{13}C=^{18}O$ labels are from the COO$^-$ bend of glutamic acid (Glu, E) and aspartic acid (Asp, D), the NH$_2$ bend of asparagine (Asn, N) and glutamine (Gln, Q), and the C=NH$_2$ mode of arginine (Arg, R). The frequencies of these bands are given in Table 6.1 in both H$_2$O and D$_2$O. These five amino acids have oscillator strengths that are comparable to the amide I stretch and thus will have similar intensities as an isotope-labeled peptide. By a judicious choice of either H$_2$O or D$_2$O for the solvent, one can minimize these side-chain absorbances. For Glu and Asp residues whose absorbances do not depend on solvent, one can ^{18}O exchange their side-chains, which shifts their frequencies far from the $^{13}C=^{18}O$ label [145]. Other side-chains also absorb in the region of the $^{13}C=^{18}O$ label, but their oscillator strengths are weaker. As we have pointed out before, 2D IR spectra scale as $|\mu|^4$ whereas linear spectra scale as $|\mu|^2$. Thus, $^{13}C=^{18}O$ labels appear more prominent in 2D than 1D spectra, and those weaker bands are not so much an issue.

We note that H$_2$O cannot usually be used in IR transmission studies of proteins because the H$_2$O bend obscures the amide I band, so D$_2$O is used instead. But for membrane systems either H$_2$O or D$_2$O can be used because the amount of excess water can be kept small. As the amide I vibration also involves to a certain extent an N–H bend vibration, deuteration has a small effect on the amide I frequency as

Table 6.1 *Approximate local mode frequencies for isotope-labeled amide I modes and side-chain absorbance in H_2O and D_2O solvent. Absorbance of the COO^- side chain of Glu and Asp when ^{18}O labeled is also given.*

Amide I isotope label	Local mode frequency	
$^{12}C^{16}O$	1645 cm^{-1}	
$^{13}C^{16}O$	1620	
$^{13}C^{18}O$	1590	
Amino acid side-chain	H_2O solvent	D_2O solvent
Glutamic acid (Glu,E) COO– asym	1550–1590 cm^{-1}	1550–1590 cm^{-1}
Aspartic acid (Asp,D) COO– asym	1550–1590	1550–1590
Glu and Asp $C^{18}O^{18}O$	<1550	<1550
Asparagine (Asn,N) NH_2 bend	1580–1625	<1550
Glutamine (Gln,Q) NH_2 bend	1580–1625	<1550
Arginine (Arg,R) $C=NH_2$ mode	1630–1640	1585–1610

well (≈ 5 cm^{-1}). A deuterated amide vibration is often labeled with a prime, e.g. amide I'. For small peptides, it is straightforward to exchange the labile protons, but for proteins with a rigid 3D structure, the accessibility of the hydrophobic core is limited and so care must be taken to fully H–D exchange the protein.

Isotope labeling changes the diagonal, not the off-diagonal, elements of the Hamiltonian. Thus, it does not alter the magnitude of the coupling. But by introducing an energy gap, it effectively does decouple the labeled residue from the rest of the peptide if the energy gap is large enough. How large must the gap be to ensure that the isotope labels can be considered as their own Hamiltonian? It depends. Intensities are more sensitive to coupling than frequency shifts (see Problem 6.14). In some peptides, the integrated intensity of the $^{13}C=O$ absorption band relative to the unlabeled band is not proportional to the number of labeled residues, which may indicate that coupling is still important. Furthermore, do not forget that most excitonic states of protein secondary structures are IR inactive (see the dispersion curves above), so that unobserved modes can still couple to the labels. In both these regards, $^{13}C=^{18}O$ labeling is better localized.

6.4.1 Strategies for isotope labeling

There are several general strategies for extracting bond-specific structural information using isotope labeling. First, the 2D linewidths of isotope labels are good probes of environment, such as the heterogeneous environment caused by

6.4 Isotope labeling

membranes. Using lineshape analysis, one can probe the secondary structure and orientation of peptides in lipid bilayers [194]. Second, one can monitor the coupling between isotope labels on different peptides. This approach has proven useful in amyloid fiber studies of peptides that self-assemble into β-sheets [109, 151, 167]. If the labels are in-register when folded, they will form linear exciton chains, thereby giving information on the register of the stacked peptides. Third, one can look at the cross-peaks between the isotope-labeled and the unlabeled absorption features. Such an approach has been used on di- and tripeptides to obtain dihedral angle information [190]. If the label is in, or adjacent to, an α-helix or β-sheet excitonic band, then these cross-peaks can also provide information on the secondary structure of the residue [173]. Fourth, two residues can be labeled and the coupling measured between them using 2D IR spectroscopy or using linear spectroscopy by identifying peak shifts [9, 53]. This approach was used to study the dimerization of the transmembrane peptide Glycophorin A [54]. Fifth, hydrogen/deuterium exchange can be measured with residue-specificity using isotope labeling by looking at the cross-peak between the label and the amide II band, which is not possible with linear IR spectroscopy [13]. Finally, one can express uniformly ^{13}C labeled proteins so that protein–protein or domain-specific studies can be performed [140].

The above ideas probe structure using the spectroscopic signatures of the label itself, either its frequency, lineshape, or cross-peaks. Another approach is to use the isotope labels to create defects in excitonic bands. For example, if one or more isotope labels are placed into a delocalized vibrational mode, such as in an α-helix, then the unlabeled absorption band will be shifted because the exciton will become more localized. Labels placed into the middle of the exciton band will have a larger effect than isotope labels on the edges [39]. One can also create defects that alter the cross-peaks in the unlabeled residues. For instance, by placing isotope labels in between β-sheet and β-turn structures, one effectively decouples them, which suppresses the cross-peaks between the two [173].

To illustrate how isotope labeling can be used to probe structure either directly through couplings or indirectly through defects, we present two sets of simulations shown in Fig. 6.10. These simulations are of a ^{13}C=^{18}O isotope-labeled α-helical peptide. Although these spectra are simulated, they very closely match the experimental spectra after which they are modeled (when ^{13}C=O isotopes are simulated as was used in the experiments) [9, 39]. The simulations include both diagonal and off-diagonal disorder. The off-diagonal disorder is generated by using an NMR distribution of structures from a comparable helical structure from the protein data bank. The first set of simulations is of the α-helix labeled with pairs of ^{13}C=^{18}O isotopes spaced by one, two and three residues (Fig. 6.10a), as well as a singly labeled helix. Notice that the $i/i+1$ simulation has a peak that is

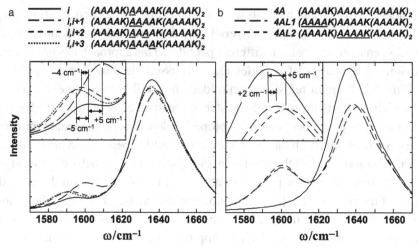

Figure 6.10 Simulated linear IR spectra of a $^{13}C = {}^{18}O$ isotope-labeled α-helical peptide. (a) Single and double isotope labels. Inset shows the isotope labeled features. (b) Unlabeled and four-residue stretches (one turn) of the helix labeled. The inset shows the shift in the unlabeled amide I peak.

+5 cm^{-1} higher than the single label, the $i/i + 2$ pair appears at -5 cm^{-1}, and the $i/i + 3$ pair appears about -4 cm^{-1} lower than the single label (see inset). How do we interpret these frequencies? If the label residues are not affected by the unlabeled ones (which can be tested in the simulations) then they can be simulated independently using a 2×2 Hamiltonian using the frequency of the single label as the diagonal matrix elements. Thus, upon diagonalization, there should be two observed modes that are shifted by $\pm\beta_{i,i+n}$ from the local mode frequency. In the simulations (and the experiments), only one band is observed because the transition dipoles are nearly parallel (see Fig. 6.1), but the frequency of that band gives the coupling strength. Thus, one can measure from the spectrum the coupling constants of an α-helix, which are found to be $\beta_{i,i+1} = +5$, $\beta_{i,i+2} = -5$ and $\beta_{i,i+3} = -4$. The actual couplings used in the simulations are +6, −4 and −4.5, respectively. The observed spectra are close to the predicted splitting, but are not perfect because of the disorder in the Hamiltonian and the broad linewidths. Thus, fits and simulations of experimental data are recommended when working with experimental data.

The second set of simulations also comes from an experiment in which four consecutive residues were isotope-labeled so that an entire turn of an α-helix could be resolved. Figure 6.10(b) contains simulations of the unlabeled helix (4A), the helix with the first four residues labeled (4AL1), and the second set of four residues labeled (4AL2). In this case, the isotope labeled bands are nearly all identical, because it is a well-formed helix in these regions (in the experiment, the N-terminus is capped with an acetyl group to prevent fraying). However, notice that the frequency of the unlabeled helix shifts upon labeling. When the first four residues are labeled, the unlabeled band shifts by about 2.5 cm^{-1}, and it shifts an additional 2.5 cm^{-1} when the middle four residues are labeled instead. Why does this happen? The answer lies in how the isotope labels affect the ratio of the nearest neighbor to long-range couplings. Recall Eq. 6.34 that gives the eigenstate energy of the IR observable exciton in helices. It is a summation of all the couplings in the helix and the nearest neighbor coupling ($\beta_{i,i+1}$ is positive while all the others are negative). When four residues are labeled, the length over which the excitons can delocalize on the unlabeled portion of the helix is smaller, which decreases the number of long-range couplings to offset the frequency shift of the nearest neighbors. Thus, the absorption band shifts to higher frequencies. The largest effect is when the labels are placed directly in the middle of the peptide, which creates two very short helical peptides (from an exciton perspective), and thus shifts the absorption band even higher. Thus, one can probe structures by creating defects in exciton modes and observing its effect on the direction and magnitude of the unlabeled absorption bands.

6.5 Local mode transition dipoles

In Section 2.1 we defined the transition dipole moment, but we did not discuss its origin, which we do now in the context of local modes. One-photon transitions between $|v\rangle$ and $|v'\rangle$ can only occur if $\langle v|\hat{\mu}|v'\rangle \neq 0$. Since the vibrational states depend on the local mode coordinate q, the q-dependence of the permanent dipole $\hat{\mu} \rightarrow \hat{\mu}(q)$ will influence the integral. Thus, we Taylor expand $\hat{\mu}(q)$

$$\langle v|\hat{\mu}(q)|v'\rangle = \mu(q_e)\langle v|v'\rangle + \left(\frac{\partial \mu}{\partial q}\right)_{q_e} \langle v|\hat{q}|v'\rangle + \cdots \qquad (6.36)$$

where q is the local mode coordinate at equilibrium. Since the vibrational eigenstates are orthogonal, the first term is zero. $\langle v|\hat{q}|v'\rangle$ of the second term gives the selection rules of $\Delta v = \pm 1$, which are found when q is written as ladder operators (Appendix B). $(\partial \mu/\partial q)$ gives the transition dipole strength.

Figure 6.11 illustrates the transition dipole for the amide I mode of one residue in a small polypeptide. What is shown is the difference in the charge density, which can be calculated from an electronic structure program like *Gaussian* for the amide I mode stretched and compressed (i.e. ∂q). One can see that the transition charge density does not simply consist of two opposite charges separated by a small distance. That is, it does not look like a simple dipole. The transition charge

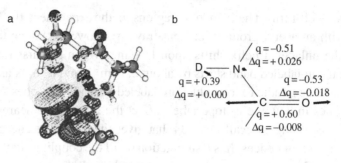

Figure 6.11 (a) Transition charge density of one amide I vibration in a short α-helical segment. Adapted from Ref. [139] with permission. (b) The nuclear displacements, partial charges and charge flow of the amide I normal mode [84].

extends not only over the C=O group, but also covers the C–N bond and to a certain extent the C_α atoms as well. The amide I vibration is accompanied by a charge flow between the carbonyl oxygen and the nitrogen which is responsible for about half of the total transition dipole moment of the amide I mode, which is why the transition dipole does not lie along the carbonyl bond (Fig. 6.5).

How do we know how to displace the molecular bonds since the local coordinates q are not just the carbonyl stretches? We use the normal mode of a model monomer, such as N-methyl acetamide (NMA), which we compute from *ab initio* quantum chemistry calculations (Fig. 6.11b). We then use the displacements of the normal mode as our q for each of the peptide "local" modes.

6.6 Calculation of coupling constants

Although being very intuitive, the transition dipole–dipole coupling model (Eq. 6.16) includes two approximations that often limit its applicability in quantitative modeling: (a) the limitation of the dipole approximation and (b) the neglect of through-bond effects such as charge flow or mechanical couplings. Regarding the first limitation, we saw in Fig. 6.11(a) that the *transition charge density* is more complicated than a simple dipole. Moreover, it extends over a length of about 2.5 Å, which is comparable to the distance between adjacent peptide units in a folded polypeptide. Thus, when residues are far apart, transition dipole–dipole coupling is adequate, but for nearest neighbors the dipole approximation is less accurate. The second limitation is that, since the charge density extends to the C_α atoms, nearest neighbors will contain mechanical effects as well. Transition dipole coupling can only account for electrostatic effects. However, there are a number of more sophisticated coupling models that can account for electrical and mechanical couplings, which we outline below.

6.6 Calculation of coupling constants

One can use molecular orbital calculations to take into account electrostatic and electrodynamic contributions to the coupling. To this end, we take a local mode point of view. Consider a dipeptide that has two amide I "local" modes. To compute the coupling, we displace the atoms of each peptide unit along the coordinates of the local amide I modes, q_1 or q_2, and calculate the potential curvature to get the coupling constant

$$\beta_{12} = \frac{\partial^2 V(q_1, q_2)}{\partial q_1 \partial q_2} \tag{6.37}$$

where the energies $V(q_1, q_2)$ are obtained from single-point *ab initio* calculations and the second derivatives are computed as finite differences. This approach implicitly takes into account the electrostatic interaction between the charge densities of the two peptide units, as well as the charge flow between units.

If charge flow between local modes is unimportant, then one can use the transition charge density of the model compound (such as in Fig. 6.11) for each of the two coupled local modes and integrate the electrostatic interaction energy between the two. Such an approach has been referred to as the *transition charge density model* [139]. A numerically simpler approach is to first condense the transition charge density to Mulliken point charges, in which case the integration is replaced by a summation over the atoms. That leads to the *transition point charge model*, which can be very efficiently parameterized [84], and which still avoids some of the limitations of the dipole approximation. In the transition point charge model, a point charge e_n and point charge flow de_n are assigned to each atom n in peptide unit i, with position $r_n(q_i)$ (Fig. 6.5b). The vibration of one peptide unit gives rise to an oscillating electric field which interacts with that of a second peptide unit according to

$$\beta_{12} = \frac{1}{4\pi\epsilon_0} \frac{\partial^2}{\partial q_1 \partial q_2} \sum_{n,m} \frac{(e_n + de_n q_1)(e_m + de_m q_2)}{|r_n(q_1) - r_m(q_2)|} \tag{6.38}$$

where q_1 and q_2 once again are the normal mode coordinates of the amide I mode of peptide units 1 and 2, respectively.

None of the approaches discussed so far include mechanical coupling. In particular the –CH bending mode of the C_α atom connecting nearest neighbor peptide units vibrates a little when either mode is stretched. Thus, it mechanically couples the amino acids to one another. In order to include this effect, one can perform a full normal mode calculation and deduce the coupling from a Hessian matrix reconstruction [73, 78]. That is, one uses Eq. 6.17 as an ansatz for the sub-Hamiltonian of the amide I modes and deduces the mixing angle directly from the normal mode eigenvectors, using the C=O distance as a measure of the normal

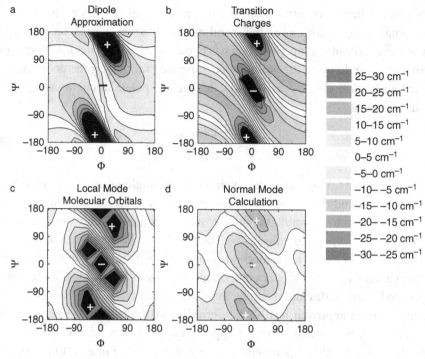

Figure 6.12 Coupling maps for the glycine dipeptide as a function of the central (ϕ, ψ) dihedral angles for (a) the transition dipole coupling [112, 178], (b) transition charge coupling [84], (c) local mode molecular orbital calculation [81] and (d) the full normal mode calculation [73]. Adapted from Refs. [73, 81] with permission.

mode displacement. With Eq. 6.20, one may then calculate the coupling and the local mode frequencies.

6.6.1 Comparison of the models and coupling maps

Figure 6.12 compares the couplings predicted by the various coupling models for the nearest neighbor amide I vibrational modes in the glycine dipeptide. The calculations were performed as a function of the central (ϕ, ψ) dihedral angles. All of the coupling models predict the same sign for the coupling, although transition dipole coupling does not give the correct coupling value near $(\phi, \psi) = (0, 0)$. The angular dependence of the three more sophisticated models all agree qualitatively, although they differ on the overall strength of the coupling. To our knowledge, there are very few quantitative experimental tests of these calculations.

It turns out that the coupling maps in Fig. 6.12 do not depend very much on the types of amino acids, the size of the proteins, or the presence of solvent (according to theory). Hence, one may use a precalculated map such as Fig. 6.12(d) for nearest neighbor coupling, and the transition charge model for the coupling between more distant peptide units, for which through-bond effects are not expected to play a significant role. This is what we will use in the case study in Section 10.3, because it is believed to be reasonably accurate and at the same time computationally efficient.

With present computer power, a full *ab initio* normal mode calculation is in principle feasible for mid-sized peptides with 10–20 amino acids (strongly depending on the level of theory used). However, in order to account for inhomogeneous broadening, a realistic simulation of 2D IR spectra will always require an average over an ensemble of structures, which makes calculations on all but the smallest peptides prohibitively expensive. Moreover, to search for a structure that matches an experimental spectrum requires some sort of optimization algorithm, which iteratively varies the conformation of the molecule, calculates a set of coupling constants for each configuration and compares them with the experimental data. Since this will be a high-dimensional optimization problem, it will once again be prohibitively expensive. That is why the models discussed above are so important, because they enable fast analyses while also providing an intellectually insightful method for interpreting the spectra.

6.7 Local versus normal modes

The vibrational eigenstates of molecules are, in principle, obtained by solving the nuclear Schrödinger equation. A common method to solve the Hamiltonian is to write it in the basis set of normal modes. The mathematical derivation of normal modes is not discussed here, but is contained in many introductory spectroscopy textbooks (see Appendix E). For peptides and polymers with repeating patterns, we often use the local mode basis, which was the focus of the preceding sections. The aim of this section is to compare these two descriptions to better understand when one provides a closer approximation to the true eigenstates and to provide deeper insight into the interpretation of 2D IR spectra.

The decision of whether to use local or normal modes depends upon which most easily describes the potential energy surface. Normal modes are exact when the potential energy surface is harmonic, but when there is anharmonicity, the Hamiltonian in the normal mode basis set is not diagonal. So, the vibrational potential is expanded in the normal mode basis

$$V(Q_1, Q_2) = \frac{1}{2} \sum_i \left.\frac{\partial^2 V}{\partial Q_i^2}\right|_0 Q_i^2 + \frac{1}{6} \sum_{ijk} \left.\frac{\partial^3 V}{\partial Q_i \partial Q_j \partial Q_k}\right|_0 Q_i Q_j Q_k$$
$$+ \frac{1}{24} \sum_{ijkl} \left.\frac{\partial^4 V}{\partial Q_i \partial Q_j \partial Q_k \partial Q_l}\right|_0 Q_i Q_j Q_k Q_l + \cdots \quad (6.39)$$

where Q_n are the normal mode coordinates. Because we do not care about the offset and are evaluating the potential at the bottom, the zeroth- and first-order terms are zero (not shown), as are the bilinear Q_iQ_j terms by construct of the normal mode transformation (also not shown). Therefore, the first nonzero terms are the quadratic terms, Q_i^2, as well as the cubic and quartic terms $Q_iQ_jQ_k$ and $Q_iQ_jQ_kQ_l$. These higher-order terms introduce anharmonicity. It turns out that the anharmonicity shifts Δ_{ij} have two contributions. The first scales quadratically with the cubic expansion coefficients, while the second scales linearly with the quartic expansion coefficients, which is why both have to be considered.

Each term can be translated into ladder operators B_i and $B_i^\dagger$ that operate on the normal mode basis functions, just like we used ladder operators b_i and $b_i^\dagger$ in the local mode basis (Eq. 6.4). The summation over $Q_iQ_jQ_k$ contains terms that couple normal modes of different quanta, such as $B_1B_1B_1^\dagger$ (e.g. $|00\rangle$ with $|10\rangle$) and $B_1^\dagger B_1 B_2$ (e.g. $|10\rangle$ with $|11\rangle$). The fourth-order summation also has terms that couple states of different quanta such as $B_1^\dagger B_2^\dagger B_2 B_2^\dagger$ (e.g. $|00\rangle$ with $|11\rangle$) as well as "diagonal" terms like $B_1 B_1^\dagger B_2 B_2^\dagger$ (e.g. $|10\rangle$ with $|10\rangle$) that accounts for anharmonicity along a single coordinate. To illustrate the locations of these anharmonic terms, we schematically write them in the same form as the Hamilton matrix:

$$\begin{pmatrix} B_iB_i^\dagger B_jB_j^\dagger & \text{h.c.} & \text{h.c.} \\ B_i^\dagger B_jB_j^\dagger & B_iB_i^\dagger B_jB_j^\dagger & \text{h.c.} \\ & B_i^\dagger B_jB_jB_j^\dagger & \\ B_iB_i^\dagger B_j^\dagger B_j^\dagger & & B_iB_i^\dagger B_jB_j^\dagger \\ B_i^\dagger B_j^\dagger B_jB_j^\dagger & B_i^\dagger B_jB_j^\dagger & B_i^\dagger B_jB_jB_j^\dagger \\ & & B_iB_iB_j^\dagger B_j^\dagger \end{pmatrix} \cdot \quad (6.40)$$

Degenerate perturbation theory can be used to remove the coupling between the blocks of this Hamiltonian, which gives the diagonal elements that are described by the *Dunham expansion*, which is an expression commonly used for the energies of the normal mode eigenstates

$$\begin{aligned} E_{n_1,n_2} = \langle n_1n_2|H|n_1n_2\rangle &= \hbar\omega_1(n_1+1/2) + x_{11}(n_1+1/2)^2 \\ &+ \hbar\omega_2(n_2+1/2) + x_{22}(n_2+1/2)^2 \\ &+ x_{12}(n_1+1/2)(n_2+1/2) \end{aligned} \quad (6.41)$$

where ω_i and x_{ii} are the normal mode frequencies and anharmonicities, x_{ij} is the *intermode* anharmonicity, and n_i are the quantum numbers of the normal modes [94]. From this equation, one can derive the 0–1 transition frequency of mode 1, for example, as:

$$\nu_{0\to 1}^{(1)} = E_{1,n_2} - E_{0,n_2} = \nu^{(1)} + x_{12}n_2, \quad (6.42)$$

6.7 Local versus normal modes

where $\nu^{(1)} = \omega_1 + 2x_{11} + x_{12}/2$. Hence, the mixed-anharmonic constant x_{12} causes a frequency shift in mode 1 when mode 2 is excited. That is, the 0–1 frequency of mode 1 is ν_1 if the other mode is in its ground state $n_2 = 0$, while it is $\nu_1 + x_{12}$ if the other mode is vibrationally excited.[3] In fact, x_{12} is related to the off-diagonal anharmonic shift by $x_{12} = \Delta_{12}$ so that it can be read directly from a 2D IR spectrum by the splitting of the cross-peak pairs (Fig. 1.5), which is unlike β from the local mode basis which must be deduced by fitting the eigenstates. Thus, normal modes give a complementary view of molecular couplings. The calculation of the anharmonic constants x_{ij} is part of modern quantum chemistry programs such as Gaussian [58] (see Section 10.2 for a case study).

It is important to note that there are still off-diagonal elements that have not been considered, the most important of which are $B_i B_i B_j^\dagger B_j^\dagger$. Since they couple nearly degenerate states, perturbation theory does not apply for them. These terms are called the *Darling–Dennison* coupling terms, γ, which are given by [123]

$$\langle n_1 + 2, n_2 - 2 | H | n_1 n_3 \rangle = \frac{1}{2}\gamma[(n_1+1)(n_1+2)n_2(n_2-1)]^{1/2} \quad (6.43)$$

$$\langle n_1 - 2, n_2 + 2 | H | n_1 n_3 \rangle = \frac{1}{2}\gamma[n_1(n_1-1)(n_2+1)(n_2+2)]^{1/2}. \quad (6.44)$$

To relate normal modes to local modes, we consider as a simple example a normal mode Hamiltonian in which the following holds [135]:

$$x_{11} = x_{22} = x_{12}/4 = x/2 = -\Delta/4$$
$$\omega_1 = \nu + \beta$$
$$\omega_2 = \nu - \beta$$
$$\gamma = x \quad (6.45)$$

where ν is the local mode *transition* frequency (e.g. $\nu = \omega_{01} - 2x$) and Δ is the local mode anharmonic shift, which we used above in Eq. 6.10. When the zero-point energy is subtracted ($E_{00} = 0$), then the normal mode Hamiltonian becomes

$$H = \begin{pmatrix} 0 & & & & \\ & \hbar\omega - \beta & & & \\ & & \hbar\omega + \beta & & \\ & & & 2\hbar\omega - 2\beta - \frac{1}{2}\Delta & -\frac{1}{2}\Delta \\ & & & -\frac{1}{2}\Delta & 2\hbar\omega + 2\beta - \frac{1}{2}\Delta \\ & & & & & 2\hbar\omega - \Delta \end{pmatrix}. \quad (6.46)$$

---

[3] For carbonyl vibrations x_{12} is negative and so a downshift in the frequency is observed.

Notice that the normal mode Hamiltonian is already diagonalized, except for the Darling–Dennison coupling terms that create the off-diagonal elements in the two-quantum block of states. When this normal mode Hamiltonian is fully diagonalized, one obtains the same eignestates as for the local mode Hamiltonian (Eq. 6.13). Thus, the molecular eigenstates can be described equally well (or poorly) by either Morse oscillator local modes coupled by β, or Dunham style normal modes coupled by the Darling–Dennison terms [135]. In other words, it is β that couples the local modes and Δ that couples the normal modes. Which basis set is better for describing the true molecular vibrations? It depends on the relative magnitudes of β and Δ. If $\beta \gg \Delta$, then the normal modes more closely represent the true molecular eigenstates because Δ is just a perturbation. If $\Delta \gg \beta$, then β acts as a perturbation to the local modes. This is the case when "forbidden" transitions become weakly allowed (Fig. 6.4b). If the potential energy surface happens to be exactly described by Eq. 6.45 then either basis set works equally well. But generally one basis set will naturally lie closer to the true molecular eigenstate energies, which is generally the preferable one because then the inaccuracies will be minimized.

In symmetric systems, normal modes are often preferable for the one-quantum states, but we usually prefer the local mode description for repeating systems like peptides and proteins. In systems like these, local modes are preferable because (a) we can build a Hamiltonian using the amino acids as the basis set, (b) disorder is structurally intuitive and (c) cubic and higher-order anharmonic terms are mostly local. The third point expresses the fact that a C=O vibration is more accurately described by a Morse oscillator rather than a harmonic oscillator. The leading coupling term of two local modes, $q_1 q_2$, is bilinear, thus still harmonic, whereas mixed higher-order terms such as $q_1^2 q_2$ are smaller in a local mode basis (and we usually neglect them). In contrast, in a normal mode basis (Eq. 6.39) these mixed higher-order terms generally all contribute.

6.8 Fermi resonance

Another example of intramolecular coupling is a *Fermi resonance*. Overtone and combination bands are usually very weak in linear spectroscopies, but if an overtone or combination band is accidentally in resonance with a bright fundamental transition to which it is coupled, it will gain intensity by stealing it from the bright state. Fermi resonances often appear unexpectedly in infrared spectra and so cause confusion unless they are recognized for what they are. With linear spectroscopy, isotope labeling is about the only means of testing for Fermi resonances. 2D IR spectroscopy can help assign Fermi resonances by the peculiar peak patterns that they cause and through their polarization dependence.

6.8 Fermi resonance

A spectroscopic Hamiltonian of a Fermi resonance between two otherwise harmonic oscillators would read:

$$H = \hbar\omega_1 \left(b_1^\dagger b_1 + \frac{1}{2}\right) + \hbar\omega_2 \left(b_2^\dagger b_2 + \frac{1}{2}\right) + \frac{\Delta}{2}\left(b_1 b_2^{\dagger 2} + b_1^\dagger b_2^2\right). \quad (6.47)$$

An on-site anharmonicity could be added in analogy to Eq. 6.10, but is skipped here for clarity. The coupling term $b_1 b_2^{\dagger 2} + b_1^\dagger b_2^2$ is related to a $q_1 q_2^2$ term in a Taylor expansion of the potential energy surface, in analogy to Eq. 6.4. In a Fermi resonance, we have in addition $\omega_1 \approx 2\omega_2$ as an accidental resonance condition. Then a coupling term $b_1 b_2^{\dagger 2}$, which de-excites oscillator #1 by one quantum, and at the same time excites oscillator #2 by two quanta, will strongly couple these two modes. As a result, resonant states $|10\rangle$ and $|02\rangle$, as well as $|20\rangle$, $|12\rangle$ and $|04\rangle$, will start to mix (Fig. 6.13a). Expanding the Hamiltonian in a site basis $\{|00\rangle, |10\rangle, |02\rangle, |20\rangle, |12\rangle, |04\rangle\}$ reveals:

$$H = \begin{pmatrix} 0 & & & & & \\ & \hbar\omega_1 & \Delta/2 & & & \\ & \Delta/2 & 2\hbar\omega_2 & & & \\ & & & 2\hbar\omega_1 & \sqrt{\frac{1}{2}}\Delta & 0 \\ & & & \sqrt{\frac{1}{2}}\Delta & \hbar\omega_1 + 2\hbar\omega_2 & \sqrt{\frac{3}{2}}\Delta \\ & & & 0 & \sqrt{\frac{3}{2}}\Delta & 4\hbar\omega_2 \end{pmatrix}. \quad (6.48)$$

The Hamiltonian is again block-diagonal, and can be diagonalized [51]. In linear spectroscopy, one only measures the $\{|1, 0\rangle, |0, 2\rangle\}$ manifold of states, and so one would see a doublet. Both states have oscillator strength due to mixing; we say the

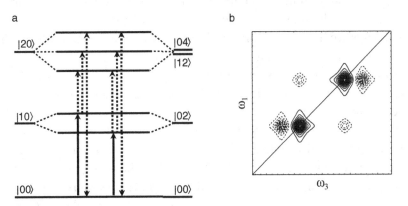

Figure 6.13 (a) Level scheme of a perfect Fermi resonance with $\omega_1 = \omega_2$ (Eq. 6.47), and (b) the resulting purely absorptive 2D IR spectrum. The solid and dashed contour lines depict positive and negative signs, respectively, of the corresponding bands.

dark state $|0, 2\rangle$ borrows oscillator strength from the bright state $|1, 0\rangle$. Of course, a doublet can also be caused by standard coupling between two bright states, like we have been discussing throughout this chapter, which is what causes the confusion. However, transitions to the next higher manifold of states, $\{|2, 0\rangle, |1, 2\rangle, |0, 4\rangle\}$, causes a very distinct peak pattern in the 2D IR spectrum that is unlike the standard spectrum of two coupled bright states (Fig. 6.13b). Note that the apparent diagonal anharmonicity is actually inverted in sign for one of the two sets of diagonal peaks [51]. Moreover, one can check the polarization dependence of the spectra. If the mode is indeed a Fermi resonance, then it must have a transition dipole direction equal to that of the mode from which it is stealing intensity since the only oscillator strength it has comes from the bright state. Thus, no matter what polarization is measured, the relative intensities of the 2D IR peaks should not change.

Exercises

6.1 With the rules from Appendix B, verify that the Hamiltonian 6.2 translates into the Hamilton matrix 6.5.

6.2 Verify that the Hamiltonian 6.10 translates into the Hamilton matrix 6.11.

6.3 Show that the terms $b_i^\dagger b_i^\dagger b_i b_i$ are related to $V_{1122} q_1^2 q_2^2$ in a Taylor expansion of the potential energy surface. Why do we retain only the term $b_i^\dagger b_i^\dagger b_i b_i$, and not, e.g. $b_i b_i b_i b_i$?

6.4 Show that the Hamiltonian in Eq. 6.5 would no longer be block-diagonal if we were to include cubic anharmonicity, e.g. terms of the sort $V_{122} q_1 q_2^2$ in a Taylor expansion of the potential energy surface. Would these coupling terms create cross-peaks?

6.5 With the rules from Appendix B, verify that we obtain for the transition dipoles of an harmonic oscillator: $\langle 1|\hat{\mu}|2\rangle = \sqrt{2}\langle 0|\hat{\mu}|1\rangle$. To that end, keep in mind that we have for a vibrational transition dipole operator $\hat{\mu} = d\mu/dq \cdot \hat{q}$, where $d\mu/dq$ is the change of the molecular dipole with coordinate q.

6.6 Diagonalize the two-exciton matrix of the Hamiltonian in Eq. 6.11 with $\omega_1 = \omega_2 \equiv \omega$ to get Eq. 6.13. Hint: Do a coordinate transformation into a basis set defined by $|2\pm\rangle = 1/\sqrt{2}(|02\rangle \pm |20\rangle)$, which will create a block-diagonal two-exciton Hamiltonian that can then be analytically diagonalized.

6.7 Calculate the dispersion relation of an exciton in a 3_{10}-helix which has 3.2 residues per turn and $\beta_{i,i+1} = 0.9$ cm^{-1}, $\beta_{i,i+2} = -2.5$ cm^{-1}, $\beta_{i,i+3} = -2.8$ cm^{-1}, $\beta_{i,i+4} = -0.8$ cm^{-1} [128], and vanishing coupling for larger distances. Calculate the expected splitting between the A and E modes.

6.8 Consider two coupled oscillators in a plane, like the two carbonyl groups in Fig. 6.1(a). Using transition dipole coupling, determine the sign of the coupling and the intensities of the two peaks in a linear infrared spectrum when one carbonyl is rotated out of the plane. Do the same for a 2D IR spectrum, predicting both the diagonal and cross-peak intensities.

6.9 The transition dipole strength of a vibrational mode can be determined from its infrared spectrum. The integrated absorption coefficient A is the integral of the molar absorption coefficient, ϵ, over the band [6]

$$A = \int \epsilon(\nu) d\nu \qquad (6.49)$$

which is related to the transition dipole by

$$|\mu_{01}|^2 (m^2 C^2) = \frac{3\hbar n c \varepsilon_0 \ln 10}{\pi \nu_0 N_A} A = 1.02 \times 10^{-61} \times \frac{A(M^{-1} cm^{-2})}{\nu_0 (cm^{-1})} \qquad (6.50)$$

where the units are given in parenthesis, ε_0 is the permutivity of free space, n is the index of refraction, N_A is Avogadro's number, and $\nu_0 = \omega_0/2\pi$ is the center frequency of the band. (a) Do a unit analysis and verify the constant (constants are given in Appendix C). (b) Given that a protein with 18 residues has a maximum ϵ of 8000 M^{-1} cm^{-1} and a spectral width of 40 cm^{-1} for the amide I band, estimate the transition dipole strength $|\mu_{01}|$ for a single residue in debye.

6.10 When calculating transition dipole strengths and couplings, units must be done properly, which is not always straightforward. For instance, the transition dipole moment, μ_{01} (debyes (D), Eq. 6.36), is often written [112]

$$\mu_{01} = \left(\frac{\hbar}{8\pi^2 c \nu_{01}}\right)^{1/2} \frac{\partial \mu}{\partial q} = \left(\frac{4.1058}{\nu_{01}^{1/2}}\right) \frac{\partial \mu}{\partial q} \qquad (6.51)$$

where ν_{01} (cm^{-1}) is the observed frequency, $\partial \mu / \partial q$ is in units of D Å^{-1}u$^{-1/2}$ where u is the reduced mass, and the constant has units of Å$u^{1/2}$ $cm^{-1/2}$. q is written in units of Å$u^{1/2}$ so that the coordinate system is mass-weighted, which is often used in normal mode analysis [174]. Derive the constant in Eq. 6.51 by solving the integral for the transition dipole moment $\mu_{01} = d\mu/dq \langle 0|\hat{q}|1\rangle$. Hint: Use the following wavefunctions for a harmonic oscillator to evaluate the integral: $\psi_0(q) = (\frac{\alpha}{\pi})^{1/4} e^{-\alpha q^2/2}$ and $\psi_1(q) = (\frac{\alpha}{4\pi})^{1/4} 2\alpha^{1/2} q e^{-\alpha q^2/2}$ where $\alpha = \sqrt{ku/\hbar^2}$ (in SI units). Replace the force constant, k, with a conversion to ν_{01}.

6.11 The amide I mode of N-methylacetamide has $\partial \mu / \partial q = 3.466$ D Å$u^{1/2}$. What is its transition dipole strength? Hint: Use the conversion factor from Problem 6.10.

6.12 In some theory papers it is common to report couplings between vibrational groups in units of mdyn/(Å u) where u is the reduced mass. These units are used because $\beta_{12} = \partial^2 V / \partial q_1 \partial q_2$, and so should have units of a

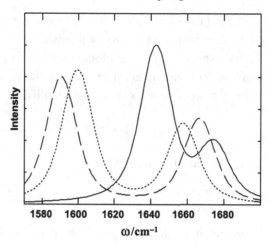

Figure 6.14 Simulation of two coupled oscillators without (solid) and with one $^{13}C = ^{18}O$ isotope labels (dashed and dotted).

force constant (N/m) in mass-weighted coordinates ($1/u^{1/2}$). For example, in one particular geometry Torii and Tasumi reported a coupling constant of $\beta = 0.02$ mdyn/(Å u) between two amide I stretches of trialanine [179]. Compute the coupling term in the Hamiltonian, namely $\beta_{12}q_1q_2$, in cm^{-1}. Use the conversion factor in Problem 6.10 to convert β into units of mdyn Å, which is energy. Then convert to cm^{-1}. (Overview of units in Appendix C.)

6.13 Generate the normal mode Hamiltonian in Eq. 6.46 using the Dunham expansion and the Darling–Dennison coupling terms.

6.14 In Chapter 1 we stated that 2D IR spectroscopy is not needed to measure the coupling if one can isotope-label the local modes. Given the 3 linear spectra of two coupled oscillators in Fig. 6.14, two of which have one oscillator isotope labeled, (a) find the unknowns ω_1, ω_2 and β. (b) Find the relative angles between the transition dipoles. (c) Explain why the intensities of the isotope-labeled peaks are different.

7
2D IR lineshapes

So far, we have implicitly assumed that the transition frequency ω_{01} of a vibrational mode is infinitesimally sharply defined and does not vary as a function of time. In an actual sample, this will not be the case because the solvent molecules will push and pull at the molecule, thereby deforming the molecular potential energy surface of the vibrational transition under study and hence modulating its transition frequency ω_{01} (Fig. 7.1). The time dependence of the transition frequency leads to pure and inhomogeneous dephasing. So far, we have considered pure dephasing by just including a phenomenological T_2 damping term whenever the system is in a coherent state. In what follows, we will develop a microscopic theory that explains dephasing and relates it to the microscopic motion of the solvation shell or the molecule itself. Measuring dephasing processes turns out to be a powerful tool to study the dynamics of molecular systems in the solution phase.

7.1 Microscopic theory of dephasing

The theory we outline was originally formulated by Kubo to describe the dephasing of NMR transitions [116], but has also proved adequate for describing dephasing of vibrational transitions. Kubo's stochastic theory of lineshapes leads to a microscopic theory of dephasing. It treats the vibrational transitions quantum mechanically, and the solvent classically (it is therefore sometimes called a semiclassical theory of dephasing).[1]

We start out with the time evolution of the density matrix of a single molecule during a coherence time

$$\dot{\rho}_{01}(t) = -i\omega_{01}(t)\rho_{01}(t) \qquad (7.1)$$

---

[1] Kubo's stochastic theory of lineshapes describes dephasing of vibrational transitions accurately unless the linewidth is larger than $k_B T$. It fails for electronic transitions, for which the full quantum counterpart of Kubo's stochastic theory, the Brownian oscillator model, becomes necessary (see Ref. [141]).

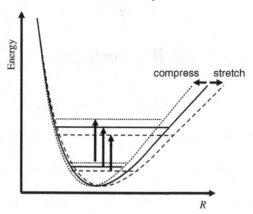

Figure 7.1 A Morse potential describing a chemical bond, at which the solvent is pushing and pulling.

which follows directly from the Liouville–von Neumann equation (see the derivation of Eq. 3.29)

$$\dot{\rho} = -\frac{i}{\hbar}[\hat{H}_0(t), \rho(t)]. \tag{7.2}$$

Note that we have not included the dephasing term here since this is what we ultimately want to derive. In contrast to the previous chapters, we now take into account that the molecular Hamiltonian $\hat{H}_0(t)$, and hence the transitions frequency $\omega_{01}(t)$, is time dependent due to fluctuations in the bath structure. Integration of Eq. 7.1 gives for a single molecule:

$$\rho_{01}(t) \propto \exp\left(-i \int_0^t \omega_{01}(\tau)d\tau\right) \tag{7.3}$$

which reduces to the expression we have used many times previously

$$\rho_{01}(t) \propto e^{-i\omega_{01}t} \tag{7.4}$$

if the vibrational frequency ω_{01} is constant in time.

We recall that the very concept of a density matrix was introduced to describe statistical ensembles. In a realistic experiment, we will have a large number of molecules in our laser focus, each of which is sitting in a different environment with a different instantaneous frequency $\omega_{01}(t)$. We therefore average the density matrix over all individual molecules:

$$\rho_{01}(t) \propto \left\langle \exp\left(-i \int_0^t \omega_{01}(\tau)d\tau\right)\right\rangle \tag{7.5}$$

where $\langle ... \rangle$ denotes the ensemble average, like in previous derivations throughout this book.

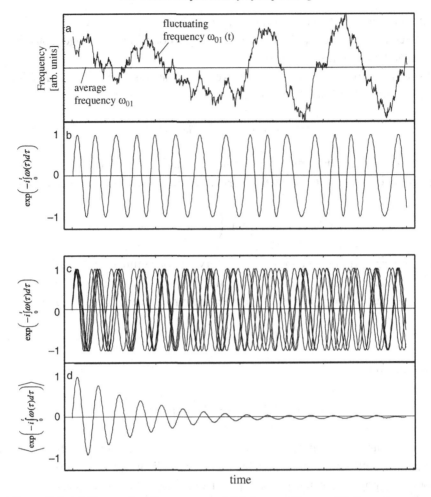

Figure 7.2 (a) Fluctuating frequency and (b) the resulting oscillation where the oscillation periods vary (Eq. 7.3). Panel (c) overlays five such trajectories, which are initially in phase, but run out of phase quickly. Panel (d) shows an average over 1000 trajectories (Eq. 7.5).

This expression illustrates the central concept of dephasing. As a result of changes in the environment structure, the instantaneous frequency $\omega(t)$ fluctuates around its average ω_{01} (Fig. 7.2a). If we had just a single molecule, the change of frequency causes the oscillation period to change as a function of time (Fig. 7.2b). That is, whenever the instantaneous frequency $\omega(t)$ happens to be low, the oscillation period is long, and vice versa. Nevertheless, the amplitude of the oscillating off-diagonal density matrix $\rho_{01}(t)$ would still be constant (Eq. 7.3) if we had just a single molecule. Due to ensemble averaging, however, the various oscillating terms will eventually become out of phase (Fig. 7.2c), so that the amplitude of the ensemble averaged density matrix $\langle \rho_{01}(t) \rangle$ will decay in time (Fig. 7.2d). We will see in Section 7.3 that Eq. 7.5 includes homogeneous and inhomogeneous dephasing, as well as spectral diffusion.

As a first step to simplify Eq. 7.5, we write the time dependent transition frequency of the vibrational transition as

$$\omega_{01}(t) = \omega_{01} + \delta\omega_{01}(t) \tag{7.6}$$

with a time independent mean frequency ω_{01} and a fluctuating part $\delta\omega_{01}(t)$. The time average $\langle \delta\omega_{01} \rangle = 0$ vanishes since we subtracted off the average. We then get:

$$\rho_{01}(t) \propto e^{-i\omega_{01}t} \left\langle \exp\left(-i \int_0^t \delta\omega_{01}(\tau)d\tau\right) \right\rangle. \tag{7.7}$$

Using this procedure, we can translate every response function we have developed so far by replacing each term of the form $e^{\pm i\omega_{01}t}e^{-t/T_2}$ with Eq. 7.7 or its complex conjugate, respectively. For example, the linear response function, Eq. 4.3,

$$R^{(1)}(t_1) = i\mu_{01}^2 e^{-i\omega_{01}t_1} e^{-t/T_2} \tag{7.8}$$

is modified into

$$R^{(1)}(t_1) = i\mu_{01}^2 e^{-i\omega_{01}t_1} \left\langle \exp\left(-i \int_0^{t_1} d\tau \delta\omega_{01}(\tau)\right) \right\rangle. \tag{7.9}$$

Note that this approach replaces only pure dephasing, but not the population relaxation contribution to dephasing. The formalism we will derive does not account for population relaxation T_1, but its contribution to dephasing could be added phenomenologically along the lines of Section 4.2.

With Eq. 7.9, we implicitly make the *Condon approximation* that assumes that the transition dipole moment μ_{01} is time-constant and thus can be pulled out of the ensemble average. This approximation is valid when a transition dipole moment does not vary as the transition frequency ω_{01} changes. Or, said another way, it does not depend on the bath coordinates. The Condon approximation is reliable when the solvent influences the nuclear potential, but not so much that it alters the electronic structure appreciably.

An example for which the Condon approximation does not apply is the OH stretch vibration of water [161]. When an OH group forms a hydrogen bond to a surrounding water molecule, it changes its vibrational frequency (since hydrogen bonding red-shifts the OH stretch vibration) and oscillator strength (since hydrogen bonding polarizes the OH group) at the same time. Hence, the transition frequency and strength are correlated, in which case one has to evaluate:

$$R^{(1)}(t_1) = i e^{-i\omega_{01} t_1} \left\langle \mu_{01}(0) \mu_{01}(t_1) \exp\left(-i \int_0^{t_1} d\tau \, \delta\omega_{01}(\tau)\right) \right\rangle. \quad (7.10)$$

Orientational motions of water are also correlated to the transition frequency and strength [92, 119], but we neglect this effect here (see Chapter 5).

7.2 Correlation functions

It turns out that the ensemble average in Eq. 7.7 is slowly converging since the exponential function is highly oscillatory. Therefore, the expression is commonly evaluated using the *cumulant expansion* truncated after second order, which also makes it possible to write analytical expressions for the integral. To this end, we expand the exponential function in powers of $\delta\omega_{01}$:

$$\left\langle \exp\left(-i \int_0^t d\tau \, \delta\omega_{01}(\tau)\right) \right\rangle = 1 - i \int_0^t d\tau \, \langle \delta\omega_{01}(\tau) \rangle \quad (7.11)$$

$$- \frac{1}{2} \int_0^t \int_0^t d\tau' d\tau'' \, \langle \delta\omega_{01}(\tau') \delta\omega_{01}(\tau'') \rangle + \cdots$$

Furthermore, we postulate that we can write this expression in the following form:

$$\left\langle \exp\left(-i \int_0^t d\tau \, \delta\omega_{01}(\tau)\right) \right\rangle \equiv e^{-g(t)} = 1 - g(t) + \frac{1}{2} g^2(t) + \cdots \quad (7.12)$$

and expand $g(t)$ in powers of $\delta\omega_{01}$:

$$g(t) = g_1(t) + g_2(t) + \cdots \quad (7.13)$$

where $g_1(t)$ is of order $O(\delta\omega_{01})$, $g_2(t)$ of order $O(\delta\omega_{01}^2)$, and so on. Inserting Eq. 7.13 into Eq. 7.12 and ordering the terms in powers of $\delta\omega_{01}$ gives:

$$e^{-g(t)} = 1 - (g_1(t) + g_2(t) + \cdots) + \frac{1}{2}(g_1(t) + g_2(t) + \cdots)^2 + \cdots \quad (7.14)$$

The linear term in Eq. 7.12 is the average of $\delta\omega_{01}$, which vanishes by construct and so $g_1(t) = 0$. Hence, the leading term in the so-called *lineshape function* is $g_2(t)$:

$$g(t) = \frac{1}{2} \int_0^t \int_0^t d\tau' d\tau'' \, \langle \delta\omega_{01}(\tau') \delta\omega_{01}(\tau'') \rangle. \quad (7.15)$$

This equation can be simplified because we are calculating the lineshape for a molecule at equilibrium. As a result, the absolute times at which we measure the frequency fluctuations does not matter, just their time difference, which is to say that the frequency correlation function is *stationary*. Thus we can shift the time origin

$$\langle \delta\omega_{01}(\tau')\delta\omega_{01}(\tau'')\rangle = \langle \delta\omega_{01}(\tau' - \tau'')\delta\omega_{01}(0)\rangle. \tag{7.16}$$

Moreover, because the time-correlation function is stationary, it is also even (see Problem 7.1):

$$\langle \delta\omega_{01}(\tau' - \tau'')\delta\omega_{01}(0)\rangle = \langle \delta\omega_{01}(\tau'' - \tau')\delta\omega_{01}(0)\rangle. \tag{7.17}$$

These two equations allow us to rewrite Eq. 7.15 as

$$g(t) = \int_0^t \int_0^{\tau'} d\tau' d\tau'' \langle \delta\omega_{01}(\tau' - \tau'')\delta\omega_{01}(0)\rangle \tag{7.18}$$

$$= \int_0^t \int_0^{\tau'} d\tau' d\tau'' \langle \delta\omega_{01}(\tau'')\delta\omega_{01}(0)\rangle \tag{7.19}$$

the exact derivation of which is outlined in Problem 7.2. The function $\langle \delta\omega_{01}(\tau)\delta\omega_{01}(0)\rangle$ is called the *frequency fluctuation correlation function* (FFCF). Therefore, within the approximation of the cumulant expansion truncated after second order, we have:

$$\left\langle \exp\left(-i \int_0^t d\tau \, \delta\omega_{01}(\tau)\right)\right\rangle = e^{-g(t)}. \tag{7.20}$$

Plugging $g(t)$ into Eq. 7.9 the linear response function simplifies to:

$$R^{(1)}(t) = i\mu_{01}^2 e^{-i\omega_{01}t} e^{-g(t)} \tag{7.21}$$

and we obtain for the linear absorption spectrum

$$S(\omega) \propto \Re \int_0^\infty e^{i(\omega-\omega_{01})t} e^{-g(t)} dt. \tag{7.22}$$

The cumulant expansion is an intelligent way of reordering terms in different powers of $\delta\omega_{01}$. The cumulant expansion effectively shifts the averaging from $\langle e^{-i\int \cdots}\rangle$ to $e^{-\int \langle \cdots \rangle}$, which is simple to calculate when one has a functional form for the correlation function, and is also simpler when one is analyzing frequency trajectories (see Section 10.1).

One can show that the cumulant expansion truncated after second order is exact when the distribution of $\delta\omega_{01}$ is Gaussian [184], which is often the case to a very good approximation. The *central limit theorem* states that when a stochastic process is the result of many independent stochastic processes, then the sum will

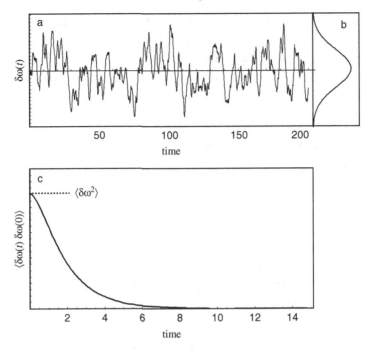

Figure 7.3 (a) An example frequency trajectory deduced from a Langevin simulation (in arbitrary units), with (b) its Gaussian distribution. (c) Resulting frequency fluctuation correlation function.

be Gaussian distributed, regardless of whether or not the statistics of the individual contributions is Gaussian. In the context of solvation, many solvent molecules typically contribute to the frequency fluctuations of a vibrational transition, hence the central limit theorem often applies. For example, carbonyl vibrations are affected by the electrostatic fields originating from molecules 10 Å away [194].

Figure 7.3 illustrates the concept of how frequency fluctuations are described by a correlation function. Shown in Fig. 7.3(a) is an example frequency trajectory of a single molecule deduced from a random walk in a harmonic potential, whose distribution is Gaussian (Fig. 7.3b). If the system is ergodic, we can use a sliding time average rather than an ensemble average to calculate the frequency fluctuation correlation function:

$$\langle \delta\omega_{01}(t)\delta\omega_{01}(0)\rangle = \frac{1}{T}\int_0^T \omega_{01}(\tau)\omega_{01}(\tau+t)d\tau \quad (7.23)$$

where T is the total length of the trajectory. The initial value of the frequency fluctuation correlation function at $t = 0$ is the variance of the frequency fluctuations, $\langle \delta\omega_{01}^2\rangle$. As time increases, the molecule's frequency changes, and eventually is no longer correlated from where it started. The decay time of the frequency fluctuation correlation function reveals the characteristic time-scale of the fluctuations, thus, the time-scale of the memory in the system (Fig. 7.3c). In the context of solvation of a vibrational mode, the frequency fluctuation correlation function directly reveals the time-scale of the motion of the solvent molecules to which the mode is coupled.

7.3 Homogeneous and inhomogeneous dynamics

Kubo introduced an exponential ansatz for the frequency fluctuation correlation function [116]:

$$\langle \delta\omega(\tau)\delta\omega(0)\rangle = \Delta\omega^2 e^{-\frac{|\tau|}{\tau_c}} \quad (7.24)$$

which includes two parameters, the fluctuation amplitude $\Delta\omega$ and the correlation time τ_c. It decays to zero as we expect from the last section, but is unphysical at $t = 0$ because it is discontinuous. Integrating the correlation function twice reveals the *Kubo lineshape function*:

$$g(t) = \Delta\omega^2 \tau_c^2 \left[e^{-\frac{t}{\tau_c}} + \frac{t}{\tau_c} - 1 \right]. \quad (7.25)$$

When the frequency fluctuations are small or very rapid, as defined by $\Delta\omega \cdot \tau_c \ll 1$, then we say that we are in the *fast modulation limit* or the *homogeneous limit*. In this limit, the Kubo lineshape function simplifies to:

$$g(t) = \Delta\omega^2 \tau_c t \equiv t/T_2^* \quad (7.26)$$

with a pure dephasing time $T_2^* = (\Delta\omega^2 \tau_c)^{-1}$. One obtains this expression since $e^{-t/\tau_c} \to 0$ and $t/\tau_c \gg 1$. When this occurs, the frequency fluctuation correlation function can be approximated as a δ-function, $\langle \delta\omega(\tau)\delta\omega(0)\rangle = \delta(t)/T_2^*$. Moreover, in this limit, the lineshape function creates a Lorentzian line with width T_2^{*-1} (see Eq. 4.9):

$$A(\omega) \propto \Re \int_0^\infty e^{i(\omega-\omega_{01})t} e^{-g(t)} dt = \Re \int_0^\infty e^{i(\omega-\omega_{01})t_1} e^{-t_1/T_2^*}$$

$$= \frac{1/T_2^*}{(\omega - \omega_{01})^2 + 1/T_2^{*2}}. \quad (7.27)$$

If this limit applies, the correlation time is much smaller than the pure dephasing time, $\tau_c \ll T_2^*$, which we see from the definition of $T_2^* = (\Delta\omega^2 \tau_c)^{-1}$ and $\Delta\omega \cdot \tau_c \ll 1$. Notice that, in the fast modulation limit, the linewidth $(1/T_2^*)$ becomes narrower than the actual frequency distribution $\Delta\omega$. We call this phenomenon *motional narrowing* (see Fig. 7.4). The linewidth narrows since it scales linearly

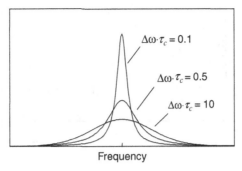

Figure 7.4 Simulation illustrating motional narrowing. Shown is the transition of a Kubo lineshape from the inhomogeneous limit $\Delta\omega \cdot \tau_c \gg 1$ into the homogeneous limit $\Delta\omega \cdot \tau_c \ll 1$. The width of the distribution, $\Delta\omega$, was kept constant in this model, while the correlation time τ_c was varied.

with the correlation time τ_c (since $T_2^{*-1} = \Delta\omega^2 \tau_c \ll \Delta\omega$). In principle, $1/T_2^* \to 0$ as $\tau_c \to 0$. Narrowing occurs because the frequency of the molecules fluctuates so quickly that one only sees the average.

The *slow modulation* or *inhomogeneous limit* is the opposite limit with $\Delta\omega \cdot \tau_c \gg 1$. In this case, the frequency fluctuation correlation function can be approximated as constant, $\langle \delta\omega(\tau)\delta\omega(0) \rangle = \Delta\omega^2$, and the lineshape function reduces to

$$g(t) = \frac{\Delta\omega^2}{2} t^2 \tag{7.28}$$

which can be seen when expanding the exponential function in Eq. 7.25 for $t/\tau_c \ll 1$ (see Problem 7.3). The lineshape function now is independent of the correlation time τ_c so that it resembles the static distribution of frequencies, and the absorption spectrum

$$A(\omega) \propto \Re \int_0^\infty e^{i(\omega-\omega_{01})t} e^{-g(t)} dt = \Re \int_0^\infty e^{i(\omega-\omega_{01})t} e^{-\frac{\Delta\omega^2}{2}t^2} dt$$

$$\propto e^{-\frac{(\omega-\omega_{01})^2}{2\Delta\omega^2}} \tag{7.29}$$

yields a Gaussian line with bandwidth $\Delta\omega$. Since we used the cumulant expansion, the Kubo model in the slow modulation limit necessarily leads to a Gaussian lineshape, but in reality the lineshape will just give the frequency distribution, whether or not it is Gaussian (see Problem 7.4).

The frequency fluctuation correlation function does not have to decay monoexponentially. Multi-exponential decays could be caused by multiple-component systems. For example, membrane peptides exhibit fast frequency fluctuations due to water dynamics and slower fluctuations caused by the lipids [194]. And even

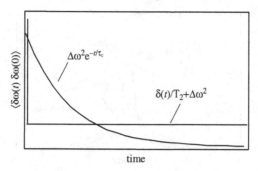

Figure 7.5 Frequency fluctuation correlation function for a Kubo lineshape and a Voigt lineshape.

single-component systems typically exhibit two time-scales [97]: the so-called *inertial* component on a 100 fs time-scale and a slower *diffusive* component in the picosecond range (see, e.g. the model calculation on water in Section 10.1). One often approximates the resulting bi-exponential decay by a lineshape function:

$$\langle \omega_{01}(t)\omega_{01}(0) \rangle = \delta(t)/T_2^* + \Delta\omega^2$$
$$g(t) = t/T_2^* + \frac{\Delta\omega^2}{2}t^2 \qquad (7.30)$$

assuming that the inertial component is in the motional narrowing limit (e.g. Lorentzian lineshape), and that the diffusive component is quasi-static on the time-scale of the experiment (Fig. 7.5). In this case, the absorption spectrum

$$A(\omega) \propto \Re \int_0^\infty e^{i(\omega-\omega_{01})t} e^{-t/T_2^*} e^{-\frac{\Delta\omega^2}{2}t^2} dt$$

$$\propto \frac{1}{(\omega-\omega_{01})^2 + 1/T_2^{*2}} \otimes e^{-\frac{(\omega-\omega_{01})^2}{2\Delta\omega^2}} \qquad (7.31)$$

becomes a Voigt profile, which is a convolution of a Lorentzian line with the Gaussian frequency distribution. In this limit, we say that we have *Bloch dynamics*.

Oftentimes neither the fast nor slow modulation limits apply for vibrations. For typical solution phase systems at room temperature, we find the homogeneous dephasing time is about $T_2 \approx 0.5$ ps to ≈ 2 ps as is the solvent correlation time. Hence, typical solution phase systems are in between the homogeneous and inhomogeneous limits. In this situation, the molecules gradually shift their frequencies, which causes what is known as *spectral diffusion*, because the frequency shifts are slow enough that they can be measured. With linear spectroscopy, lineshapes caused by spectral diffusion are very difficult to distinguish from Voigt lineshapes that result from Bloch dynamics. 2D IR spectroscopy can measure the time-scale of the dynamics and thus make the distinction quite clear.

7.4 Nonlinear response

The formalism can be extended to nonlinear spectroscopy in a straightforward manner. Recall the response function when we treat dephasing phenomenologically (Section 4.2):

$$R_{1,2}(t_1, t_2, t_3) \propto i\mu_{01}^4 e^{+i\omega_{01}t_1} e^{-t_1/T_2} e^{-i\omega_{01}t_3} e^{-t_3/T_2}$$
$$R_{4,5}(t_1, t_2, t_3) \propto i\mu_{01}^4 e^{-i\omega_{01}t_1} e^{-t_1/T_2} e^{-i\omega_{01}t_3} e^{-t_3/T_2} \quad (7.32)$$

and replace terms of the sort:

$$e^{\pm i\omega_{01}t} e^{-t/T_2} \quad (7.33)$$

with

$$e^{\pm i\omega_{01}t} \left\langle \exp\left(\pm i \int_0^t d\tau \, \delta\omega_{01}(\tau)\right) \right\rangle. \quad (7.34)$$

Hence, we obtain for the third-order response functions of a two-level system:

$$R_{1,2} = i\mu_{01}^4 e^{-i\omega_{01}(t_3-t_1)} \left\langle \exp\left(+i\int_0^{t_1} \delta\omega_{01}(\tau)d\tau - i\int_{t_2+t_1}^{t_3+t_2+t_1} \delta\omega_{01}(\tau)d\tau\right) \right\rangle$$
$$R_{4,5} = i\mu_{01}^4 e^{-i\omega_{01}(t_3+t_1)} \left\langle \exp\left(-i\int_0^{t_1} \delta\omega_{01}(\tau)d\tau - i\int_{t_2+t_1}^{t_3+t_2+t_1} \delta\omega_{01}(\tau)d\tau\right) \right\rangle$$
(7.35)

or, when applying the cumulant expansion truncated after second order (again exact for Gaussian statistics):

$$R_{1,2} = i\mu_{01}^4 e^{-i\omega_{01}(t_3-t_1)} e^{-g(t_1)+g(t_2)-g(t_3)-g(t_1+t_2)-g(t_2+t_3)+g(t_1+t_2+t_3)}$$
$$R_{4,5} = i\mu_{01}^4 e^{-i\omega_{01}(t_3+t_1)} e^{-g(t_1)-g(t_2)-g(t_3)+g(t_1+t_2)+g(t_2+t_3)-g(t_1+t_2+t_3)} \quad (7.36)$$

which is obtained after a tedious but straightforward calculation [141]. As before, the T_1 contribution to dephasing could be added phenomenologically to these equations. The nonlinear response function can be reduced to the same lineshape function $g(t)$ (Eq. 7.19), and hence the same frequency fluctuation correlation function $\langle \delta\omega(t)\delta\omega(0)\rangle$. If the goal is to determine the frequency fluctuation correlation function from an experiment (which tells a lot about the dynamics of the solvation shell around a molecule), it might appear one could just use linear spectroscopy, and invert Eqs. 7.19 and 7.22. While this is possible mathematically, it is ill-conditioned and works very poorly in the presence of experimental noise [82]. Nonlinear spectroscopy is a far better method for measuring the frequency fluctuation correlation function.

7.4.1 2D lineshape of a two-level system

Figure 7.6 shows rephasing, non-rephasing and purely absorptive 2D spectra of a two-level system, assuming a Kubo lineshape function with $\tau_c = 1$ ps, $\Delta\omega = 5$ ps^{-1} and $t_2 = 0.1$ ps. As expected, rephasing and non-rephasing spectra differ in their phase-twist. Moreover, the rephasing spectrum is significantly narrower in the antidiagonal direction, which we say is *line narrowed*. The purely absorptive 2D IR spectrum does not differ dramatically from the rephasing 2D spectrum, because the non-rephasing spectrum is weak. It is weak because the inhomogeneous distribution is not rephased (see Section 7.5).

These parameters produce an elongated lineshape which is about twice as long in the diagonal direction as it is in the antidiagonal direction. Line narrowing occurs because the system retains a memory of the ω_1-frequency during the excitation process when the ω_3-frequency is read out. In other words, the two frequencies are correlated.

However, we cannot say from a single 2D IR spectrum whether the transition follows Bloch dynamics or whether spectral diffusion is occurring. Spectral diffusion can be identified by taking a time series of 2D IR spectra with the population time t_2 varied and looking for changes of peak-shape. If the system follows Bloch dynamics (Eq. 7.30), then Eq. 7.36 reduces to

$$R_{1,2} \propto i e^{-i\omega_{01}(t_3-t_1)} e^{-(t_1+t_3)/T_2} e^{-\Delta\omega^2 (t_1-t_3)^2/2}$$
$$R_{4,5} \propto i e^{-i\omega_{01}(t_3+t_1)} e^{-(t_1+t_3)/T_2} e^{-\Delta\omega^2 (t_1+t_3)^2/2} \quad (7.37)$$

(see also Eq. 2.72 for an alternative derivation of this equation). The response function is independent of the population time t_2 and so the 2D IR spectra do not change in shape as the population time t_2 increases, because there is no spectral diffusion. In this limit, the antidiagonal width gives the homogeneous linewidth and the diagonal width gives the total linewidth.

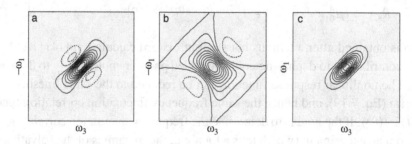

Figure 7.6 (a) Rephasing (real part), (b) non-rephasing and (c) purely absorptive 2D spectra of a two-level system, assuming a Kubo lineshape function with a correlation time $\tau_c = 1$ ps and a fluctuation amplitude $\Delta\omega = 5$ ps^{-1}. The population time t_2 was set to 0.1 ps. Solid and dotted contour lines depict positive and negative contribution to the spectra, respectively.

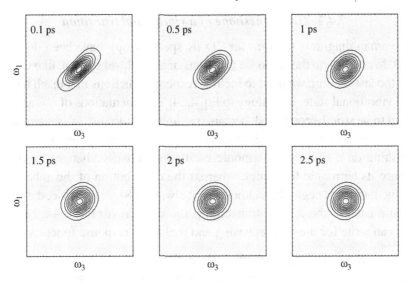

Figure 7.7 Purely absorptive 2D spectra of a two-level system, assuming a Kubo lineshape function with a correlation time $\tau_c = 1$ ps and a fluctuation amplitude $\Delta\omega = 5$ ps^{-1}. The population time t_2 was varied from 0.1 ps to 2.5 ps.

If the system exhibits spectral diffusion, then the 2D IR lineshape changes with time t_2 in Fig. 7.7, where we simulated the spectra using a correlation function that decays on a 1 ps time-scale. As the memory of the initial frequency is lost as time proceeds, the antidiagonal linewidth broadens to create round 2D IR lineshapes at delay times $t_2 \gg \tau_c$. The round shape indicates that there is no longer any correlation between ω_1 and ω_3 for times $t_2 \gg \tau_c$.

An intuitive way to think about a 2D lineshape is in terms of a two-time-point joint probability density:

$$R(\omega_1, t_2, \omega_3) \approx p(\omega_3, t_2|\omega_1, 0) \qquad (7.38)$$

i.e. the conditional probability of finding the system at frequency ω_3 at time t_2, provided it has been at frequency ω_1 at time 0. That is, at early times, frequencies ω_1 and ω_3 are correlated and the 2D IR lineshape is elongated along the diagonal, while at late times, when the correlation is lost, we get for the joint probability:

$$p(\omega_3, t_2|\omega_1, 0) \xrightarrow{t_2 \to \infty} p(\omega_3)p(\omega_1) \qquad (7.39)$$

which is just the product of the 1D frequency distribution and is why the lineshape becomes round. One can show that Eq. 7.38 is exact in the inhomogeneous limit, when the time-scale of the fluctuations is slow compared to the inverse linewidth (Section 7.3) [79]. This relation is very general and holds even for non-Gaussian statistics when the cumulant expansion does not apply.

7.4.2 2D IR lineshape of a vibrational transition

The Feynman diagrams relevant for 2D IR spectroscopy have been discussed in Fig. 4.3. In addition to the response functions of a two-level system, like we considered in the last section, we need to include response functions that reach the second excited vibrational state, in analogy to Eq. 4.24. The fluctuations of ω_{01} and ω_{12} are assumed to be strictly correlated, because the anharmonic shift Δ is much less than the frequency. This assumption is motivated by Fig. 7.1, according to which pulling and pushing on a slightly anharmonic oscillator primarily changes its curvature, and hence its harmonic frequency, whereas the contribution of the anharmonicity to the fluctuations is negligible, although not always [68]. In this case, the lineshape function related to the $0 \rightarrow 1$ transition is the same as for the $1 \rightarrow 2$ transition, and we can write for the non-rephasing and rephasing response functions:

$$R_{1,2,3} = 2i\mu_{01}^4 \left(e^{-i\omega_{01}(t_3-t_1)} - e^{-i((\omega_{01}-\Delta)t_3-\omega_{01}t_1)} \right)$$
$$\cdot e^{-g(t_1)+g(t_2)-g(t_3)-g(t_1+t_2)-g(t_2+t_3)+g(t_1+t_2+t_3)}$$

$$R_{4,5,6} = 2i\mu_{01}^4 \left(e^{-i\omega_{01}(t_3+t_1)} - e^{-i((\omega_{01}-\Delta)t_3+\omega_{01}t_1)} \right)$$
$$\cdot e^{-g(t_1)-g(t_2)-g(t_3)+g(t_1+t_2)+g(t_2+t_3)-g(t_1+t_2+t_3)} \qquad (7.40)$$

where the second terms are for the anharmonically shifted overtone band. Figure 7.8 shows a series of 2D IR spectra with varying population time t_2, using

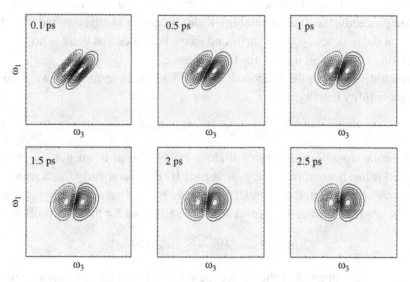

Figure 7.8 Purely absorptive 2D spectra of a two-level system, assuming a Kubo lineshape function with a correlation time $\tau_c = 1$ ps, a fluctuation amplitude $\Delta\omega = 5$ ps^{-1} and an anharmonicity $\Delta = 5$ ps^{-1}. The population time t_2 was varied from 0.1 ps to 2.5 ps. Dashed and solid contour lines mark the different signs of the two contributions.

a Kubo lineshape function with the same parameters as Fig. 7.7. The anharmonicity was chosen to be $\Delta = 5$ ps^{-1}, so that the $0 \to 1$ and $1 \to 2$ contributions (which have opposite signs) overlap and partially cancel. If necessary, one can build in analytical solutions that account for differences in fluctuations of ω_{01} and ω_{12} [68].

7.4.3 Extracting spectral diffusion times from 2D IR spectra

Various methods have been suggested to extract the frequency fluctuation correlation function from a t_2-series of 2D spectra. The most involved procedure is to fit the experimental 2D IR spectra. Fitting involves constructing a physically motivated model for the frequency fluctuation correlation function, running it through the corresponding response function and the Fourier transform, and varying the parameters in the frequency fluctuation correlation function until good agreement is obtained. One can instead measure certain features of the 2D lineshape that resemble the frequency fluctuation correlation function. For example, one often measures the tilt of the *nodal line* between the 0–1 peak and the 1–2 peak. If the anharmonicity is smaller than or equal to the linewidth of the transition, i.e. when both peaks overlap and partially cancel, then this tilt is a measure of the amount of inhomogeneity [157]. For a very inhomogeneous line, it is essentially 45° at early t_2-times, and as spectral diffusion occurs, it rotates to vertical at late t_2-times (see Fig. 7.8). Plotting the tilt as a function of time t_2, one obtains a qualitative measure of the frequency fluctuation correlation function.

The procedure is demonstrated in Fig. 7.9. Shown are 2D IR spectra of ^{13}C^{15}N$^-$ dissolved in D$_2$O (Fig. 7.9a), which at early time is inhomogeneous to a certain extent, reflecting the various hydrogen bond environments in which the ^{13}C^{15}N$^-$ ions are sitting [111]. Figure 7.9(b) shows a global fit to the experimental data, using a bi-exponential ansatz for the frequency fluctuation correlation function:

$$\langle \delta\omega(t)\delta\omega(0)\rangle = \Delta\omega_1^2 e^{-\frac{t}{\tau_1}} + \Delta\omega_2^2 e^{-\frac{t}{\tau_2}} \qquad (7.41)$$

and varying $\Delta\omega_1$, $\Delta\omega_2$, τ_1, and τ_2 as free fit parameters. Figure 7.9(c) compares the tilt of the nodal plane as a function of time t_2 with the resulting correlation function in Fig. 7.9(d). The tilt measure resembles the correlation function to a close approximation. From global fitting we find that the correlation time of the fast inertial component is $\tau_1 = 200$ fs. This number puts the dynamics close to the motional narrowing limit with $\Delta\omega_1\tau_1 = 0.4 < 1$, which is why the extracted amplitude of this component is too small in the tilt measure (Fig. 7.9c).

For a two-level system, or, if the anharmonicity is so large that the 0–1 peak and the 1–2 peak are separated in the 2D IR spectrum, one may evaluate the ellipticity:

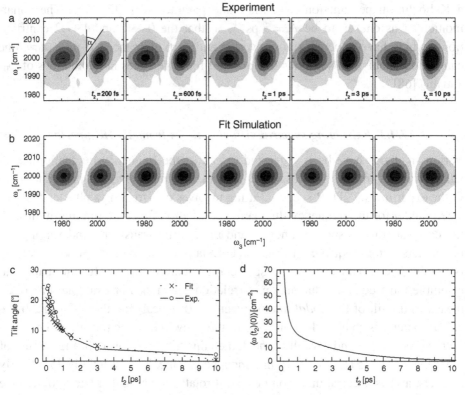

Figure 7.9 (a) A series of purely absorptive 2D IR spectra of $^{13}C^{15}N^-$ dissolved in D_2O. (b) Global fit to the data using lineshape theory discussed in this paragraph. (c) Tilt angle α of the nodal line between the 0–1 and the 1–2 peaks, as indicated in panel (a), from both the experimental data and the simulation results. (d) Frequency fluctuation correlation function that resulted from the fit. Adapted from Ref. [111] with permission.

$$\epsilon = \frac{\Delta\omega_d^2 - \Delta\omega_a^2}{\Delta\omega_d^2 + \Delta\omega_a^2} \quad (7.42)$$

where $\Delta\omega_d$ and $\Delta\omega_a$ are the diagonal and antidiagonal width of the line, respectively [121]. At early t_2-times, when the correlation between ω_1 and ω_3 still persists, so that the 2D lineshape is elongated along the diagonal, the parameter ϵ is large (with 1 being an upper limit). At long delay times, it decays to zero, as the memory gets lost and the 2D lineshape becomes a circle. Note that the main axis of the ellipse is always 45° for an isolated transition, so it is the ratio of the diagonal and antidiagonal width that changes as spectral diffusion occurs, but not the tilt of the peak. Other measures of the frequency fluctuation correlation function are summarized in Ref. [157].

7.5 Photon echo peak shift experiments

2D IR spectroscopy can measure the dynamics of individual vibrational transitions even if the spectrum is congested. But if one has a spectrally isolated vibrator, like in the example above, then the experimentally simpler method to measure its frequency fluctuation correlation function is the photon echo peak shift experiment in which the signal is homodyne detected and analyzed in the time rather than the frequency domain.

To explain this technique we consider a transition that follows Bloch dynamics (Eq. 7.37):

$$R_{1,2} \propto i e^{-i\omega_{01}(t_3-t_1)} e^{-(t_1+t_3)/T_2} e^{-\Delta\omega^2(t_1-t_3)^2/2}$$
$$R_{4,5} \propto i e^{-i\omega_{01}(t_3+t_1)} e^{-(t_1+t_3)/T_2} e^{-\Delta\omega^2(t_1+t_3)^2/2}. \quad (7.43)$$

In either set of response functions, the signal consists of a product of two terms, $e^{-(t_1+t_3)/T_2}$ and $e^{-\Delta\omega^2(t_1\mp t_3)^2/2}$. Figure 7.10 shows the contributions of these two terms to the overall rephasing and non-rephasing signals. In the rephasing diagrams, the $e^{-\Delta\omega^2(t_1-t_3)^2/2}$ term acts as if it were a gate, and light is emitted predominantly after a time that is equal to the time separation between the two incident pulses t_1. In other words, an echo is emitted in very much the same way as what we constructed in Fig. 2.12(b). The appearance of a photon echo is demonstrated in Fig. 7.11 for the asymmetric stretch vibration of azide (N_3^-) in an ionic glass, which is very strongly inhomogeneously broadened.

In the non-rephasing diagram, the gate $e^{-\Delta\omega^2(t_1+t_3)^2/2}$ is conceptually shifted to negative times, but of course there is no signal at negative times due to causality (the signal is causal because it is generated by the laser pulses). Therefore, the signal is smaller in the non-rephasing diagrams because it is just the tail of the

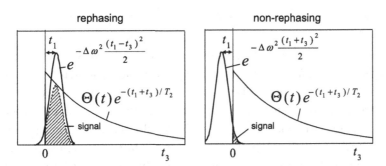

Figure 7.10 Contributions to the rephasing and non-rephasing signals (Eq. 7.37) as a function of time t_3 for a fixed time t_1. Shown are the homogeneous damping term (thick solid lines), the Gaussian term (thin solid lines) and the product of the two (dotted line and shaded area).

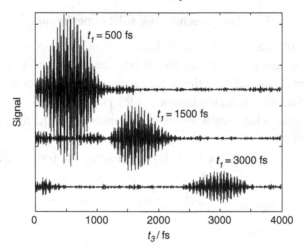

Figure 7.11 Photon echo of the asymmetric stretch vibration of azide in an ionic glass. As the separation between the first two pulses t_1 is increased, a photon echo is emitted at time $t_3 = t_1$.

Gaussian term that causes to the signal. Consequently, in the presence of inhomogeneous broadening the emitted signal is significantly larger in the rephasing direction than it is in the non-rephasing direction.

One typically uses a box-CARS geometry to measure photon echoes. It is essentially the same as what is used for 2D IR spectroscopy, except that the signal is homodyne rather than heterodyne detected. Consequently, the square-law detector will measure the intensity of the emitted field, integrated over its duration, which is the area shaded in Fig. 7.10:

$$S(t_1, t_2) = \int_0^\infty \left| E^{(3)}_{sig}(t_3; t_1, t_2) \right|^2 dt_3. \qquad (7.44)$$

Because the photon echo is integrated over its duration, this experiment is often called an *integrated photon echo*. Despite the information loss, the frequency fluctuation correlation function can be still regained, albeit very indirectly, by measuring the integrated signal intensity as a function of the first coherence time t_1. For time $t_1 = 0$, there is a certain amount of emitted light (the shaded area in Fig. 7.12a). As time t_1 increases, the Gaussian shaped gate shifts to positive times and the amount of emitted light increases, until t_1 roughly equals half the width of the Gaussian gate (Fig. 7.12b). When t_1 is increased even further, the emitted signal decreases due to the exponential damping term (Fig. 7.12c). Thus, when plotting the integrated photon echo signal as a function of t_1, it will initially rise and peak at a certain time, which is what we call the *peak shift* (Fig. 7.12d). The position of

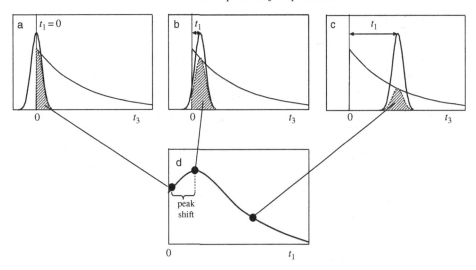

Figure 7.12 (a–c) Photon echo signals for a rephasing diagram with time t_1 varied. Thick solid lines: homogeneous damping term; thin solid line: Gaussian term; dotted line and shaded area: product of both. (d) Plotting the integrated photon echo signal (the shaded area in panels (a–c)) as a function of t_1, one obtains a peak shift.

the peak shift is an indirect measure of the ratio of homogeneous versus inhomogeneous broadening.

In solution phase systems, when spectral diffusion is taking place, inhomogeneity is not static. As we increase the population time t_2, the memory of the system gets lost, the effective inhomogeneity decreases, as does the peak shift (Fig. 7.13a). It turns out that when we plot the peak shift as a function of population time t_2, we obtain a measure that resembles to a certain extent the frequency fluctuation correlation function of the transition (Fig. 7.13b) [28, 37]). Instead, one can extract the frequency correlation function by global fitting of the data [82]. Similar to fitting the series of 2D IR spectra that we describe above, one can write a physically motivated ansatz for the frequency fluctuation correlation function, plug it into the response functions 7.40, calculate the integrated signal 7.44, and vary the parameters in the ansatz until best agreement with the experimental data is obtained.

Finally we note that the signal strength in an integrated photon echo experiment scales with the eighth power of the transition dipole as a result of homodyne detection as compared to the fourth power for a heterodyned experiment. Thus, while more complicated, 2D IR spectroscopy can usually measure the correlation function to much longer time-scales. Moreover, 2D IR spectroscopy is necessary if one needs to resolve the frequency dependence of the correlation function, as occurs when the Condon approximation breaks down.

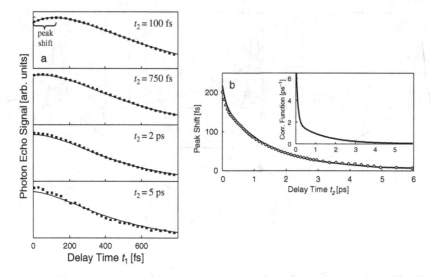

Figure 7.13 (a) Integrated photon echo response of the asymmetric stretching frequency of N_3^- in H_2O as a function of t_1 and t_2. (b) Peak shift as a function of population time t_2 and (inset) the frequency fluctuation correlation function deduced from a global fit of the data. Adapted from Ref. [82] with permission.

7.5.1 Lineshapes of excitonically coupled systems

We end by noting that the lineshape theory presented in this chapter only applies to vibrational modes that are spectrally resolved and uncoupled to any other mode. If this is not the case, isotope labeling can often be used, as was discussed in Sect. 6.4. Either way, the lineshape analysis discussed here cannot be applied quantitatively to lineshape analysis of coupled systems, such as bulk water or unlabeled proteins. In systems like these, the local mode Hamiltonian must be diagonalized, which distributes the excitation over multiple oscillators and will be affected by diagonal and off-diagonal disorder. The theory presented here applies to the disorder of a single diagonal element in the local mode Hamiltonian, not the diagonalized exciton states. Thus, one can use isotope labeling to obtain the frequency correlation function of the local modes, and then model the exciton states. If Bloch dynamics are appropriate to describe the local modes, then modeling the exciton states is quite straightforward, but explicitly including the local mode dynamics is computationally challenging [102].

Exercises

7.1 Use the stationary property of a time correlation function to show that it is an even function. That is, prove

$$\langle A(0)A(t)\rangle = \langle A(0)A(-t)\rangle. \tag{7.45}$$

7.2 Prove that

$$g(t) = \frac{1}{2} \int_0^t \int_0^t d\tau' d\tau'' \langle \delta\omega_{01}(\tau') \delta\omega_{01}(\tau'') \rangle$$

$$= \int_0^t \int_0^{\tau'} d\tau' d\tau'' \langle \delta\omega_{01}(\tau' - \tau'') \delta\omega_{01}(0) \rangle \qquad (7.46)$$

$$= \int_0^t \int_0^{\tau'} d\tau' d\tau'' \langle \delta\omega_{01}(\tau'') \delta\omega_{01}(0) \rangle. \qquad (7.47)$$

Hint: Break the integral of $\int_0^t d\tau''$ into two terms, one of which goes from 0 to τ' and the other from τ' to t, and then show that the second term is equal to the first. Hint 2: To get the final step in the derivation, do a coordinate transformation of the integral by defining $x = \tau' - \tau''$.

7.3 Starting from the Kubo lineshape function (Eq. 7.25), prove Eqs. 7.26 and 7.28 in the homogeneous and inhomogeneous limits, respectively.

7.4 (a) Show that in the slow modulation limit a lineshape is produced that matches the frequency distribution of the molecules, whether or not it is Gaussian. Hint: Start with Eq. 7.9. Rewrite in terms of $\omega_{01}(t)$ using Eq. 7.6. Let $\omega_{01}(t) = \omega_{01}(0)$. Take the Fourier transform to get the spectrum.

7.5 Imagine a carbonyl vibration on the surface of a protein. The time-scale of protein structural fluctuations is typically much slower than that of solvent fluctuation. What ansatz might you use for the frequency fluctuation correlation function?

7.6 Not all molecules have a frequency fluctuation correlation function that decays monotonically. For example, for the OH stretch vibration of water, one observes a partial recurrence at about 150 fs. Discuss what this implies for the frequency trajectory, and what could be the structural cause of it.

8

Dynamic cross-peaks

In Chapter 7 we found that a time dependent Hamiltonian (Eq. 7.2) is the source of infrared lineshapes. It also causes energy transfer between eigenstates that will alter the intensity of cross- and diagonal peaks or create completely new peaks. As for lineshapes, the time dependence of the Hamiltonian results from the fluctuations of the solvent and molecular structure. Three types of relaxation occur, all of which have been observed experimentally: population relaxation, population transfer, and coherence transfer. Figure 8.1 shows an example of a Feynman diagram for each of these cases. In contrast to the Feynman diagrams that we have seen so far, these diagrams time-propagate with more than just the usual oscillatory terms. They flip quantum states (indicated by the dotted lines and the arrows in Fig. 8.1). Population relaxation, population transfer, and coherence transfer correspond to flipping an excited population state $|1\rangle\langle 1|$ back into the ground state $|0\rangle\langle 0|$ (Fig. 8.1a), flipping an excited population state $|1\rangle\langle 1|$ into another excited state $|1'\rangle\langle 1'|$ (Fig. 8.1b), and flipping a coherence state $|0\rangle\langle 1|$ into another coherence state $|0\rangle\langle 1'|$ (Fig. 8.1c), respectively. A rigorous treatment of relaxation processes has become quite demanding, and we outline here only the simplest cases. For a more comprehensive discussion, we refer readers to Ref. [106].

8.1 Population transfer

As the simplest case, consider a dimer of coupled oscillators that is sitting in a fluctuating bath. The coupling Hamiltonian, Eq. 6.17, will then become time dependent as a whole:

$$H(t) = \begin{pmatrix} \hbar\omega_A(t) & \beta_{AB}(t) \\ \beta_{AB}(t) & \hbar\omega_B(t) \end{pmatrix}$$
$$\equiv \begin{pmatrix} \hbar\omega_A + \hbar\delta\omega_A(t) & \beta_{AB} + \delta\beta_{AB}(t) \\ \beta_{AB} + \delta\beta_{AB}(t) & \hbar\omega_B + \hbar\delta\omega_B(t) \end{pmatrix} \quad (8.1)$$

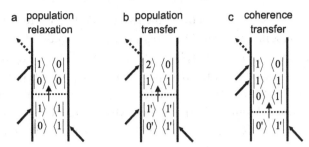

Figure 8.1 Three types of relaxation processes. (a) Population relaxation, (b) population transfer and (c) coherence transfer.

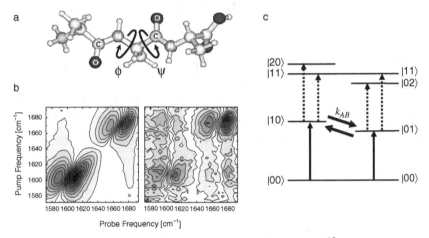

Figure 8.2 Cross-relaxation between the two amide I states of ^{13}C-isotope substituted trialanine (A^{13}C-AA) dissolved in water. (a) Structure, (b) purely absorptive 2D IR spectra at two different population times t_2, and (c) a level scheme with population transfer indicated. Adapted from Ref. [193] with permission.

with diagonal and off-diagonal disorder. Here, $\hbar\delta\omega_A(t)$, $\hbar\delta\omega_B(t)$ and $\delta\beta_{AB}(t)$ are the instantaneous deviations of the site energies and couplings from their time-average values $\hbar\omega_A$, $\hbar\omega_B$ and β_{AB} with vanishing time averages $\langle\hbar\delta\omega_A(t)\rangle = 0$, $\langle\hbar\delta\omega_B(t)\rangle = 0$ and $\langle\delta\beta_{AB}(t)\rangle = 0$. The diagonal elements fluctuate for the same reasons as they did in Chapter 7, i.e. the solvent is pushing and pulling at the local mode coordinate, thereby deforming the potential energy surface and thus the vibrational frequency (Fig. 7.1). Fluctuations of the off-diagonal element, on the other hand, bring in a new aspect. For example, consider a peptide dissolved in water (Fig. 8.2a). Peptides are flexible molecules, and will fluctuate mostly around the dihedral (ϕ, ψ) angles of each amino acid. Since the coupling β_{AB} is a function of geometry (Fig. 6.12), the coupling will become time dependent as the structure changes. We will see that this leads to transfer of population between the two states.

Figure 8.2(b) shows 2D IR spectra of the two amide I bands for trialanine [193]. Because isotope labeling was used to separate the two amide I bands by 70 cm^{-1} the cross-peaks are small (<5% relative to the size of the diagonal peaks). As population time is increased, the cross-peak intensity grows relative to the diagonal peaks ($\approx$35% of the size of the diagonal peaks; the total signal decreases due to T_1 relaxation). Figure 8.2(c) explains why. At early population times, we have a quasi-static level scheme as in Fig. 1.5 with very weak coupling, so that, e.g. the $|00\rangle \to |10\rangle$ and the $|01\rangle \to |11\rangle$ transitions appear at almost the same frequency, and largely cancel each other. However, as time t_2 goes on, population from the initially pumped $|10\rangle$ state may flow into $|01\rangle$ with rate k_{AB}, and we observe the $|01\rangle \to |02\rangle$ transition. The latter appears at a very different frequency from the initially pumped $|00\rangle \to |10\rangle$ transition, and the size of the cross-peak grows.

A rigorous treatment of Eq. 8.1 becomes quite involved in the general case [204]. However, in the weak coupling limit $|\beta_{AB}| \ll |\hbar\omega_B - \hbar\omega_A|$, Eq. 8.1 decouples into two simple and very intuitive contributions. To this end, we diagonalize the Hamiltonian Eq. 8.1 with respect to its time average part up to first order of the mixing angle (the smallness parameter, see Eq. 6.20).

$$\alpha \approx \frac{\beta_{AB}}{\hbar\omega_B - \hbar\omega_A} \tag{8.2}$$

and obtain:

$$H'(t) = H_{\text{stat}} + H_{\text{dyn}}^{(0)}(t) + H_{\text{dyn}}^{(1)}(t) + \mathcal{O}(\alpha^2) \tag{8.3}$$

with

$$H_{\text{stat}} = \begin{pmatrix} \hbar\omega_A - 2\alpha \cdot \beta_{AB} & 0 \\ 0 & \hbar\omega_B + 2\alpha \cdot \beta_{AB} \end{pmatrix}$$

$$H_{\text{dyn}}^{(0)}(t) = \begin{pmatrix} \hbar\delta\omega_A(t) & \delta\beta_{AB}(t) \\ \delta\beta_{AB}(t) & \hbar\delta\omega_B(t) \end{pmatrix}$$

$$H_{\text{dyn}}^{(1)}(t) = \begin{pmatrix} -2\alpha \cdot \delta\beta_{AB}(t) & \alpha \cdot \hbar\delta\Delta\omega(t) \\ \alpha \cdot \hbar\delta\Delta\omega(t) & 2\alpha \cdot \delta\beta_{AB}(t) \end{pmatrix}, \tag{8.4}$$

and $\delta\Delta\omega(t) = \delta\omega_A(t) - \delta\omega_B(t)$. The first term, H_{stat}, is the static contribution which is diagonal by construction, the second term, $H_{\text{dyn}}^{(0)}(t)$, is zeroth-order in the mixing angle α, and the third term, $H_{\text{dyn}}^{(1)}(t)$, is first-order in the mixing angle α. The time averages of $\langle H_{\text{dyn}}^{(0)}(t)\rangle = 0$ and $\langle H_{\text{dyn}}^{(1)}(t)\rangle = 0$ vanish.

To zeroth-order in $\alpha = \beta_{AB}/(\hbar\omega_B - \hbar\omega_A)$, the time dependence of the coupling Hamiltonian separates completely. For instance, the time dependence of the diagonal peaks is governed solely by fluctuations of the diagonal elements $\delta\omega_A(t)$ and $\delta\omega_B(t)$, and we can formulate a lineshape function of each site along the lines of

Chapter 7, as if it were a separated (uncoupled) state. Fluctuations of the diagonal elements will give rise to inhomogeneous and pure dephasing as well as spectral diffusion.

The time dependence of the cross-peaks, on the other hand, is exclusively governed by fluctuations of the off-diagonal elements $\delta\beta_{AB}(t)$. Fluctuation of off-diagonal elements $\delta\beta_{AB}$ lead to irreversible population transfer between the two eigenstates on the diagonal of the Hamiltonian H_{stat} (Eq. 8.4). To see that, we employ time dependent perturbation theory, starting from

$$H(t) = \begin{pmatrix} \hbar\omega'_A & \delta\beta_{AB}(t) \\ \delta\beta_{AB}(t) & \hbar\omega'_B \end{pmatrix} \quad (8.5)$$

with diagonal elements in the eigenstate basis $\hbar\omega'_A = \hbar\omega_A - 2\alpha\beta_{AB}$ and $\hbar\omega'_B = \hbar\omega_B + 2\alpha\beta_{AB}$ which we assume to be constant in time. We look for the solution of the time dependent Schrödinger equation:

$$i\hbar \frac{\partial \Psi(t)}{\partial t} = \hat{H}(t)\Psi(t) \quad (8.6)$$

with

$$\Psi(t) = c_A(t)\phi_A e^{-i\omega'_A t} + c_B(t)\phi_B e^{-i\omega'_B t} \quad (8.7)$$

where ϕ_A and ϕ_B are the eigenstates of the time independent part of the Hamiltonian. Substituting Eq. 8.7 in Eq. 8.6, we obtain for the time evolution of $c_A(t)$ and $c_B(t)$:

$$i\hbar \frac{\partial c_A(t)}{\partial t} = \delta\beta_{AB}(t) e^{i(\omega'_A - \omega'_B)t} c_B(t)$$

$$i\hbar \frac{\partial c_B(t)}{\partial t} = \delta\beta_{AB}(t) e^{i(\omega'_B - \omega'_A)t} c_A(t). \quad (8.8)$$

We assume that the system is initially in one state, for example A, so that $c_A(0) = 1$ and $c_B(0) = 0$. Integration of Eq. 8.8 with $c_A = 1$ gives the first-order correction to c_B:

$$c_B(T) = -\frac{i}{\hbar} \int_0^T \delta\beta_{AB}(t) e^{i(\omega'_B - \omega'_A)t} dt. \quad (8.9)$$

The probability that the system has made a transition to state B is $|c_B(T)|^2$:

$$P_{AB} = \frac{1}{\hbar^2} \int_0^T dt_2 \int_0^T dt_1 \delta\beta_{AB}(t_1) \delta\beta_{AB}(t_2) e^{i(\omega'_B - \omega'_A)(t_1 - t_2)}. \quad (8.10)$$

Finally, we average over the ensemble of molecules:

$$P_{AB} = \frac{1}{\hbar^2} \int_0^T dt_2 \int_0^T dt_1 \langle \delta\beta_{AB}(t_1) \delta\beta_{AB}(t_2) \rangle e^{i(\omega'_B - \omega'_A)(t_1 - t_2)}. \quad (8.11)$$

In equilibrium, the correlation function $\langle \beta_{AB}(t_1)\beta_{AB}(t_2)\rangle$ will depend only on the time difference $\tau \equiv t_1 - t_2$. Substituting this new time variable, we get

$$P_{AB} = \frac{1}{\hbar^2}\int_0^T dt_2 \int_{-t_2}^{T-t_2} d\tau \langle \delta\beta_{AB}(\tau)\delta\beta_{AB}(0)\rangle e^{i(\omega_B'-\omega_A')\tau}. \tag{8.12}$$

The fluctuation autocorrelation function $\langle \delta\beta_{AB}(\tau)\delta\beta_{AB}(0)\rangle$ typically decays quickly with τ, so when T is large enough, the second integral becomes almost independent of T and we can send the integration limits to $\pm\infty$. The integration with respect to t_2, on the other hand, gives just a term linear in T. Putting everything together, we obtain for the transfer rate (with $P_{AB} = k_{AB}T$):

$$k_{AB} = \frac{1}{\hbar^2}\int_{-\infty}^{\infty} \langle \delta\beta_{AB}(\tau)\delta\beta_{AB}(0)\rangle e^{i(\omega_B'-\omega_A')\tau} d\tau. \tag{8.13}$$

To zeroth order in α, the transfer rate is given by the Fourier transform of the fluctuation autocorrelation function $\langle \delta\beta_{AB}(\tau)\delta\beta_{AB}(0)\rangle$, i.e. the spectral density of the fluctuations, evaluated at a frequency which equals the energy gap $\Delta\omega \equiv (\omega_B' - \omega_A')$ between both states. When the first-order term $H_{\text{dyn}}^{(1)}(t)$ becomes important, the correlation function $\langle \delta\Delta\omega(t)\delta\Delta\omega(0)\rangle$ (with $\Delta\omega(t) = \omega_B - \omega_A$) and the cross-correlation function $\langle \delta\beta_{AB}(t)\delta\Delta\omega(0)\rangle$ would contribute as well to population transfer. However, it has been verified by MD simulations [193] that the first-order term has only negligible effect, at least in the example of Fig. 8.2.

Figure 8.3 illustrates the concept [193]. Shown is the time dependence of the central (ϕ, ψ) dihedral angles of trialanine, as obtained from an all-atom molecular dynamics (MD) simulation (Fig. 8.3a). With the help of a coupling map such as Fig. 6.12(d), the fluctuating structure can be translated into a fluctuating coupling constant $\beta_{AB}(t)$ (Fig. 8.3b), whose autocorrelation function decays on two timescales (Fig. 8.3c). The Fourier transform of the autocorrelation function reveals the spectral density of the fluctuations (Fig. 8.3c, insert). Depending on the energy gap between the two amide I states, which can be varied by varying the isotope substitution of the peptide, different relaxation rates are obtained, as is indeed observed experimentally [193].

What we have outlined here is the simplest version of relaxation theory, which applies in the high-temperature limit with $k_B T \gg \hbar\Delta\omega$. The forward and backward rates are equal and the two cross-peaks at both sides grow equally. In other words, this theory does not preserve detailed balance, which is a consequence of the coupling correlation function $\langle \delta\beta_{AB}(\tau)\delta\beta_{AB}(0)\rangle$ being real valued, revealing $k_{AB} = k_{BA}$ after the Fourier transform. Nevertheless, in most 2D IR experiments we often only consider a frequency window of 100–200 cm^{-1}, so we will automatically fulfill the condition $\hbar\Delta\omega < k_B T$.

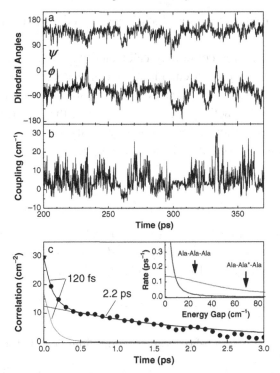

Figure 8.3 (a) Trajectory of the dihedral angles $\phi(t)$ and $\psi(t)$ of the central amino acid of trialanine as obtained from an MD simulation; (b) instantaneous coupling $\beta_{AB}(t)$ computed from the trajectory and (c) coupling autocorrelation function $\langle \delta\beta_{AB}(\tau)\delta\beta_{AB}(0)\rangle$, the Fourier transform of which determines the population transfer rate (insert). The arrows mark the frequency splittings of Ala-Ala-Ala and Ala-^{13}C-Ala-Ala, respectively. The light gray and dark gray curves represent the fast and slow components of the correlation function, respectively. Adapted from Ref. [193] with permission.

In 2D electronic spectroscopy, or in two-color 2D IR spectroscopy, when energy gaps are larger than k_BT, relaxation will preferably go downhill, and cross-peaks grow only in one off-diagonal region [22]. In this case, a full quantum version of the coupling correlation function needs to be applied, a full account of which is beyond the scope of this book (a nice discussion of the issue is given, e.g. in Ref. [120]). From the general time symmetries of quantum correlation functions, one can deduce for their Fourier transforms $k_{AB} = e^{-\hbar\Delta\omega/k_BT}k_{BA}$. However, it is often not possible to calculate an accurate quantum correlation function from an MD simulation, which is why one resorts to a classical theory with rate $k_{AB}^{(cl)}$, as we have done here, and adds to it a quantum correction factor, such as:

$$k_{AB} \approx \frac{\hbar\Delta\omega}{(1-e^{-\hbar\Delta\omega/k_BT})k_BT} k_{AB}^{(cl)} \qquad (8.14)$$

that enforces detailed balance. Note that the choice of the quantum correction factor is not unique [120].

Population relaxation is commonly calculated on the same footing. That is, it is a random force acting on a chemical bond that couples the ground and first excited state of the vibrational transition. Hence, it is the Fourier transformation of the autocorrelation function of that random force that describes the relaxation rate. However, since current laser technology limits us to only study states with $\hbar\omega_{01} \gg k_B T$, quantum correction factors have to be taken into account [120].

8.2 Dynamic response functions

Now that we know the transfer rate k_{AB}, we can reformulate the response functions (see Fig. 8.4 for the corresponding Feynman diagrams). To that end, we first need to solve the kinetic equations for the simple equilibrium:

$$A \overset{k_{AB}}{\rightleftarrows} B \tag{8.15}$$

to which we add population relaxation, assuming identical T_1 for both A and B:

$$\frac{d}{dt}\begin{pmatrix} n_A \\ n_B \end{pmatrix} = \begin{pmatrix} -k_{AB} - 1/T_1 & k_{BA} \\ k_{AB} & -k_{BA} - 1/T_1 \end{pmatrix}\begin{pmatrix} n_A \\ n_B \end{pmatrix}. \tag{8.16}$$

Its solution is:

$$n_A(t_2) = \Gamma_{AA}(t_2)n_A(t_2=0) + \Gamma_{BA}(t_2)n_B(t_2=0)$$
$$n_B(t_2) = \Gamma_{AB}(t_2)n_A(t_2=0) + \Gamma_{BB}(t_2)n_B(t_2=0) \tag{8.17}$$

with

$$\Gamma_{AA}(t_2) = \frac{k_{BA}e^{-t_2/T_1} + k_{AB}e^{-(k_{AB}+k_{AB}+1/T_1)t_2}}{k_{AB} + k_{BA}}$$

$$\Gamma_{BA}(t_2) = \frac{k_{BA}\left(e^{-t_2/T_1} - e^{-(k_{AB}+k_{AB}+1/T_1)t_2}\right)}{k_{AB} + k_{BA}}$$

$$\Gamma_{BB}(t_2) = \frac{k_{AB}e^{-t_2/T_1} - k_{BA}e^{-(k_{AB}+k_{AB}+1/T_1)t_2}}{k_{AB} + k_{BA}}$$

$$\Gamma_{AB}(t_2) = \frac{k_{AB}\left(e^{-t_2/T_1} - e^{-(k_{AB}+k_{AB}+1/T_1)t_2}\right)}{k_{AB} + k_{BA}}. \tag{8.18}$$

With these factors, one may now formulate the response functions. For the diagonal rephasing diagrams we obtain (see Eq. 7.40):

8.2 Dynamic response functions

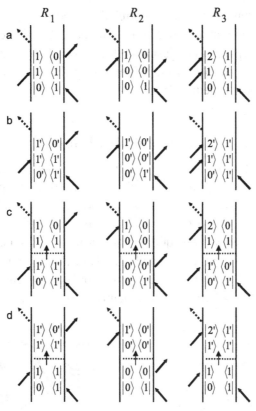

Figure 8.4 Rephasing Feynman diagrams that need to be considered in an exchange experiment; the corresponding non-rephasing diagrams are obtained by interchanging the time ordering of the first two interactions. (a,b) Feynman diagrams that stay in species A and B, and (c,d) Feynman diagrams that flip between species A and B during the population time t_2. Additional diagrams are in principle present, describing the direct coupling between both states from the static part of β_{AB} in Eq. 8.1; however, since that is often weak, it is neglected here.

$$R_{AA} = i\mu_A^4 \left(e^{-i\omega_A(t_3-t_1)} - e^{-i((\omega_A-\Delta)t_3-\omega_A t_1)}\right) \Gamma_{AA}(t_2) \cdot$$
$$\cdot e^{-g(t_1)+g(t_2)-g(t_3)-g(t_1+t_2)-g(t_2+t_3)+g(t_1+t_2+t_3)}$$

$$R_{BB} = i\mu_B^4 \left(e^{-i\omega_B(t_3-t_1)} - e^{-i((\omega_B-\Delta)t_3-\omega_B t_1)}\right) \Gamma_{BB}(t_2) \cdot$$
$$\cdot e^{-g(t_1)+g(t_2)-g(t_3)-g(t_1+t_2)-g(t_2+t_3)+g(t_1+t_2+t_3)} \quad (8.19)$$

with corresponding terms for the non-rephasing diagrams. For the off-diagonal diagrams (Fig. 8.4c,d):

$$R_{AB} = i\mu_A^2\mu_B^2 \left(e^{-i(\omega_B t_3-\omega_A t_1)} - e^{-i((\omega_B-\Delta)t_3-\omega_A t_1)}\right) \Gamma_{AB}(t_2) \cdot e^{-g(t_1)} e^{-g(t_3)}$$
$$R_{BA} = i\mu_A^2\mu_B^2 \left(e^{-i(\omega_A t_3-\omega_B t_1)} - e^{-i((\omega_A-\Delta)t_3-\omega_B t_1)}\right) \Gamma_{BA}(t_2) \cdot e^{-g(t_1)} e^{-g(t_3)}$$

$$(8.20)$$

one often assumes that any frequency correlation is destroyed during jumping from state A to B (although that does not necessarily have to be the case), hence the 2D lineshape function separates into just a product of two 1D lineshape functions. The diagonal and cross-peak overall intensities reflect the Γ-factors, from which the exchange kinetics can be deduced.

8.3 Chemical exchange

Another process that leads to time dependent cross-peaks in exactly the same manner as population transfer, but whose physical origin is completely different, is chemical exchange [107, 192, 207]. Consider phenol that weakly hydrogen bonds to benzene in CCl_4 solution (Fig. 8.5a). Depending on the concentration of the compounds, one can adjust the equilibrium constant between a hydrogen-bonded form and a dissociated form to essentially 1:1. Hydrogen bonding strongly deforms the potential energy curve of the OD bond of phenol, which causes the OD frequency to shift by ≈ -40 cm^{-1}. Consequently, two approximately equal intensity

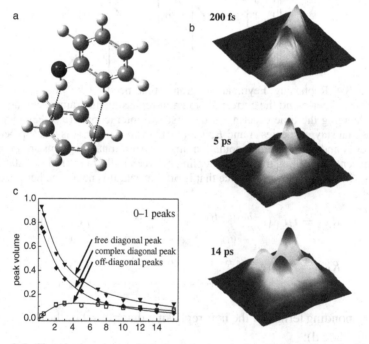

Figure 8.5 Chemical exchange between a hydrogen-bonded and dissociated form of a phenol-benzene complex dissolved in CCl_4. (b) A series of purely absorptive 2D IR spectra with increasing population time t_2 and (c) the intensity of both the diagonal and off-diagonal peaks as a function of population time t_2. Adapted from Ref. [207] with permission.

peaks are observed in the IR absorption spectrum, *despite* the fact that only one type of molecular group is present in this frequency range. Two peaks are observed since the one molecular group exists in two different chemical environments.

The phenol–benzene hydrogen bond is weak, and hence is constantly formed and broken. As a result, the rate k_{AB} with which the equilibrium is maintained lies in the picosecond regime so that we again have a kinetic exchange between two states of the molecule, just like in Eq. 8.15. Although the physical origin of that kinetic process is now distinctively different, the response functions are exactly the same as before (Eqs. 8.19 and 8.20). The exchange process can directly be observed by 2D IR spectroscopy (Fig. 8.5b) by the appearance of cross-peaks. At early times, two peaks are observed along the diagonal, representing the two states, without any cross-peaks. Cross-peaks are absent because the coupling between the two molecules is weak. As time t_2 is incremented, cross-peaks appear. In essence, we tag the molecule in one state by vibrationally exciting it. The molecular group will remain tagged for as long as that excitation lives, which is set by T_1. If the molecular group changes its spectroscopic properties due to chemical exchange, it will take its tag (i.e. its excitation) with it, and radiate with a new frequency during the detection period t_3. Consequently, a cross-peak appears whose kinetics directly reflect the exchange kinetics k_{AB}. The time window accessible to observe the exchange process is limited by the T_1 population relaxation time and the signal-to-noise ratio. Typically, one can measure spectra to a few times T_1, which is usually in the range of 1–20 picoseconds. 2D IR spectroscopy is therefore well suited to study fast exchange processes.

A basic assumption of exchange 2D IR spectroscopy is that the vibrational excitation of the molecular group by the laser pulses does not influence the way in which the molecules move in time; the excitation is just a label to follow the natural thermal motion. This assumption is certainly valid in the case of NMR spectroscopy, where the excitation energy of spins is far below $k_B T$ and hence very unlikely to influence the course of a reaction. However, in the case of an IR excitation, vibrational energy could, in principle, change the equilibrium of a thermally equilibrated reaction [170], but intra-vibrational relaxation usually dominates.

9

Experimental designs, data collection and processing

In the preceding chapters we have discussed many of the mathematical and logical underpinnings of 2D IR spectroscopy. In this chapter we focus on some of the practical aspects of implementing 2D IR spectroscopy in the laboratory and methods of processing the data. We have written this chapter assuming that our audience is already familiar with basic laser and ultrafast spectroscopic techniques and is interested in transitioning into the 2D IR field. The field of 2D IR spectroscopy is still growing and evolving. Improvements are continuously being made. Thus, some parts of this chapter may go out of date quickly, but at this time there is no commercial 2D IR spectrometer available nor is there a consensus on the best way of collecting multidimensional spectra, and so we think it prudent to discuss the design aspects of current 2D IR spectrometers, especially since the spectrometer design is tied closely with the theory.

Shown in Fig. 9.1 are some of the designs that have been invented so far for measuring 2D IR spectra. We have separated the designs into three categories, based on whether they operate in the frequency or time domains, or either.[1] We have also divided this chapter into these same three categories. And, embedded throughout this chapter, are discussions pertaining to phase stability, data workup, and other issues pertinent to deciding how to implement 2D IR spectroscopy.

9.1 Frequency domain spectrometer designs

One of the simplest methods for implementing 2D IR spectroscopy both in the ease at which the spectrometer is assembled and the simplicity of the data processing, is the narrowband pump–probe approach that Hamm, Lim and Hochstrasser used to collect the first 2D IR spectrum in 1998 [83]. A schematic of the optical layout

---

[1] There are other designs developed for visible wavelengths that are difficult to implement in the mid-IR [88, 176, 185].

9.1 Frequency domain spectrometer designs

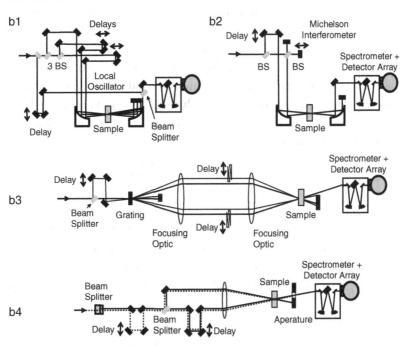

Figure 9.1 Some existing spectrometer designs separated into the categories into which they operate. (a) Frequency domain spectrometer using a pump–probe beam geometry and a Fabry–Perot filter [83]. (b) Time domain spectrometers using (b1) a four-wave mixing geometry [5], (b2) a pump–probe beam geometry with a Michelson interferometer [40], (b3) a phase stable design using diffractive optics [72], and (b4) a phase stable design using conventional optics [163]. (c) A pulse shaping method that can operate in either the time or frequency domains by shaping the pump pulse [166].

is shown in Fig. 9.1(a), or one now also has the choice of using a pulse shaper as shown in Fig. 9.1(c). Either way, femtosecond mid-IR pulses (from an optical parametric amplifier [85]) are split into an intense pump beam (typically 1–2 µJ) and two weak beams (a few nJ each) using a wedged piece of CaF_2 or ZnSe. The bandwidth of the pump beam is filtered using either a Fabry–Perot style etalon or a pulse shaper to create a narrowband pulse whose center frequency can be swept across the vibrational modes of interest. The weak beams serve as a probe and a reference (the reference is optional and not shown in the figures). The probe beam is spatially and temporally overlapped in the sample with the pump beam. The reference beam also passes through the sample but is slightly offset from the probe. It is used to normalize the shot-to-shot intensity fluctuations of the laser system.

If a pulse shaper is being used to filter the pump pulse, then the bandwidth and the temporal shape is computer adjustable (see Section 9.4.2 below). If an etalon is being used, then the optics must be designed for the particular experiment at hand. An etalon consists of two partially reflective dielectric mirrors separated by a small gap so that a femtosecond laser pulse that enters the etalon will bounce back and forth between the mirrors like a laser cavity (Fig. 9.2a). Each time the pulse hits one of the mirrors, a little light leaks out, leading to a series of femtosecond pulses with an exponential decay in their intensities. Thus, the pump pulse generated by an etalon has a Lorentzian-shaped bandwidth (because an exponential decay in time is equivalent to a Lorentzian in frequency), and can be described approximately according to the equation

$$E(\omega) = E_0(\omega) \frac{1 - R}{1 - R \cdot e^{i\frac{2d\omega}{c}}} \qquad (9.1)$$

where $E_0(\omega)$ is the electric field of the laser pulse before the etalon, R the reflection coefficient of the mirrors in the etalon, d their spacing, and c the speed of light. The transmission of the etalon will be unity whenever the phase factor in the denominator is $2d\omega/c = 2\pi n$, where the integer number n is the order of interference. The order of interference must be chosen such that only one transmission peak

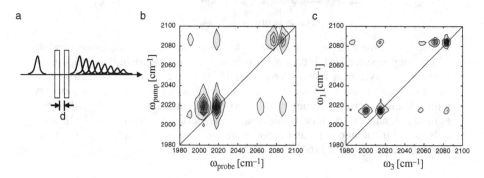

Figure 9.2 (a) A schematic of how a Fabry–Perot filter works. 2D IR spectra of a metal dicarbonyl collected with (b) a Fabry–Perot filter and (c) a time domain method. Notice that the Fabry–Perot filter elongates the peaks because the measured spectrum is a convolution of the pump bandwidth with the molecular linewidths. Adapted from Ref. [24] with permissions.

lies within the laser pulse spectrum $E_0(\omega)$ (typically $n = 5$–10). The reflection coefficient of the mirrors is then manufactured to set the desired bandwidth of the transmitted pulse (typically $R = 90\%$) and the spacing is adjusted during the experiment to scan the center frequency. The pulse emerging from the etalon has a tail with an exponential decay that stretches towards the probe pulse (Fig. 9.2a), and so it is necessary for the probe pulse to be time-delayed from the pump to avoid deleterious interferences (this is a problem that we return to in Section 9.4.2, because with a pulse shaper, the time domain shape can be designed to minimize these unwanted interferences). Piezocrystal actuators are not very reproducible, so when the gap between the mirrors is changed to shift the wavelength, the spectrum of the pump should be measured on a spectrometer and a feedback loop used to set the final spacing.

The spectral width of the pump is typically chosen so that it is comparable to the homogeneous linewidth. That way, the duration of the laser pulse matches the fast vibrational dynamics of the system and slower spectral diffusion dynamics can be measured by scanning the pump–probe delay time. If the pump bandwidth matches the homogeneous linewidth, then the antidiagonal width of the 2D IR peaks gives the homogeneous linewidth and the diagonal width is the total vibrational linewidth. Of course, the third-order emission is always a convolution of the molecular response with the laser pulse (Eq. 2.74), and so the linewidths will always be broadened by the pump pulse, at least along one frequency axis. Along the probe frequency axis, the lineshapes will more closely reflect the natural molecular widths since a femtosecond probe pulse is used in this style of data collection (see Fig. 9.2b,c).

One of the main advantages of the narrowband pump–probe method is that data collection and processing are very straightforward. One uses a chopper to block the pump laser beam at half the laser repetition rate so that a background probe spectrum is collected every other shot. Chopping largely suppresses slow drifts of the laser system, and only noise which is uncorrelated from one laser shot to the next remains. From these two spectra, the probe absorbance is calculated, which is a function of the probe wavelength as well as the center wavelength of the pump pulse which is scanned across the vibrational modes of interest

$$S(\omega_{\text{probe}}; \omega_{\text{pump}}, t_2) = -\log \frac{I(\omega_{\text{probe}}; \omega_{\text{pump}}, t_2)}{I_0(\omega_{\text{probe}})}. \tag{9.2}$$

Thus, one arrives at a 2D IR spectrum directly, without further data processing. Moreover, the spectra are absorptive and properly phased, which is a major advantage of collecting data in the frequency domain over the time domain four-wave mixing beam geometries.

Shot-to-shot normalization with a separate reference beam improves the signal-to-noise ratio by typically a factor of 5–10. To gain that improvement, it is very important that both probe and reference beams are treated as symmetrically as possible; e.g. both should be guided over the same optics, and the precise imaging of both onto the two lines of a double array detector is critical. The final signal is calculated using

$$S = -\log \frac{\sum_s I_{\text{probe,on}} \sum_s I_{\text{ref,off}}}{\sum_s I_{\text{probe,off}} \sum_s I_{\text{ref,on}}} \quad (9.3)$$

where I_{probe} and I_{ref} are the probe and reference channels, respectively, I_{on} and I_{off} is with the chopper open and closed, respectively, and where the sums are to average over a certain number of laser shots (typically 1000). Note that this is not the same as:

$$S = -\log \sum_s \frac{I_{\text{probe,on}} I_{\text{ref,off}}}{I_{\text{probe,off}} I_{\text{ref,on}}}. \quad (9.4)$$

Although the latter seems to correlate shot-to-shot fluctuations more directly, the first type of averaging is preferred (in the presence of noise, one can show that the second type of averaging leads to a signal $S > 0$ even if the pump beam is blocked).

9.2 Experimental considerations for impulsive spectrometer designs

If one needs to collect 2D IR spectra with optimum frequency and/or time resolution then a fully impulsive setup is needed in which all of the laser beams are femtosecond laser pulses. In this case, one can use any of the methods shown in Fig. 9.1(b) or (c). When choosing one of these designs, or inventing a new one, the main considerations are phase stability, accurate time delays, phase matching geometry, and detection. In what follows, we describe these considerations.

9.2.1 Phase stability

A major consideration when designing a time domain 2D IR spectrometer is phase drift. Time domain 2D IR spectroscopy must measure the oscillation in signal strength caused by coherences during the t_1 and t_3 time delays (or other delays, depending on the pulse sequence). A drift in phase will cause the oscillations at the early times of the free induction decay to destructively interfere with later oscillations in which they should normally constructively interfere. The result is often-times an asymmetric peak shape, such as shown in Fig. 9.3 (which can be found

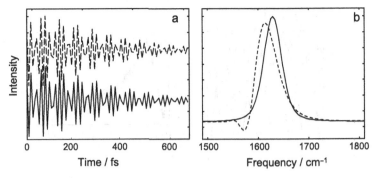

Figure 9.3 (a) Time domain signal from a simulated free induction decay in which phase drift has occurred on a time-scale close to the acquisition time. Free induction decay (solid) without and (dashed, offset) with phase drift. Notice that the oscillations are out of phase near the end of the free induction decay. (b) A slice through the simulated 2D IR spectrum (solid) without and (dashed) with phase drift, calculated from the free induction decays in (a). Notice the asymmetry caused by phase drift.

in many of the first reports of 2D IR spectra collected using the four-wave mixing geometry). Phase drift is a misnomer because it is caused by small changes in path length rather than by a change in the relative carrier phases of the femtosecond pulses. At mid-IR wavelengths of 3–6 μm, a phase change of 90° only requires a change in the beam path length of 750–1500 nm.

The designs in Fig. 9.1 suffer from varying degrees of phase drift. The design (b1) is the worst, because every laser pulse hits an entirely different set of mirrors, so that the movement of any mirror in the setup changes the signal phase. The designs (b3), (b4) and (c) are the best, because the pairs of pulses responsible for each coherence time hit the same optics. That way, if a mirror moves, the relative phase drift between pulse pairs is constant (for instance, if a coherence is generated by pulses 1 and 2, then if they hit all the same optics their relative time delay will not be affected by mirror translations). The design (b3) uses diffractive optics or gratings to generate stable pairs of pulses. The design (b4) only uses conventional optics. The pulse shaping design (c) is inherently phase stable, because pulse 1 and 2 as well as pulse 3 and the local oscillator are collinear (see Section 9.2.3).

Shown in Fig. 9.4c is the phase stability of a typical mid-IR four-wave mixing design measured every seven seconds over the course of seven hours. The phase is obtained by scanning the free induction decay across the same range of delays over and over again. Figure 9.4(a) shows the free induction decay measured at 7 s, 17 and 34 min. From these measurements it is apparent that the phase drifts on two time-scales. On the time-scale of a few minutes, the phase fluctuates rapidly with a standard deviation of 3°, corresponding to $\lambda/120$. These fast phase fluctuations are

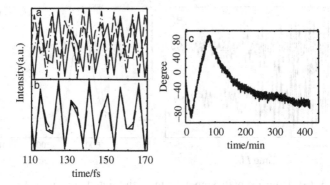

Figure 9.4 (a) A free induction decay measured at 7 s, 17 and 34 min. The phase has clearly changed during this time. (b) The free induction decays after phase correction. (c) The phase of the free induction decay measured every 7 s over 7 hours. Adapted from Ref. [45] with permission.

presumably stochastic and their effects removed from 2D IR spectra with enough averaging (each time delay in Fig. 9.4a was only averaged for 20 laser shots). The more grievous phase shift occurs over a longer time-scale of minutes to hours, where the phase rotates a full 360° within 2 hours. While any phase drift degrades the quality of the multidimensional spectra, this slow drift distorts 2D IR spectra that take longer than 5–10 minutes to collect and prohibits averaging multiple spectra together.

To remove phase drifts, the beam paths can either be actively stabilized [186] or the phase drift can be measured and corrected later on in the computer [45]. The time-scale of the slow phase shift is slow enough that it can be reliably characterized with periodic measurements every few minutes. To monitor the phase drift during a measurement, a reference time-scan (or spectral interferogram) should be collected using the same delay times for each reference scan. Linear interpolation of the data is then used to correct the phase of each data point in the 2D IR data set. Shown in Fig. 9.4(b) are the same scans as in Fig. 9.4(a) after phase correction, which makes them nearly identical. For samples in which the signal strength is so weak that a quick reference scan is not possible, one can design a sample mount to quickly translate to a high signal sample or a small pinhole (see Section 9.5.6).

To actively phase stabilize a 2D IR setup, one overlaps the infrared laser beam with a HeNe beam so that they co-propagate through the optical layout, and then, just before the sample, measure the phase of the HeNe beams. Phase drift is then compensated by feedback loops to mirrors fitted with piezoelectric transducers that make small changes to the path lengths. The HeNe method complicates the optical design through additional optics and electronics, but the result is high phase stability for as long as needed.

9.2.2 Achieving accurate time delays

Collecting data with precise time steps is imperative for achieving reliable 2D IR spectra. Inaccuracies in time steps diminish the intensity of 2D IR spectra and distort the lineshapes. The most common way of incrementing time delays in optical spectroscopy is by using a retroreflector and a translation stage. Translation stages have varying degrees of accuracy and reproducibility, but in our opinion, few are suitable for measuring 2D IR spectra. Consider that a typical 2D IR spectrum might be collected from $t_1 = 0$ to 3000 fs in 9 fs steps, for example. The most common problem with translation stages is that there are sinusoidal variations in the encoders and stepping motors that give rise to periodic variations in the actual step size taken. The effect is to alias the true frequency to a new frequency. For example, if we think the stage is taking 9 fs steps, but it is really taking 9.5 fs steps (which is within the resolution of most translation stages), then the observed frequency will be shifted by 10%. The more egregious problem are the sinusoidal variations. Shown in Fig. 9.5 is the Fourier transform of a HeNe interferogram collected with a Michelson interferometer. The dashed line in Fig. 9.5 is generated from data collected by incrementing one arm of the interferometer in 50 nm steps to cover 237 HeNe periods using a retroreflecting translation stage. For retroreflection, 150 nm = 1.0 fs of translation. A HeNe interferogram should have only a single peak at 632.8 nm, but multiple peaks appear in Fig. 9.5 (dashed) because of sinusoidal inaccuracies in the translation stage steps on the scale of 16 nm. When collecting 2D IR data sets, it is even possible to get replica 2D IR spectra when the time domain data is Fourier transformed. Distortions often appear as asymmetric lineshapes similar to those caused by phase drift (Fig. 9.3).

There are three ways to circumvent or fix the problem caused by the translation stages. First, the step sizes can be set extremely accurately with a pulse shaper (see Sect. 9.4.2). Second, HeNe tracer beams, that we discussed above,

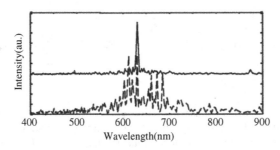

Figure 9.5 The Fourier transforms of two HeNe interferograms collected with a Michelson interferometer. The solid line is obtained using the ZnSe wedges and the dashed line is measured with a retroreflecting translation stage. Adapted from Ref. [45] with permissions.

can be used to very accurately define the time steps while they are stabilizing the phase [186]. Third, one can use ZnSe wedges instead of retroreflectors [45]. The solid line in Fig. 9.5 is another Fourier transform from the interferences of the same Michelson interferometer, except this time the translation stage increments one of two ZnSe wedges in 150 nm steps. Thus, instead of altering the path lengths, the stage changes the amount of material in one arm of the interferometer. The delay for 5° wedges is 2107 nm = 1.0 fs of translation, improving the delay accuracy by 2107/150 = 14 times. With these wedges in place, the translational accuracy (750 nm) and minimum step size (20 nm) of common translational stages correspond to an absolute time accuracy of 0.32 fs with a 0.01 fs time resolution. As the index of refraction of the ZnSe wedges depends on wavelength, the conversion factor changes in the IR, so it has to be calibrated, which could be done by comparing with a linear (FTIR) spectrum. The material also broadens the pulse widths and rotates the phase of light passing through the wedges. These effects can be corrected, but are so small as to be negligible.

9.2.3 Phase matching geometry

Perhaps the most involved consideration when deciding on a spectrometer design is choosing the phase matching geometry. This choice depends on the desired pulse sequence one wants to utilize, as well as the detection method, the type of sample, data workup, and other issues. There are primarily three commonly used phase matching geometries, which are shown in Fig. 9.6. In the equilateral triangle geometry (Fig. 9.6a), which was illustrated previously in Fig. 2.14, signals emerge symmetrically about the excitation pulses. Each direction corresponds to one of the three third-order pulse sequences, rephasing, non-rephasing, and two-quantum sequence (the last of which we will discuss in Section 11.1), and the

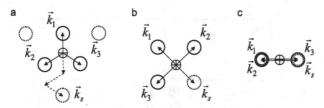

Figure 9.6 Some typical phase matching geometries, which are what we refer to as (a) equilateral triangle, (b) box-CARS and (c) pump–probe geometries. The arrows show the directions of the pulses relative to the focus at the sample, which is marked with a hatched circle. The third-order signal usually measured in these geometries is marked with dashed circles. In (a), the dashed arrows illustrate how one adds the wavevectors to get the signal direction, which in this case is $\vec{k}_s = -\vec{k}_1 + \vec{k}_2 + \vec{k}_3$.

direction that the pulse sequences appear can be chosen with the appropriate delays. This geometry is convenient when one wants to measure integrated photon echoes, because both the rephasing and non-rephasing signals can be measured simultaneously with two detectors in two different directions. However, one does not usually collect both types of 2D IR spectra at the same time because it would require two local oscillator pulses. The box-CARS geometry (Fig. 9.6b) has the three excitation pulses arranged at the three corners of a square (Fig. 9.6a). Like the equilateral geometry, the third-order signals are spatially resolved, but because the box-CARS geometry is not symmetric, only the signal at the fourth corner is used. Rather than using two detectors, one switches between the rephasing and non-rephasing signals by changing the time ordering of the first two pulses. Of the spectrometer designs illustrated in Fig. 9.1, (b1) can be set in either an equilateral or box-CARS geometry, and (b3) and (b4) use the box-CARS geometry.

The final phase matching geometry used in designs (b2) and (c) is the pump–probe or partially collinear geometry (Fig. 9.6c). In this geometry, two of the excitation pulses, $\vec{k}_1$ and $\vec{k}_2$, are collinear so that the signal is emitted in the same direction as the third excitation beam, which is often called a "probe" pulse. In the pump–probe geometry, both the rephasing and non-rephasing processes are emitted in the same direction. Thus, absorptive 2D IR spectra are automatically detected with this beam geometry. Moreover, the pump–probe geometry is convenient because the third excitation beam also serves to heterodyne the signal so that the spectra are properly phased along ω_3 since $t_3 = 0$ and $\Delta\phi = \phi_{LO} - \phi_3 = 0$ [165].

The designs shown in Fig. 9.1 are one-color 2D IR setups with all pulses originating from one IR laser pulse, in which case the lengths of all wavevectors $|\vec{k}_n|$ are identical. In principle, all these designs could be converted into a two-color 2D IR spectrometer [159], which correlates the response of a set of vibrational transitions in one frequency range with that in another frequency range that is spectrally so far away that it cannot be reached with the bandwidth of a single IR optical parametric amplifier (OPA). In that case, the pairs of pulses k_1 and k_2, as well as k_3 and the oscillator, will have different frequencies generated in two different OPAs, and one must take into account the different lengths of the wavevectors according to

$$|\vec{k}| = \frac{\omega}{c} n \qquad (9.5)$$

where ω is the frequency of the laser pulse, c is the speed of light, and n is the index of refraction. The box-CARS geometry then becomes trapezoidal shaped.

In 2D IR spectroscopy, we are often very close to the *thin sample limit*, in which the sample thickness would be of the order of the wavelength or smaller (samples are usually 6–50 µm thick). In this limit, the gratings imprinted into the sample by the incident pulses are 2D gratings rather than 3D holograms. As a result, only the projection of the k-vectors onto the sample plane is relevant, as shown in Fig. 9.6, which significantly relaxes the phase matching condition. In 2D optical or Raman spectroscopy, one typically has sample thicknesses much larger than the wavelength, and phase matching requires adding k-vectors in all three dimensions.

9.2.4 Polarization control

Polarized pulse sequences are very useful in 2D IR spectroscopies. Besides the capabilities discussed in Chapter 5, polarization control can be used to suppress unwanted signals and to increase the signal-to-noise ratio. It is much more straightforward to control the polarization of the pulses in some designs than others. Designs (b1), (b3) and (b4) allow the polarization of each pulse to be easily controlled by adding waveplates. In the pump–probe phase matching geometries of (b2) and (c), it is straightforward to control the polarization of k_1 and k_2 relative to k_3 and k_{LO} to collect $\langle ZZXX \rangle$, for example, but independent control of each pulse is more difficult. In design (b2), k_1 and k_2 can be independently controlled, but rotating their polarizations will also affect their intensities due to the 50/50 beamsplitter of the Michelson interferometer. For method (c), any type of polarized pump pulses can be generated if a polarization pulse shaper is built.

9.2.5 Signal detection and balanced heterodyne detection

One may, in principle, detect the emitted signal with a single IR (MCT) detector, in which case both the first coherence time t_1 and the local oscillator time t_3 need to be scanned, and a 2D Fourier transform is then taken in the computer. Collecting the emitted electric field directly in the frequency domain with a spectrometer and a multichannel detector is nowadays more common because data collection is significantly faster. That is, the spectrometer effectively performs a Fourier transform with respect to time t_3, and only one time dimension (the first coherence time t_1) needs to be scanned explicitly. IR MCT detector arrays are quite expensive, and are also relatively poor detectors (in terms of noise and nonlinearity), which is why some researchers also up-convert the signal beam in a nonlinear crystal and subsequently measure it in the visible spectral range [147].

Balanced heterodyne detection is a way to improve the signal-to-noise ratio (Fig. 9.7) [60]. In balanced heterodyne detection, a 50/50 beamsplitter is used to overlap the local oscillator with the emitted electric field at orthogonal angles to one another so that there are two sets of co-propagating heterodyne fields, each of which is collected on a detector. Balanced heterodyne detection can be performed either in the time domain with two single-channel detectors (Fig. 9.7b) or in the

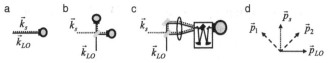

Figure 9.7 Illustration of balanced heterodyne detection. (a) In standard heterodyne detection, the emitted field and local oscillator are collinearly overlapped and detected in the time or frequency domains (not shown). (b) Time domain balanced heterodyne detection in which a 50/50 beamsplitter is used to overlap the emitted field and local oscillator into two out-of-phase beams, which can be detected in the time domain with two single-channel detectors or (c) in the frequency domain with two array detectors.

frequency domain with two array detectors (Fig. 9.7c). The signal on one detector will be the same as Eq. 4.27,

$$I_+(t_{\text{LO}}; t_1, t_2) \propto \int_0^\infty \left| E_{\text{LO}}(t_{\text{LO}} - t_3) + E_{\text{sig}}^{(3)}(t_3; t_2, t_1) \right|^2 dt_3 \quad (9.6)$$

whereas the other will be

$$I_-(t_{\text{LO}}; t_1, t_2) \propto \int_0^\infty \left| E_{\text{LO}}(t_{\text{LO}} - t_3) - E_{\text{sig}}^{(3)}(t_3; t_2, t_1) \right|^2 dt_3. \quad (9.7)$$

In this geometry, the signals on the two detectors are out of phase in order to conserve energy (when added, they must equal $|E_{\text{LO}}|^2 + |E_{\text{sig}}^{(3)}|^2$). This is convenient because if the two are subtracted, the intensity terms $|E_{\text{LO}}|^2$ and $|E_{\text{sig}}|^2$ cancel, leaving only the desired cross-term

$$I_+ - I_- \propto 2\Re \int_0^\infty \left\{ E_{\text{LO}}(t_{\text{LO}} - t_3) \cdot E_{\text{sig}}^{(3)}(t_3; t_1, t_2) \right\} dt_3. \quad (9.8)$$

With balanced heterodyne detection, $|E_{\text{LO}}|^2$ is subtracted every laser shot, whereas it would otherwise appear as a large background. Although the subtraction will not be perfect in reality since the responsivity of typical (MCT) IR detectors is not very uniform, balanced heterodyne detection improves the signal-to-noise ratio because the intensity fluctuations caused by noise are correlated on the two detectors. The two signals can be individually digitized and then subtracted on the computer, or preferably, the signals are subtracted with an analog circuit before digitization. Pre-subtraction is useful because it results in a smaller voltage range so that the analog-to-digital converter can be set to a higher resolution. We find that balanced heterodyne detection improves the signal-to-noise ratio by about four times over regular heterodyne detection.

Finally, we note that it is also possible to balanced heterodyne detect collinear signals, as they appear in the pump–probe phase matching geometry [133]. This requires that the signal and local oscillator have orthogonal polarizations, which can be achieved by measuring signals such as $\langle XYXY \rangle$. One then creates the two out-of-phase beams using a polarizer placed at 45° with respect to the other two (Fig. 9.7d).

9.2.6 Pulse intensities

The various designs of Fig. 9.1 produce different signal strengths since the intensities of the input beams are different. The signal strength of a 2D IR spectrum scales linearly with the electric field of each laser pulse. Assuming that all methods start with the same pulse energies, methods (b1) and (b4) will have the most intense pulses at the sample because there are no losses except for the surface reflections of transmissive optics. In method (b3), the intensities of the pulses are diminished by the diffraction efficiency of the custom-made gratings, which is typically about 50–80%. In the Michelson interferometer design of (b2), two of the excitation beams will have their intensities cut by 50% due to the 50/50 beamsplitter, while in the pulse shaper design of (c), there are two gratings and an acousto-optic modulator, which altogether also attenuates the intensity of the beams (we typically get a 30% throughput efficiency, although it should be possible to get >50%). It is now possible to build optical parametric amplifiers that produce >4 μJ of mid-IR or commercial ones are available that generate >10 μJ. Thus, in the long term, efficiencies will probably not be the limiting factor of the technology.

In a pump–probe geometry, the k_3 beam also acts as the local oscillator, which means that it cannot be very intense because it will saturate the detector. However, if one uses a polarization condition like $\langle XYXY \rangle$, then k_3 can be as intense as k_1 and k_2 by placing a polarizer in the signal beam oriented for Y-polarization [195]. To generate the local oscillator, the polarizer is tilted a few degrees so that it hardly alters the polarization of the signal, but allows a little k_3 to act as the local oscillator. This trick alleviates the most severe drawback of a pump–probe geometry as compared to a four-wave mixing geometry, and signal strengths as intense as those in a four-wave mixing geometry (for the same polarization condition) can be achieved, and balanced heterodyne detection via polarization is possible as well. The freedom one gives up though is the ability to measure polarization signals other than $\langle XYXY \rangle$ and $\langle XYYX \rangle$.

9.2.7 Optical densities

Another consideration is whether or not to pass the local oscillator around or through the sample before overlapping it with the emitted field. The only optical layout where there is a choice is (b1). In all the others there is no independent control over the local oscillator. If the samples to be studied will have optical densities of less than 0.2, then it does not really matter whether the local oscillator goes through the sample or not. However, if the samples have optical densities of 0.4 or larger, or the sample is being perturbed somehow (such as by heating in a transient 2D IR experiment), then we recommend passing the local oscillator through the sample so that it experiences the same distortions as the emitted signal field, which helps compensate for time and frequency distortions of the sample [197, 199]. For example, at optical densities of 0.4 or more, the emitted field is appreciably reabsorbed while leaving the sample. At optical densities below 0.2, reabsorption is usually negligible. If the emitted field is heterodyned with a local oscillator that passes around the sample, the reabsorption will appear as a valley across the 2D IR spectrum (Fig. 9.8a). But if the local oscillator passes through the sample, it will be absorbed in a similar manner to that of the emitted field. Although still present, the distortion is much less severe (Fig. 9.8b), and appears largely as a broadening along the ω_1-axis, until optical densities of about $\sim$1.2 make it infeasible to collect spectra.

9.2.8 Suppressing scatter and transient absorption signals

Scattering is a critical issue in 2D IR spectroscopy. It occurs when light from one of the exciting pulses is scattered off the sample into the direction of the emitted field. Since most 2D IR experiments performed so far are one-color experiments, all beams have the same frequency, and the scatter is heterodyne detected by the local oscillator:

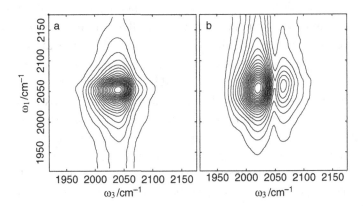

Figure 9.8 Absolute-value 2D IR spectra of azide in D_2O with an optical density of (a) 0.1 and (b) $\sim$0.8. Notice that the high OD sample has a valley at 2050 cm^{-1}, which is the fundamental frequency of azide.

$$S_{\text{scatter}} = |E_{\text{LO}} + \alpha E_n|^2$$
$$= |E_{\text{LO}}|^2 + \alpha^2 |E_n|^2 + \alpha |E_n E_{\text{LO}}| \qquad (9.9)$$

where $\alpha \ll 1$ is a measure of the amount of scatter. Even in weakly scattering samples (homogeneous solutions), minor scratches on the cuvette windows, finger prints, etc., render the third term so strong that it easily reaches the level of the actual signal, $|E_{\text{sig}}^{(3)} E_{\text{LO}}|$. Since the third term is an interference term, it is phase sensitive as well.

The other problematic contribution is transient absorption signals. That is, while the desired signal in a four-wave mixing geometry originates from the interaction of all input pulses k_1, k_2, and k_3 with the sample, each of these pulses will also independently generate an ordinary pump–probe signal, i.e. all three field interactions coming from one of these pulses. In designs (b3) and (b4), the local oscillator then acts as a probe pulse, and phase matching of a pump–probe process $+k_n - k_n + k_{\text{LO}}$ dictates that the field is emitted into the direction of the local oscillator. The same is true for the first two pulses k_1 and k_2 in a time domain pump–probe setup (design b2 and c). In this case, the transient absorption background can be as large as the desired signal itself. In design (b1), when the local oscillator is guided around the sample, there is no transient absorption signal. There are various ways to suppress scatter and transient absorption signals, which we will discuss now.

In design (a), scatter and not background transient absorption is the issue. The easiest way to suppress scatter is to measure the signal twice with shifting the population time t_2 by half the laser carrier frequency (10 fs at 1600 cm^{-1}), thereby flipping the sign of the scatter signal.

One often uses a mechanical chopper to separate the desired signal, which is emitted only when all three laser pulses k_1, k_2, and k_3 are present in the sample, from the transient absorption and scatter signals that originate from the individual pulses. However, it matters which beam is chopped. It is best to put the chopper into the beam which is not scanned. That way, the scatter and transient absorption from the beams that are moving is subtracted by the chopper, whereas these signals from the stationary beam appear at zero frequency in the 2D IR spectrum. Thus, unless one strongly under-samples (see Section 9.5.2), or is working in a rotating frame (see Section 9.3.3), it will not affect the 2D IR spectrum.

Alternatively, one can also use polarization to suppress the undesired signals, which is particularly interesting for a pump–probe geometry (design b2 and c) [195]. In this case, one may measure the $\langle XYXY \rangle$ signal by using orthogonally polarized k_1 and k_2 pulses and a polarizer after the sample. The $\langle XYXY \rangle$ polarization condition removes the transient absorption background from each laser pulse because it is not allowed in the dipole approximation. To understand this statement, consider the transient absorption signal from k_1 acting on the sample twice. The signal passing through the polarizer would have to be $\langle XXXY \rangle$, which is zero (see Chapter 5). Scatter from beams k_1 and k_3, on the other hand, is suppressed in this configuration by the polarizer, whereas scatter from beam k_2 is again suppressed by the Fourier transform, if beam k_1 is the one whose time is scanned.

Phase cycling, which we will discuss in the following section, offers another very elegant way to suppress undesired signals.

9.3 Capabilities made possible by phase control

Active control over the individual pulse phases can be used to improve 2D IR spectroscopy significantly. Phase cycling enables scatter subtraction, chopping, background subtraction, selection of Feynman pathways and the use of the rotating frame. The largest flexibility is provided by pulse shaping, which has been introduced in both 2D IR [166] and 2D VIS spectroscopies [75, 146, 181] in connection with a pump–probe geometry (design c) and is used in other configurations as well [176, 185]. With pulse shapers, nearly arbitrary control of the temporal, phase and polarization of the pulse train is possible. Moreover, pulse shaping eliminates moving parts in the spectrometer for faster data collection. Alternatively, photoelastic modulators or fast wobbling Brewster windows can be used for phase cycling. We will return to the technical realization of phase cycling devices in Section 9.4, and discuss in the following how phase control can be used to improve 2D IR spectroscopies.

9.3.1 Removing scatter and transient absorption background by phase cycling

As discussed above in Section 9.2.8, scatter and transient absorption background is superimposed on the desired signal. In the previous section, we outlined how chopping or polarization can be used to remove this background, while here we discuss how the same can be achieved by phase cycling. To design a phase cycling scheme, one starts by comparing the phase dependence of the desired and unwanted signals. For example, the phases of the desired third-order signals depend on $\Delta\phi_{12}$ whereas the background transient absorption from pulse k_1 and k_2 are independent of the pulse phases. Thus, for each time delay, we collect the signal with $\Delta\phi_{12} = 0$ and $\Delta\phi_{12} = \pi$, which we subtract, e.g. $I(\Delta\phi_{12} = 0) - I(\Delta\phi_{12} = \pi)$. The subtraction procedure is demonstrated in Fig. 9.9 for NMA in D_2O, where we show the data for each of the signals separately. The oscillation when $\Delta\phi_{12} = \pi$ appeared out of phase compared to the oscillation when $\Delta\phi_{12} = 0$, while the backgrounds, which cause the signal offsets, remain the same. Therefore, when the two scans with $\Delta\phi_{12} = 0$ and π are subtracted, the backgrounds are removed, leaving only

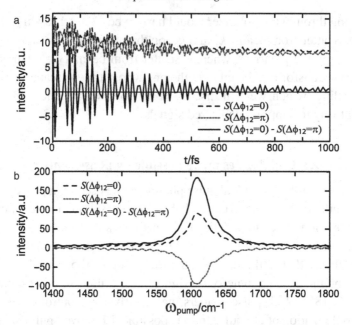

Figure 9.9 Removing backgrounds from transient absorption using a phase cycling scheme for the amide I mode of NMA in D_2O. (a) Time scans at $\omega_3 = 1625$ cm^{-1} when $\Delta\phi_{12} = 0$ (dashed) and $\Delta\phi_{12} = \pi$ (dotted), which are 180° out of phase, and the subtracted product of the two scans (solid). (b) Fourier transforms of the time scans in (a). These are equivalent to the slices along ω_1 of 2D IR spectra. Adapted from Ref. [165] with permission.

the desired oscillatory part (solid). Figure 9.9(b) displays the resultant spectra after Fourier-transformation giving a peak with the opposite sign for $\Delta\phi_{12} = 0$ (dashed) and π (dotted). Thus, the signal is collected for every laser shot so that there is no loss of repetition rate, and furthermore, the signal size is doubled.

Scatter from beams k_1 and k_2, on the other hand, carry phases ϕ_1 and ϕ_2, respectively. The phase cycling sequence that we find to be the most useful in a pump–probe geometry (designs b2 and c) is:

$$+S(\phi_1 = 0, \phi_2 = 0) - S(\phi_1 = 0, \phi_2 = \pi)$$
$$+S(\phi_1 = \pi, \phi_2 = \pi) - S(\phi_1 = \pi, \phi_2 = 0). \quad (9.10)$$

It subtracts out the unwanted scatter and transient absorption signals at the same time, while the desired third-order signal is additive for all four pulse signals. In a four-wave mixing geometry (designs b1, b3, and b4), where scatter from beam k_3 can contribute as well, a good scheme is [15]:

$$+S(\phi_1 = 0, \phi_2 = 0, \phi_3 = 0) - S(\phi_1 = 0, \phi_2 = 0, \phi_3 = \pi)$$
$$+S(\phi_1 = \pi, \phi_2 = \pi, \phi_3 = 0) - S(\phi_1 = \pi, \phi_2 = \pi, \phi_3 = \pi) \quad (9.11)$$

9.3 Capabilities made possible by phase control

with two independent phase modulators. In a frequency domain 2D IR spectrometer (designs a and c), one may rotate the phase of the pump pulse to eliminate scatter.

9.3.2 Phase cycling to extract the rephasing and non-rephasing spectra

A pump–probe beam geometry naturally gives rise to purely absorptive lineshapes because both the rephasing and non-rephasing spectra are generated collinearly. Absorptive spectra are frequently used in multidimensional experiments because they have the highest frequency resolution and lack phase twist which can distort spectra. But sometimes it is preferable to analyze the rephasing and non-rephasing spectra separately, such as when analyzing the joint frequency fluctuations between coupled chromophores or the dephasing time of coupled systems. As described in Sect. 2.8, phase cycling can be used to extract the rephasing and non-rephasing spectra from an absorptive signal [38, 176, 185]. One does so by their relative phase differences, which are $R_1 \propto e^{i(-\phi_1+\phi_2+\phi_3)}$ and $R_4 \propto e^{i(+\phi_1-\phi_2+\phi_3)}$. However, the two linear combination used in Section 2.8 requires that both the real and imaginary components of R_1 and R_4 are known. There are two ways of obtaining both components.

Consider the actual measured signal $E_{\text{sig}}^{(3)}$ rather than just the polarizations. In the first method, two phases of the local oscillator are used to measure $E_{\text{sig}}^{(3)}$, one of which gives $\Re$ and the other $\Im$. Thus, the heterodyned signal that we measure is

$$I(\phi_1, \phi_2, \phi_3, \phi_{\text{LO}}) \propto \Re\{E_{\text{LO}} \cdot (E_1^{(3)} + E_4^{(3)})\}$$
$$= \Re\{R_1 e^{i(-\phi_1+\phi_2+\phi_3-\phi_{\text{LO}})} + R_4 e^{i(+\phi_1-\phi_2+\phi_3-\phi_{\text{LO}})}\}$$
$$\equiv \Re\{R_1 e^{i(-\phi_{1,2}+\phi_{3,\text{LO}})} + R_4 e^{i(+\phi_{1,2}+\phi_{3,\text{LO}})}\}. \quad (9.12)$$

Remembering that the response functions are complex (e.g. $R_n = \Re R_n + i\Im R_n$), one gets four terms from this equation, which are

$$I(\Delta\phi_{1,2}, \Delta\phi_{3,\text{LO}}) \propto \quad (9.13)$$
$$+\Re R_1 \cos(-\Delta\phi_{1,2} + \Delta\phi_{3,\text{LO}}) - \Im R_1 \sin(-\Delta\phi_{1,2} + \Delta\phi_{3,\text{LO}})$$
$$+\Re R_4 \cos(+\Delta\phi_{1,2} + \Delta\phi_{3,\text{LO}}) - \Im R_4 \sin(+\Delta\phi_{1,2} + \Delta\phi_{3,\text{LO}}).$$

There are four unknowns in these equations ($\Re R_1, \Im R_2, \Re R_4, \Im R_4$) and so we need to measure four observables, which we choose to be:

$$I(\Delta\phi_{1,2}=0, \Delta\phi_{3,\text{LO}}=0) = \Re R_1 + \Re R_4$$
$$I(\Delta\phi_{1,2}=\pi/2, \Delta\phi_{3,\text{LO}}=0) = \Im R_1 - \Im R_4$$
$$I(\Delta\phi_{1,2}=0, \Delta\phi_{3,\text{LO}}=\pi/2) = -\Im R_1 - \Im R_4$$
$$I(\Delta\phi_{1,2}=\pi/2, \Delta\phi_{3,\text{LO}}=\pi/2) = \Re R_1 - \Re R_4. \quad (9.14)$$

Thus, linear combinations of these four measurements give the individual response functions

$$R_1 = \Re R_1 + i\Im R_1 = I(0,0) + I(\pi/2, \pi/2) + i\{I(0, \pi/2) - I(\pi/2, 0)\}$$
$$R_4 = \Re R_4 + i\Im R_4 = I(0,0) - I(\pi/2, \pi/2) - i\{I(0, \pi/2) + I(\pi/2, 0)\}.$$
(9.15)

Thus, a combination of four signals must be taken, rather than just the two in Section 2.8. The summations can be performed on either the time or frequency domain data.

The above method works when the phase of the local oscillator can be independently adjusted, but in a pump–probe style 2D IR experiment, the probe acts as both k_3 and k_{LO} so that $\phi_{3,LO}$ is always zero. However, one can regain the missing information with a simple trick [146, 165]. One takes the inverse Fourier transform of the spectral interferogram, zeros the negative times, and then Fourier transforms back to the frequency domain. For example, suppose that the interferogram is $\cos(\omega t_0)$. The Fourier transform is $\frac{1}{2\sqrt{2\pi}}(\delta(t+t_0) + \delta(t-t_0))$ (see Eq. A.11 in Appendix A). If we eliminate $\delta(t+t_0)$, and then take the inverse Fourier transform, we get $e^{-i\omega t_0}$. Thus, by eliminating the negative times, one enforces *causality* on the Fourier transformed signal so that the resulting interferogram is now complex. Response functions are causal because the field is emitted after the laser pulse. Thus, if one can only measure $I(\phi_{12}=0, \phi_{3,LO}=0)$ and $I(\phi_{12}=\pi/2, \phi_{3,LO}=0)$, then this process can be used to obtain the other two necessary signals in Eq. 9.14. Shown in Fig. 9.10 are the results of applying the method to N-methylacetamide. Notice that the rephasing and non-rephasing spectra have the opposite phase twist and that the non-rephasing spectra are weaker than the rephasing spectra, which is consistent with the known vibrational dynamics. It is interesting to note that one can even calculate all four of the necessary phases from a single absorptive 2D IR spectrum without any phase cycling by enforcing causality on one and then the other frequency axis.

9.3.3 Rotating frame

So far we have considered phase cycles that switch between fixed values such as $\pm\pi/2$. Continuous phase shifts make it possible to shift the observed frequency of the emitted field so that the signal is measured in the rotating frame, and offers other capabilities. The phase of the measured signal depends upon both the time delays and the relative phases of the laser pulses. To see that, we write the laser pulses as

$$E_n(\tau) = E_n^{(0)}(\tau_n)e^{i(\vec{k}_n \cdot \vec{r} - \omega\tau_n + \phi_n)} \tag{9.16}$$

9.3 Capabilities made possible by phase control 195

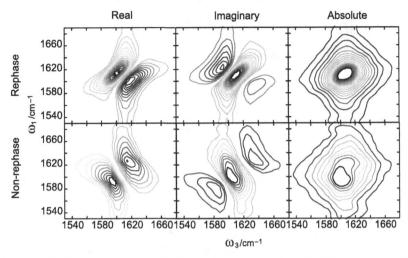

Figure 9.10 Extracting the rephasing and non-rephasing 2D IR spectra for the amide I mode of NMA in D_2O. The top row shows components of the rephasing signal. The bottom row displays the non-rephasing spectra. Adapted from Ref. [165] with permission.

with $\vec{k}_{LO} = -\vec{k}_1 + \vec{k}_2 + \vec{k}_3$ (for a rephasing diagram). Plugging this into Eq. 4.25 and then 4.27, we find for the measured signal in the impulsive limit and in this particular phase matching direction:

$$I_+ \propto \Re \left(e^{i(\Delta\phi_{1,2} - \Delta\phi_{3,LO})} e^{-i\omega(t_1 - t_3)} R'_1(t_1, t_2, t_3) \right) \quad (9.17)$$

where $t_1 = \tau_2 - \tau_1$ and $t_3 = \tau_{LO} - \tau_3$ are the time separations between pulses 1 and 2, and pulse 3 and the local oscillator, respectively, and $R'_1(t_1, t_2, t_3)$ is the envelope of the response function with the oscillatory term $e^{-i\omega(t_1 - t_3)}$ written separately. We are most interested in the envelope of the response function, because it is the envelope that contains information on the frequency spread of the vibrators, the linewidths, etc. If we move t_1 by half the period of the fundamental frequency, we change the envelope of $R'_1(t_1, t_2, t_3)$ and flip the phase of the emitted signal because of the $e^{-i\omega(t_1 - t_3)}$ term. The latter can be compensated partially or wholly by changing $\Delta\phi_{12}$ at the same time. For example, if when t_1 is incremented by half the period, we also set $\Delta\phi_{12} = \pi/2$, then the measured signal does not appear to oscillate, although its intensity will still decay due to $R'_1(t_1, t_2, t_3)$. Thus, the envelope of the emitted signal is measured, but not the fundamental frequency. It is as if the fundamental frequency has been shifted to zero. Since the fundamental frequencies are already known from the linear spectrum, it is not important to measure their absolute values with 2D IR spectroscopy, but only the differences in frequencies which are what create the envelope. Of course one does not have to shift the observed fundamental frequency all the way to zero, but can partially downshift it

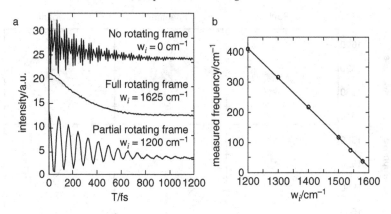

Figure 9.11 The rotating frame demonstrated with the amide I mode of NMA by incrementing the relative phase, $\Delta\phi_{12}$, in proportion to τ. (a) Time scans at $\omega_3 = 1625$ cm^{-1} by using $\omega_i = 0$, 1200 and 1625 cm^{-1} for no, partial and full rotating frame, respectively. (b) The measured frequency after Fourier-transforming the time scans similar to those in panel (a) plotted with various ω used for partial rotating frames. Adapted from Ref. [165] with permission.

(or upshift it). This process of shifting the fundamental frequency is equivalent to making the measurements in a rotating frame.

Figure 9.11 demonstrates the rotating frame with time scans measured for N-methylacetamide (NMA) in D_2O [165]. When no rotating frame is used, $\Delta\phi_{12} = 0$ and the 2D IR signal oscillates at ω_0 along t_1, which is 21 fs for NMA (top). To rotate the frame, $\Delta\phi_{12}$ is incremented with the time delay, i.e. $\Delta\phi_{12} = -\omega_i t_1$ where ω_i is a constant. The phase increment has the effect of shifting the frequency of the emitted signal to $\omega_0 - \omega_i$. For example, in Fig. 9.11, bottom, the signal has a period of 78 fs, which corresponds to $\omega_i = 1200$ cm^{-1}. And in the extreme case when $\omega_i = \omega_0$, the observed signal no longer oscillates (middle). The time scans reported in Fig. 9.11 were measured using several hundred points to clearly illustrate the frequency shift, but in practice only slightly more than two points are needed per period according to the Nyquist frequency (see Section 9.5.2). Hence, the rotating frame provides an alternative and convenient means of under-sampling.

Any signal with a phase dependence can be frequency shifted. For example, the rotating frame can also be used to suppress scatter. By rotating $\phi_1 = \phi_2 = \omega_i t_1$, the signal remains at the same frequency because $\Delta\phi_{12} = 0$, while the scatter from beams k_1 and k_2 is shifted to $\omega_0 - \omega_i$ where ω_0 is the center frequency of the incident laser pulses. Figure 9.12 illustrates this capability on a sample of aggregated fibrils from human Islet amyloid polypeptide (hIAPP) in D_2O. These fibrils are ≈ 100 nm in diameter and up to several microns in length, and hence are extremely strongly scattering, causing a broad line along the spectrum diagonal (Fig. 9.12a).

9.4 Phase control devices

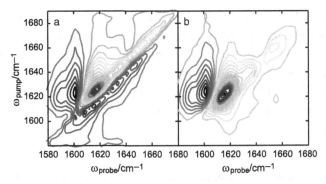

Figure 9.12 Removing scatter by incrementing the absolute phase of the pump pulses. Purely absorptive 2D IR spectra of hIAPP fiber in D_2O were collected with (a) no phase increments and (b) $\omega_i = 1800\,\text{cm}^{-1}$. The scatter which appears along the diagonal in (a) is now absent in (b). Adapted from Ref. [166] with permission.

When $\omega_i = 1800\,\text{cm}^{-1}$ is used to increment both ϕ_1 and ϕ_2, the scatter from the stationary pump pulse (k_2) moves up to $\omega_1 = 1800\,\text{cm}^{-1}$, while the scatter from the moving pump pulse (k_1) shifts to $\omega_1 = 1620–1800\,\text{cm}^{-1} = 180\,\text{cm}^{-1}$, leaving the desired 2D spectrum between $\omega_1 = 1580$ and $1700\,\text{cm}^{-1}$ scatter free (Fig. 9.12b). The general procedure for designing a phase cycling scheme is to write down the phase dependence of all the signals, and then look for linear combinations that will subtract the unwanted ones and leave the desired signals.

9.4 Phase control devices

Most optical devices cannot control the phase of a pulse independent of its envelope.[2] To change the phase without altering the envelope, one must alter the group versus the phase velocity by passing the pulse through some material or by using a pulse shaper (see Section 9.4.2). Instead, time delays are often used to approximate phase shifts. For example, at 6 μm a 10 fs delay is all that is needed to make a π phase shift. Since a 60 fs mid-IR pulse at 6 μm has about three periods during its duration (at FWHM), the envelope does not change too much and (hopefully) the response function does not either.

Thus, one may shift pulse phases by just incrementing the delays like normally. However, the true power of phase cycling comes about only when switching phases

---

[2] We have written our equations using a *carrier envelope phase* (CEP) ϕ. Standard optical parametric amplifiers (OPAs) cannot generate pulses with the same ϕ for every laser shot. For 2D IR spectroscopy it does not matter, because what counts are phase differences so even if the CEP fluctuates, the phase difference of $\Delta\phi_{i,j}$ will stay constant.

very quickly, such as from one laser shot to the next. For example, scattering is caused (in part) by small particles diffusing in the sample solution, so it is usually not constant in time. Hence, if using phase cycling to reduce scatter, the phase should be cycled as fast as possible. It ideally takes only 4 ms at 1 kHz laser repetition rate to measure all four signals of, e.g. Eq. 9.11, and the scatter is essentially constant during this time. This calls for fast phase switching devices, which we discuss below.

9.4.1 Photoelastic modulator and wobbling Brewster windows

For fast phase modulation, one may employ a photoelastic modulator or a fast wobbling Brewster window [15]. Both these devices change the optical path length, and hence also the time delay of the pulse envelopes (see the discussion above). In the photoelastic modulator, resonant driving of a ZnSe window by two piezoactuators excites a fundamental vibration of the window that effectively stretches and compresses its thickness. Photoelastic modulators were originally designed for fast polarization modulation, which is needed for vibrational circular dichroism (VCD) spectroscopy, but if its optical axis is aligned parallel to the laser polarization, it effectively modulates the index of refraction. One practical difficulty with the photoelastic modulator is the fact that the mechanical resonance frequency is extremely sharp and factory-set, so one has to synchronize the laser to the modulator and not the other way around. It operates at 50–70 kHz, depending on the size of the ZnSe window. We synchronize the laser repetition rate to the photoelastic modulator frequency by dividing the latter by $n + 1/2$, where n is an integer to achieve a laser repetition rate that is close to where the laser is designed to operate (1 kHz, for example). By using this factor, the phase of the photoelastic modulator is flipped every other laser shot.

Alternatively, one may use Brewster windows mounted, along with a small magnet, on a flexure bearing or simply a stainless steel rod, and make it wobble by driving it with a solenoid. To achieve a phase shift of $\pm\pi/2$ at 1600 cm^{-1}, a 1.5–3 mm ZnSe window has to be modulated by $\pm 0.1 - 0.05°$. Without difficulty, one can reach wobbling frequencies up to 500–750 kHz. Frequencies into the multi-kHz range may also be possible [23]. These approaches use resonant driving as well, however, the resonance frequency is far less sharp than in the case of the photoelastic modulator, so one can tune the frequency in a certain range (several Hz) and furthermore easily change the resonance frequency by adding masses. For some phase cycling schemes we need just one modulator, but if one synchronizes two of them in the same setup one can achieve four phase cycle sequences, as in Eq. 9.11.

9.4.2 Pulse shaper

Pulse shapers offer more flexibility in phase cycling, and in particular allow one to control the phase independently of the pulse envelope. Figure 9.13 presents the basic design for a pulse shaper that operates in the mid-IR. The pulse shaper (gray area enclosed with a dashed line) consists of two gratings and two cylindrical mirrors in a $4f$ geometry, as is often used in visible pulse shapers, and an acousto-optic modulator made of germanium. The first grating and cylindrical mirror serve to disperse and collimate the frequency components of the incident mid-IR pulses along the length of the germanium crystal. The pulse is shaped by an acoustic wave that propagates along the length of the crystal, whose time dependent amplitude and phase can be set with an arbitrary waveform generator in an attached computer. Since the acoustic wave propagates slowly through the crystal as compared to the femtosecond mid-IR pulse, the wave acts as a programmable grating, diffracting the desired mid-IR frequencies with specified intensity and phase. The second grating and cylindrical mirror serve to transform the diffracted frequency

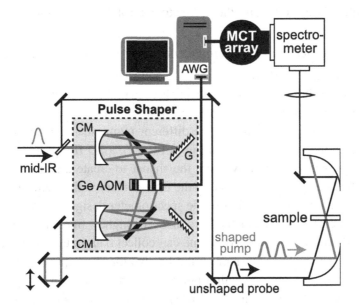

Figure 9.13 The experimental setup of a 2D IR spectrometer based on a mid-IR pulse shaper (gray area enclosed with a dashed line). The pulse shaper consists of two gratings (G) and cylindrical mirrors (CM) in a $4f$ geometry as well as a germanium acousto-optic modulator (Ge AOM) controlled by an arbitrary waveform generator (AWG) equipped on a computer. The shaped beam serves as a pump beam (thick gray line) and a small portion of the unshaped beam is used as a probe beam (solid black line). The probe spectrum is measured with a mercury cadmium telluride (MCT) array detector equipped to a spectrometer.

components back into the time domain. Thus, to create a desired optical pulse in the time domain, one just needs to calculate its Fourier transform and feed it to the acousto-optic modulator as an acoustic wave (several calibration steps are also necessary). Mathematically, the electric field of the shaped pulse in frequency is the product of the incident electric field and a mask function of the acousto-optic modulator:

$$E_{out}(\omega) = E_{in}(\omega) \cdot M(\omega) \qquad (9.18)$$

with the laser pulse electric fields $E(\omega)$. The mask $M(\omega)$ is a complex function in frequency which reflects the ω-dependent diffraction efficiency of the grating written by the acoustic wave. In this manner, all that is needed is some simple programming to delay an input pulse in time, to create two pulses or an entire pulse train from a single input pulse, to mimic an optical component such as an etalon, to control the pulse phases, or to create any frequency/time domain pulse within the limits of the shaper resolution (the shaper resolution is typically 200–500 individual frequency elements). Several reviews exist on this technology [134, 165]. It is also straightforward to modify the design to control the polarization as well [133].

Basic methods of collecting 2D IR spectra with pulse shaping

Using a pulse shaper, 2D IR spectra can be collected using either the frequency or time domains. Either way, the pulse shaper is used to create the desired pump pulses and then a probe is used to monitor the third-order response. Figure 9.14 illustrates 2D IR spectra collected with several different pump shapes using the antisymmetric stretch mode of $W(CO)_6$. In Fig. 9.14(a–c), a 2D IR spectrum of $W(CO)_6$ is shown that was collected by using the shaper to create a pair of femtosecond pulses so that the data are collected in the time domain with a semi-impulsive pair of collinear pump pulses, similar to the spectrometer design of a Michelson interferometer (design b2). This method produces the highest frequency and time resolution 2D IR spectra. Alternatively, one can collect data in the frequency domain by using the pulse shaper to narrow the bandwidth of the pump pulse and scan its center frequency, such as with a pulse shape that mimics a Fabry–Perot filter (Section 9.1), which results in the 2D IR spectrum shown in Fig. 9.14(d–f). Notice that the spectrum becomes distorted at small t_2 delays because the pump and probe pulses begin to overlap, which is a practical limitation to the Fabry–Perot filter. But the pulse shaper is not limited to these two "standard" pump shapes. One can minimize the overlap with the pump and probe pulses so that the "etalon" pulse has the same shape in the frequency domain but is reversed in the time domain so that the tail extends away from probe. A "reverse" etalon gives a nice 2D IR spectrum even for short t_2 delays. Another possible pulse shape is a Gaussian. An etalon

9.5 Data collection and data workup

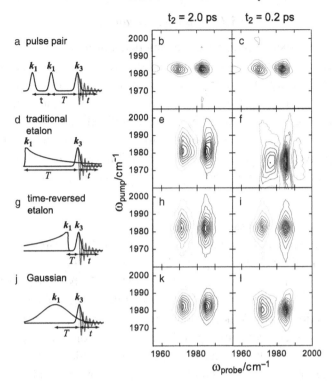

Figure 9.14 Various (left column) pulse shapes used to collect purely absorptive 2D IR spectra of W(CO)$_6$ at $t_2 = 2$ ps and 0.2 ps delays (middle and right columns, respectively). (a–c) A femtosecond pulse pair. (d–f) A pulse shaped to mimic an etalon. (g–i) A "time-reversed" etalon. (j–l) A Gaussian shaped pump pulse. Adapted from Ref. [166] with permission.

creates a Lorentzian shaped pump spectrum, which has long tails that can be problematic in congested spectra. The tails are reduced when using a Gaussian shape which enables small t_2 delays as well. One can envisage many other possible pulse shapes that might be used to enhance or suppress 2D IR features. Pulse shapes can also be designed to optimize higher vibrational states as well [171]. One especially nice feature is that absorptive spectra are automatically obtained in the pump–probe geometry and, because $t = 0$ is set perfectly, the spectra do not have to be phased, which is a topic that we address below.

9.5 Data collection and data workup

What has been discussed so far concerns the optical hardware on the laser table. There are also important issues concerning the way in which data are collected and processed, which we discuss below.

9.5.1 Frequency resolution

Generating 2D IR spectra requires collecting data during two coherence evolution times. While the "direct" time (meaning the emitted field coherence such as during t_3) can be measured in the frequency domain using a spectrometer, all other dimensions are "indirect" in that they must be collected in the time domain. For any time domain dimension, whether direct or indirect, one must decide the step size and duration over which the delay is to be measured. These parameters are related to the Nyquist frequency and the resolution, respectively (an overview of Fourier transforms is given in Appendix A). The Nyquist frequency is given by

$$\frac{\omega_N}{2\pi} = \frac{1}{2\Delta t} \tag{9.19}$$

where Δt is the step size. The Nyquist frequency is the highest-frequency sine wave that can be sampled for a given step size. For example, carbonyl groups absorb near 1600 cm^{-1}, and so have a vibrational period of about 20 fs. Thus, the Nyquist frequency is 10 fs, so that step sizes of 9 fs would be just adequate to fully sample their free induction decay. When going close to the Nyquist frequency, one must be sure that no higher-frequency signal is present, to avoid the possibility that it is aliased back, which we discuss in the next section on under-sampling. The resolution of a spectrum, on the other hand, is determined by the total acquisition time (t_{max}), according to $\Delta\omega = 1/t_{max}$ (see also the discussion of zero-padding below). Thus, to increase the resolution of a spectrum, one can either record additional data points or take larger step sizes. However, if the signal has already decayed below the signal-to-noise ratio, than increasing t_{max} increases the noise in the spectrum without improving the linewidths (see Section 9.5.7).

9.5.2 Under-sampling

It can be time consuming collecting a data set when more than one dimension must be measured in the time domain, even for high-signal samples, because so many points must be collected to fully sample the coherence evolution times. It is not atypical to collect 150 data points per dimension. Thus, two time dimensions (which is necessary when collecting 2D IR data completely in the time domain or any 3D IR spectrum) require 22 500 data points (150 × 150), which can take between 30 minutes and a couple of hours to collect depending on the amount of signal averaging. There are two ways to decrease acquisition time by collecting fewer points. One way is to lower the Nyquist frequency by working in the rotating frame, as discussed above in Section 9.3.3. Another way is to simply collect fewer points than dictated by the Nyquist frequency, which is called *under-sampling*. The Nyquist theorem requires that more than two time points must be collected

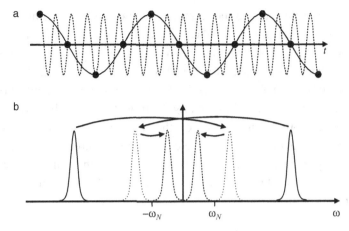

Figure 9.15 Under-sampling: (a) Sine waves with $\omega = 3.5\omega_N$ (dotted line) that appear as $\omega = 0.5\omega_N$ (solid line) due to the long time step (dots). (b) The spectrum is folded twice until it appears in the Nyquist frequency range. Since we will measure real-valued data in a real experiment, the frequency spectrum is symmetric with a positive and negative contribution.

per period in order to obtain a unique frequency. If fewer data points are collected, then frequencies higher than the Nyquist frequency will be *aliased* (potentially a couple of times) until it eventually appears inside the Nyquist frequency window (Fig. 9.15). However, since we know the range of IR frequencies that we are measuring (it is set by the mid-IR laser pulse), it is straightforward to convert the aliased spectrum, ω, into the proper frequencies, ω_0, using the relationship:

$$\omega_0 = \omega + m \cdot \omega_N \quad (9.20)$$

where m is an even integer equal to the number of times the signal has been aliased. For example, if 18 fs steps were used instead of 9 fs steps for a carbonyl mode, the spectra would be under-sampled with $m = 2$. Alternatively, if m is an odd number, we obtain:

$$\omega_0 = -\omega + m \cdot \omega_N. \quad (9.21)$$

When under-sampling many times (large m), the useable frequency range is severely reduced, so one must make sure that the frequency of the laser pulse sits in the middle of that frequency range in order to avoid aliasing effects. The condition to achieve this is

$$\Delta t = \frac{2\pi}{\omega_0}\left(\frac{m}{2} + \frac{1}{4}\right). \quad (9.22)$$

When m is an odd number, the sign of the frequency axis is inverted.

The objective of working in the rotating frame or undersampling is to decrease the amount of laboratory time it takes to fully sample the coherence times. The signal-to-noise ratio scales as the square-root of the number of laser shots, whether or not those laser shots are used to collect more time delay positions or to better average fewer time delays (at least if the laser noise is uncorrelated shot-to-shot). Hence, if there is no dead time in scanning the time delays (such as by using a pulse shaper or a continuous scanning device), the signal-to-noise ratio simply depends upon how long one averages. But if the experimental method wastes a lot of time by moving delay stages, than undersampling or the rotating frame may be advantageous (because the translation stages need time to settle, motor moving time does not scale linearly with step size).

9.5.3 Fourier transform: Initial value

The first time domain data point should be scaled by 0.5 during data analysis, for the reasons that follow. As we measure data on a discrete, equidistant time grid $t_k = k \cdot \Delta t$, with $k = 0, \ldots, N$ and a step size Δt, we replace the Fourier transform by a discrete Fourier transform (see Appendix A):

$$R(\omega) = \int_0^\infty R(t) e^{i\omega t} dt$$

$$\approx \sum_{k=0}^{N-1} R(t_k) e^{i\omega t_k} \cdot \Delta t. \qquad (9.23)$$

Response functions are single sided due to causality, i.e. they are strongly discontinuous at time $t = 0$ when the signal is large at $R(t = 0)$ but zero $R(t) = 0$ for $t < 0$. In this case, approximating the integral by a simple sum is a rather poor approximation. The much better numerical integration uses the trapezoidal rule [152], in which the first and the last summands are multiplied by 1/2. Since the signal decays, the last summand is supposed to be zero anyway, so it does not matter, but if one fails to multiply the first summand $R(t = 0)$ by 1/2, spurious backgrounds and cross-peaks appear in the 2D IR spectrum [148]. The smaller the number of time points N, the larger would be the artifacts.

9.5.4 Fourier transform: Zero-padding

Adding zeros to a given set of time domain data is called zero-padding. As this effectively increases the number of time points, and thereby the total acquisition time, t_{max}, zero-padding increases the effective frequency resolution. It has been shown in NMR spectroscopy that one should zero-pad by a factor 2 [11]. The reasoning is the following: Due to causality, response functions are single sided and

9.5 Data collection and data workup

nonzero only for positive times. When we measure data, we therefore typically take them only from $0 \leq t \leq t_{max}$, since we know that no response will be present for times $t < 0$ anyway.[3] However, we ignore causality when we neglect the zeros at negative times. After Fourier transformation into the frequency domain, causality implies that the real and the imaginary part of a spectrum are related to each other through a Kramers–Kronig relation. In other words, knowing the real part of a spectrum, we can recalculate its imaginary part, and vice versa. However, if one neglects the zeros at negative times, the real and imaginary parts carry information that is independent of each other. One can reinforce causality by adding the same number of zeros to a set of data points. Due to the implicit periodicity of the data (see Appendix A), it does not matter whether we add this block of zeros before or after the data points. This procedure effectively increases the frequency resolution by a factor of 2. Furthermore, the real and imaginary parts are now no longer independent, but are connected through a Kramers–Kronig relation. Consequently plotting one of the two displays the maximum amount of information contained in the data.

Figure 9.16 illustrates the concept. Figure 9.16(a) shows the exact Fourier transform of a spectrum with two closely spaced lines, and Fig. 9.16(b) shows a Fourier transform without zero-padding with a total acquisition time $t_{max} = 1/\Delta \omega$ that is too short to directly resolve the two lines. However, with zero-padding by a factor 2 (Fig. 9.16c), the two lines are in fact resolved without the need to measure

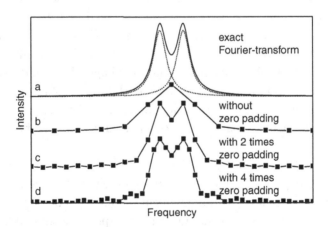

Figure 9.16 (a) Exact Fourier transform of a spectrum with two closely spaced lines. (b) Fourier transform without zero-padding, (c) with zero-padding two times and (d) with zero-padding four times.

---

[3] In reality, often, there will be a signal at times $t < 0$, however, since this changes the time ordering of pulses, this signal originates from a different set of Feynman diagrams compared to the one at times $t > 0$. For a given set of Feynman diagrams, the response functions are indeed zero for times $t < 0$.

additional data. Zero-padding by more than a factor 2 (Fig. 9.16d) does not add any new information, but just leads to a trigonometric interpolation of the data.

9.5.5 Spectral interferometry: Detection in the frequency domain

One way to dramatically decrease the data acquisition time is to use a spectrometer and an array detector to measure t_3 directly in the frequency domain across all of the relevant frequencies simultaneously. In this way, the full frequency range of ω_3 is measured simultaneously and only t_1 must be incremented. In this case, the spectrometer performs a Fourier transform of the signal and local oscillator, which is followed by an intensity measurement on an array detector:

$$I_+(\omega_{\text{LO}}; t_1, t_2) \propto \left| \int_0^\infty \left(E_{\text{LO}}(t_{\text{LO}} - t_3) + E_{\text{sig}}^{(3)}(t_3; t_2, t_1) \right) e^{i\omega_{\text{LO}} t_3} dt_3 \right|^2$$

$$= I_{\text{LO}}(\omega) + I_{\text{sig}}^{(3)}(\omega) + 2\Re \left(\int_0^\infty E_{\text{LO}}(t_{\text{LO}} - t_3) e^{i\omega_{\text{LO}} t_3} dt_3 \right.$$

$$\left. \cdot \int_0^\infty E_{\text{sig}}^{(3)}(t_3; t_2, t_1) e^{i\omega_{\text{LO}} t_3} dt_3. \right) \quad (9.24)$$

(Notice that the integral is calculated before the intensity, because the gratings come before the detector.) Thus, one measures a spectrum of the local oscillator, a spectrum of the homodyne emitted field, and a cross-term that is the spectral interferogram that we want. A chopper can be used to subtract off the I_{LO} term, or balanced heterodyne detection can be implemented by using a dual array. The homodyne signal is usually small enough to ignore. The cross-term that one wants is modulated by a periodicity that is inversely proportional to the relative time delay between the signal and local oscillator; the larger the time delay the smaller the spacing of the fringes. In the mid-IR, the Hamm and Zanni research groups set $t_{\text{LO}} = 0$ so that there is no periodicity and the 2D IR spectrum is generated by Fourier transforming the signal of each array pixel as a function of the indirect time delay (e.g. t_1).

Another approach chosen by some researchers is to make the local oscillator time delay very large so that the cross-term is highly oscillatory in the frequency domain [124]. For example, setting the local oscillator time delay large but negative is another way to suppress the transient absorption background. In this case, one can extract E_{sig} from the other terms by Fourier transforming the interferogram, filtering, and inverse Fourier transforming back. The 2D IR spectrum is then generated in the same manner as before. The process is illustrated in Fig. 9.17 and is frequently used in visible spectroscopies, but we do not think it is very useful for mid-IR spectroscopies. For the approach to work, the fringe spacing must be very small so that after Fourier transformation the cross-term is well-resolved. However,

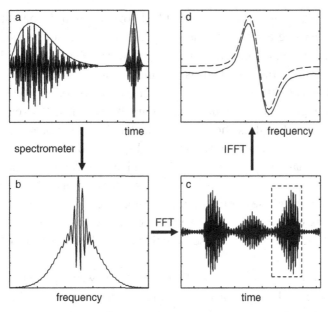

Figure 9.17 Method by which E_{sig} can be extracted using spectral interferometry. (a) The time domain signal and local oscillator (LO is shown at 1/4 intensity). Envelopes of the signal and local oscillator are shown dashed. (b) The measured signal using a spectrometer. (c) Fourier transform of the measured signal after having subtracted off the local oscillator spectrum. The data within the dashed box is what is used to retrieve the signal. (d) Spectrum of the (solid) recovered signal and (dashed) signal calculated directly from the time domain signal in (a) for comparison. The extracted electric field contains some wiggles, which is due to truncation error (see Section 9.5.7) and the relative intensity of the fundamental and overtone transitions is not quite right. These errors are caused by the interferogram overlapping with the zero-frequency components in (c). A larger local oscillator delay would increase the separation, although the delay is limited by spectrometer resolution.

the maximum time delay that one can use is determined by the resolution of the interferogram according to the spectrometer resolution and the pixel width of the array detector. In the visible, CCD cameras have many hundreds of pixels, but mid-IR arrays typically have ≤ 128 pixels and most commonly only 32 pixels. Thus, we prefer the former method.

9.5.6 Phasing of 2D IR spectra

2D IR spectra collected with any of the time domain methods, with the exception of the pulse shaper design, must be phased. "Phasing" corrects for phase errors which result, for example, from inaccurate time delays. The time delays t_1 and t_3

as well as the phases $\phi_{1,2}$ and $\phi_{3,\text{LO}}$ must be known *absolutely* to within a fraction of a light cycle (note again that time delay and phase are not necessarily equivalent). This requirement is in addition to having accurate step sizes and good phase stability. As a rule of thumb, the phase should be known with an accuracy of $\lambda/50$ to $\lambda/100$, which corresponds to 50–100 nm or 0.2–0.4 fs for a mid-IR experiment. In the optical regime, the absolute delays must be known with even higher accuracy since the wavelength is shorter. To understand why the phase and time delay of the various pulses is relevant, we return to Eq. 9.17. If the relative phases of the pairs of pulses is $\phi_{1,2} = \phi_{3,\text{LO}} = 0$ and the relative time delays can be set precisely to $t_1 = t_3 = 0$, then the peaks in the 2D IR spectra will have the correct phases. If either the relative phases or time-zeros are not precisely set, then absorptive and dispersive parts will intermix and one will get skewed peaks after taking the real part in Eq. 9.17. That is why these quantities must be known to a fraction of a wavelength.

Note that it is the phases and time delays between pulses k_1 and k_2, as well as between pulse k_3 and the LO that matter, which is when the system is in a coherent state. The phase between pulses k_2 and k_3, when the system is in a population state, is not important for these issues. There are primarily two methods for phasing 2D IR spectra: (i) post-processing in which the unknown phase is determined after data collection is complete by comparing a projection of the 2D IR spectrum with a broadband pump–probe spectrum (projection slice theorem [62]), and (ii) measuring the phase while taking the data and/or setting it to zero. In the following subsections, we outline these two methods.

Post–processing: Projection slice theorem

In a broadband pump–probe experiment, one excites the system with a single short laser pulse and probes the response of the sample. Due to the phase matching condition resulting from the pump–probe geometry, the first two field interactions come from the same pump pulse, so that t_1 and $\Delta\phi_{1,2}$ are both zero. Moreover, since the same pulse acts as k_3 and the LO at the same time, t_3 and $\Delta\phi_{3,\text{LO}}$ are both zero. The important point is that a pump–probe experiment is not inherently phase sensitive and so can be used as a calibration. Thus, if the 2D IR spectrum is properly phased, then its projection onto the ω_3 (probe) frequency axis (which is identical to setting $t_1 = 0$, see Eq. 4.39) should match the pump–probe spectrum. If it does not, then a phase factor $e^{i\phi}$ is added to manually adjust the phase until agreement is reached:

$$I_{\text{pu-pr}}(\omega_3, t_2) = \Re \int_{-\infty}^{\infty} d\omega_1 \left(\int_0^{\infty} \int_0^{\infty} dt_1 dt_3 R(t_3, t_2, t_1) e^{i\phi} e^{i\omega_1 t_1} e^{i\omega_3 t_3} \right). \tag{9.25}$$

The phase ϕ is often called a zero-order phase. Only a zero-order phase is needed if the time delays t_1 and t_3 are indeed zero. Sometimes it is difficult to determine whether the delays are really at zero or some incremental integer n of the wavelength (i.e. $t_1 + n/\omega$), especially for longer pulses that have many periods per envelope. If a period other than $n = 0$ is used, then there will be a frequency dependent phase, in which peaks at different frequencies in the 2D IR spectra require different phase factors. In this case, first-order phase corrections are necessary that vary linearly with ω_1 and ω_3 [3]. Finally, we note that it is only the time delays involved in coherences that alter the phase, not those in population times like t_2, although the population time should be the same for both the 2D IR spectrum and the pump–probe calibration spectrum since spectral diffusion could alter the lineshapes.

In order to obtain purely absorptive 2D IR spectra, rephasing and non-rephasing diagrams need to be measured independently and added. To this end, one typically switches the time-ordering between pulses k_1 and k_2. Nevertheless, if these two measurements are performed sequentially so that the instrument has not changed between measurements, their phase errors are related. That is, if the rephasing signal has a phase error of ϕ, then the non-rephasing signal will have a phase error of $-\phi$. Thus, when correcting the phase, both rephasing and non-rephasing signals can be corrected at the same time with the same set of correction parameters.

We make two caveats regarding the projection slice method. First, it is relatively insensitive to phase errors. A small phase error that is hardly noticeable in the pump–probe projection will often have sizeable effects in the 2D IR spectrum. Second, in some of the spectrometer designs, the pump–probe calibration signal is weaker than the 2D IR spectrum, and thus may be difficult to measure. One way around this latter problem is to determine the phase using a reference sample with a strong signal.

Presetting the phase

The second method is to measure the relative phase at the sample and thereby set the phase without using a calibration spectrum. We note again that the time delays must be known to within 0.2–0.4 fs to obtain a phase accuracy of $\lambda/50$ to $\lambda/100$. In a collinear geometry like for design (b2), one can achieve this accuracy with a simple interferogram. In a non-collinear geometry, such as a box-CARS design, it might seem impossible to achieve such accuracy since the beams cross at an angle, causing a time-sweep of as much as ≈60 fs across the beam profile for a typical beam waist of 100 μm and crossing angle of 10°. The remedy lies in the fact that it is the phase difference $\Delta\phi_{1,2} - \Delta\phi_{3,LO}$ that matters, not the absolute phases of each pulse. So, while the crossing angle causes the phase of each pulse to vary across its profile, the fact that we are measuring the signal in a particular phase matching direction means that the relative phases are conserved. In other words, the phase difference $\Delta\phi_{1,2} - \Delta\phi_{3,LO}$ does not depend on the position within the beam profile where we measure it, whereas the individual phases do. Thus, the goal is to measure the interference patterns of the pairs of beams directly in the focal plane of the sample.

Figure 9.18 shows the essential idea. The four beams are set into a box-CARS geometry with the local oscillator $\vec{k}_{LO}$ aligned collinearly to the phase matching direction of the emitted field, $\vec{k}_{LO} = -\vec{k}_1 + \vec{k}_2 + \vec{k}_3$. Each pair of beams forms interference fringes in the focal plane. Since we are interested in the relative phases, we block two beams and measure the interference of the remaining two. The interference will have fringes spaced according to the angle between the beams, i.e. $\Delta x = 1/|\vec{k}_2 - \vec{k}_1| = 1/|\vec{k}_{LO} - \vec{k}_3|$. Time delays $\Delta t_{1,2}$ and $\Delta t_{3,LO}$ cause the interference pattern to shift up or down. By changing the time delays to make the interference patterns of beams 1 and 2 match that of beams 3 and the LO, then the phase difference $\phi_{1,2} - \phi_{3,LO}$ is zero. One should make sure that it is the fringe associated with the maximum overlap of the envelopes so that there is no first-order phase correction needed later on.

Two methods have been proposed to measure these interference fringes. In the visible spectral range, one can image a replica of the focal plane using a beamsplitter (a piece of glass), a microscope objective, and an inexpensive CCD camera [21] or use a fiber optic cable. In the mid-IR, where 2D cameras are prohibitively expensive, one may scan a small pinhole laterally in the focal plane

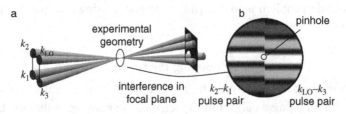

Figure 9.18 (a) The typical box-CARS geometry used in heterodyne detected 2D IR spectroscopy. (b) Graphical illustration of the interference fringes of beams k_1 and k_2 (left half) and beams k_3 and k_{LO} (right half) in the focal plane. The left and right halves, respectively, are measured independently in two subsequent scans by blocking the corresponding other pulse pair. Adapted from Ref. [7] with permission.

9.5 Data collection and data workup

and thereby map out the interference pattern [7]. If the pinhole is small enough (ca. 5 μm), it scatters light like a point source so that the scatter into the detector is independent of the input beam directions.

9.5.7 Window functions

Window functions serve as a useful tool for eliminating truncation artifacts, improving the signal-to-noise ratio of 2D IR spectra and enhancing the spectral resolution. However, they must be used carefully. If misused, they can lead to spurious peaks and bad conclusions. These data manipulations are also known as apodization and filtering.

Truncation artifacts arise when the emitted signal has not fully decayed to the baseline at the end of the scan. To illustrate these artifacts, consider the free induction decay shown in Fig. 9.19, which was calculated using the vibrational dynamics of N-methylacetamide. This type of decay (closely exponential) is often found in the indirectly detected dimension, t_1. The Fourier transform of this signal is given in Fig. 9.19. If the emitted signal is not sampled all the way to the end, then the Fourier transform gives a lineshape with side-bands. These truncation artifacts are a result of multiplying the true time domain data with a

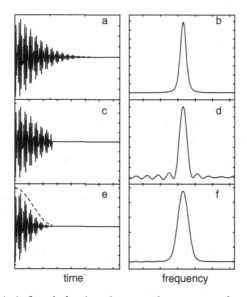

Figure 9.19 (a,b) A free induction decay and spectrum of a single absorption band. (c,d) Truncation of the free induction decay because the data were not collected until the signal fully decayed and the effect on the spectrum. (e,f) Application of a Hamming window function to force the decay to zero and its effect on the spectrum.

square function, whose Fourier transform is the sinc function. Truncation artifacts should be minimized so that they are not mistaken as cross-peaks. If the signal cannot be sampled until it completely decays, then a window function can be applied to smoothly transition the signal to zero or close to zero.[4] Shown in Fig. 9.19 is the signal that results from applying a Hamming window function, given by

$$s(t) = 0.54 + 0.46 \cos(\pi t / t_{max}). \qquad (9.26)$$

The sinc oscillations are now gone, albeit at the expense of a broadened linewidth.

Window functions may also be used to improve the signal-to-noise ratio of 2D IR spectra. That is, if the signal is decayed below the noise level at some time, then continuing to increase time will only increase the noise of the frequency domain data, but no longer enhance their spectral resolution. It is therefore advisable to again apodize the data, this time not to avoid truncation effects but to reduce noise. This is illustrated in Fig. 9.20 for the same free induction decay as above but with some simulated noise added. Multiplying the data with a Gaussian function, the signal-to-noise ratio is significantly improved, with only a slight increase of the linewidth (Fig. 9.20c,d).

Finally, one may also narrow the linewidth of bands by enhancing the signal from large over small time delays, albeit at a loss in signal-to-noise ratio. A useful filter function that is easily adjustable is a shifted sine-bell, given by

$$s(t) = \sin\left(\pi \frac{t + t_0}{t_{max} + t_0}\right) \qquad (9.27)$$

where $t_0 = \theta \cdot t_{max}/(\pi - \theta)$ and θ is an adjustable parameter in degrees. Setting $\theta = 20°$ enhances the signal from the middle of the time scan, which narrows the linewidth. Shown in Fig. 9.20(e,f) is a sine-bell applied to the data with $\theta = 20°$. In principle, one can obtain extremely narrow peaks by using two successive window functions. The first window function is a *growing* exponential with a time constant that matches the signal decay so that, at least in principle, the resulting time domain data do not decay in intensity. The second window function reduces truncation artifacts by bringing the signal to zero, such as by using a Kaiser filter. Of course, this process requires extremely high signal-to-noise spectra and it may not be possible to match the growing exponential to a complicated signal.

---

[4] Not all window functions go to zero. For example, the Hamming function is designed to minimize the first side-lobes in the truncated data.

9.5 Data collection and data workup 213

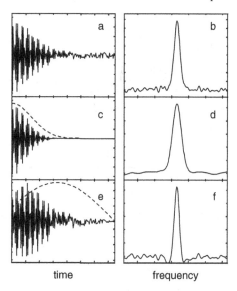

Figure 9.20 (a,b) Same free induction decay as in Fig. 9.19 with noise added. (c,d) Application of a Gaussian window function. (e,f) Resolution enhancement using a shifted sine-bell window function.

9.5.8 Cross-peak intensity and extracting accurate anharmonicities

It can be quite difficult extracting accurate anharmonicities from 2D IR spectra. One of the most important pieces of information that a 2D IR spectrum provides over a linear IR spectrum (like a FTIR spectrum), is the frequencies of the overtone and combination bands. It is these eigenstates that are necessary to determine the coupling constants that are related to molecular structure. However, to extract this information from the most commonly used 2D IR pulse sequences (the 2Q pulse sequence in Section 11.1 is an exception), one must accurately measure the frequency of the positive peaks in the spectra, which are the ones created by Feynman pathways that include transitions up to the overtone and the combination bands (see peaks B and D in Fig. 4.11, for example). These peaks are well-resolved when the anharmonicities are larger than the linewidths, such as for the carbonyl stretches in metal-carbonyl compounds when dissolved in hexane or another non-interacting solvent. But most often, the vibrational linewidths are smaller than the anharmonicities, which causes them to partially cancel each other. Shown in Fig. 9.21 is a slice through a pair of peaks from a simulated 2D IR spectrum for a fixed anharmonic shift (14 cm^{-1}) and four different linewidths. Notice that the apparent frequency separation scales with the linewidth. Thus, the measured frequency separation is always an upper bound to the anharmonic shift. Extracting the anharmonic shift is further complicated by the fact that the

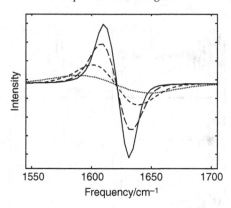

Figure 9.21 Simulated pair of out-of-phase peaks with an anharmonicity of 14 cm^{-1} and linewidths of 17, 27, 42, 70 cm^{-1}. The peak separation appears larger for broader linewidths.

combination or overtone band is usually broadened more than the fundamental due to a faster T_2 time, which is why the positive peak in Fig. 9.21 is not as intense as the negative peak. Figure 9.21 also illustrates why the cross-peaks are so much smaller than the diagonal peaks, which is a result of the smaller off-diagonal anharmonicities, causing more cancellation, and leading to weaker cross-peaks. Thus, to accurately extract the anharmonicity, one needs to know the lineshapes [203]. The ground state lineshape can be approximated by the linear spectrum. The excited state can be approximated by assuming that it has the same frequency fluctuations as the ground state (which is generally true since a change in the potential will influence the frequency more than the anharmonicity, Fig. 2.5) and using a T_1 population relaxation time measured using pump–probe or 2D IR spectroscopy.

9.6 Experimental issues common to all methods

We end this chapter by commenting on a few issues common to all the techniques. First of all, we need an IR light source. Currently, amplified Ti:sapphire laser systems are used, the pulses of which are converted into the IR in using a BBO optical parametric amplifier (OPA) followed by difference frequency mixing in a AgGaS$_2$ crystal [85]. In this way, pulses tunable between $\approx$1000 and 3500 cm^{-1} can be obtained with pulse energies of 1–10 µJ, a typical pulse duration of 50–100 fs (corresponding to a spectral width of 150–300 cm^{-1}), and at repetition rates of 1–10 kHz. Technological developments are currently underway aimed at increasing the repetition rate, the accessible frequency range, the pulse energy, and producing shorter, ultrabroadband pulses.

Second, mid-IR beams can be difficult to align because they are not visible to the naked eye. One can use pyroelectric detectors that are sensitive to roughly 200 nJ, but are sometimes inconvenient because they are bulky and the probe beams are usually too weak to measure. For coarse alignment, one can overlap the mid-IR beams with a HeNe tracer pulse by using either a flip mirror or an optic that transmits IR but reflects red light, such as a piece of germanium or a dielectric mirror. By overlapping a HeNe with the mid-IR immediately after the OPA, one can then align the remaining optics all the way to the sample (make sure to use the HeNe reflections from the same face of the beamsplitters as the mid-IR). One usually uses as many reflective optics as possible in 2D IR spectrometers (curved mirrors instead of lenses, for example) to make the beampath as wavelength insensitive as possible and also to remove dispersion. For example, a 100 fs mid-IR pulse that travels through a $f = 5$ cm CaF_2 lens is broadened by more than 50 fs. HeNe beams can be overlapped well enough that the mid-IR beams will be partially overlapped at the sample, which greatly aids in finding signals during alignment.

Third, one needs to temporally overlap the laser pulses at the sample. To do this, we usually start by performing a pump–probe experiment using a thin piece of germanium (0.5–1 mm) and adjusting the time delays with retroreflectors. Germanium has a bandgap of 0.8 eV, so two or more photons will produce free electron carriers that absorb infrared light. The lifetime of the carriers is very long (tens of picoseconds), and so one will measure a transient absorption even if the time-zeros are very far off. The time-zero can then be determined to about 50 fs using the half-rise-time of the signal. We recommend first determining time-zero with germanium and then optimizing the spatial overlap of the beams. For a more accurate time-zero, one can then switch to a $AgGaS_2$ crystal. Finally, after optimizing the spatial and time overlap, we switch to a very strong molecular absorber to optimize the signals once again. For experiments in which the relative phases of the pulses must be determined, one can use a pinhole placed at the sample position, as described in Section 9.5.6 (this has to be the last step in the alignment).

We place the sample between CaF_2 windows separated with a teflon spacer that is usually 25 or 50 μm thick. Near 5 μm wavelengths, we use azide dissolved in water at a concentration to get an OD of about 0.4. Near 6 μm, we use acetic acid in chloroform, which forms dimers to give an extremely strong transition dipole at 1715 cm^{-1}. For 4 μm, one could use carbonated water. The trickiest part of the alignment is making sure that the pump beam does not scatter into the probe beam direction, which will cause interference in the signal. To minimize scatter, make sure there are no scratches on the mirrors and place adjustable irises in various parts of the spectrometer.

Exercises

9.1 Design a 3D IR spectrometer.

9.2 Imagine you have a purely absorptive real-valued 2D IR spectrum $I(\omega_1, \omega_2)$. Write out the equations to mathematically extract the rephasing and non-rephasing spectra.

9.3 Think of a phase cycling sequence that separates the two Feynman diagrams of Fig. 9.22, assuming k_2 and k_3 are collinear (k_1 is the first pulse in the diagrams).

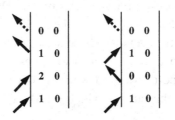

Figure 9.22

10
Simple simulation strategies

One of the strengths of 2D IR spectroscopy is the ability to quantitatively link experimental results to computer simulations, be it molecular dynamics simulations, quantum chemistry calculations, or ideally a combination of both on the level of mixed quantum mechanics/molecular mechanics (QM/MM) calculations. In the present chapter, we outline how such simulations are performed and present some examples with computer code that can be reproduced on a personal computer. We also describe more sophisticated models that have been developed. The motivation of the chapter is *not* to get the most accurate agreement with experiment, but to outline the essential concepts with working examples.

In this chapter we use the molecular dynamics simulation package *Gromacs* 3.3 [183] (which can be downloaded for free from http://www.gromacs.org), the quantum chemistry program *Gaussian09* for electronic structure calculations [58], and simple *Mathematica* or C codes (together with *Numerical Recipes* routines [152]). All the relevant computer programs in this chapter can be downloaded from the book webpage (http://www.2d-ir-spectroscopy.com), so the reader has operational programs to start with which can then be modified at will. For each of the *Mathematica* programs, *Matlab* versions are available on the book webpage as well.

10.1 2D lineshapes: Spectral diffusion of water

Perhaps the most accurate quantities that can currently be modeled are 2D IR lineshapes. As an example, we look at water for which many 2D IR experiments have been carried out [2, 33, 49, 198]. More precisely, we simulate the 2D IR lineshape of an isolated –OH vibration of isotope-diluted HOD in D_2O. This example contains the essential concepts of lineshape theory from Chapter 7 and the methods can be applied to other transitions, such as the amide I vibration in peptides. The –OH stretching frequency is a sensitive probe of the hydrogen bonding to the immediately surrounding water molecules. Hydrogen bonding shifts the vibration to lower frequencies. Hydrogen bonding in liquid water is highly fluctional. 2D IR spectroscopy can time-resolve these dynamics through lineshape analysis.

10.1.1 MD simulation

Download the following files from the book's webpage (http://www.2d-ir-spectroscopy.com):

- spc.top
  contains the topology file, i.e. the geometry of the water molecule as well as the charges and Lennard-Jones parameters of the oxygen and hydrogen atoms, respectively.
- spc.gro
  contains a box of 1019 SPC water molecules at experimental density which was equilibrated at 300 K.
- md.mdp
  contains parameters for the MD simulation, such as the simulation time, time step, etc.
- atoms.ndx
  is an index that *Gromacs* uses to know for which atoms to write out the coordinates and forces for the calculation of the frequency fluctuations. These atoms are the oxygen and one of the hydrogens of one particular water molecule; that OH group will be our test chromophore, whose frequency fluctuations we are going to evaluate.

Three successive commands will generate a trajectory of coordinates and forces for the one OH group we chose as our test chromophore:

- grompp -f md.mdp -c spc.gro -p spc.top -o spc.tpr
  starts the preprocessor that stores all parameters needed for the job in file spc.tpr.
- mdrun -s spc.tpr -o spc.trr &
  starts the actual MD simulation, which will take 1–2 h on a single processor.
- g_traj -f spc.trr -s spc.tpr -ox coord.xvg -of force.xvg -n atoms.ndx -noxvgr
  extracts coordinates and forces from the MD trajectory for the atoms specified in atoms.ndx.

After running these programs, you will obtain two files, coord.xvg and force.xvg, which can be downloaded from the book webpage for comparison and which contain the coordinates of the test chromophore, as well as the forces acting

10.1 2D lineshapes: Spectral diffusion of water 219

Figure 10.1 Water simulation box.

on it. This information will be needed in the next step to calculate a frequency trajectory. Figure 10.1 shows a snapshot of the simulation box.

10.1.2 Relating structure to a vibrational frequency

To generate a 2D IR spectrum from a molecular dynamics simulation, we need to generate a frequency trajectory analogous to the one we showed in Fig. 7.3. Thus, for each step in the molecular dynamics simulations, we need to calculate a frequency for the OH bond. At first sight, one might think one could just take a sliding Fourier transform of the bond length itself, but it turns out that this approach is not feasible for various reasons:

- A Fourier transform requires a time window of a certain length to define a vibrational frequency with a certain accuracy (time–frequency uncertainty). What we will need is the instantaneous frequency $\omega_{01}(t)$ at a given time, not the average over a longer time window. One should think of the instantaneous frequency $\omega_{01}(t)$ more as the time dependent curvature of the OH stretch potential (which is a quantity we can define on an infinitesimally short time window), rather than an actual vibration.
- In most common water models, including the one we use here (SPC), the OH bond is constrained, so it does not vibrate. This is to save computational time

(the OD/OH vibration would be the fastest motion, and would require very short integration time steps).
- Molecular dynamics force fields have not been designed for spectroscopic accuracy, but to reproduce properties such as its density and diffusion constant.
- Classical simulations only sample potentials to within kT, which is well below the zero-point energy of an O–H stretch, let alone the first excited state at 3400 cm^{-1} or the overtone and combination bands near 7800 cm^{-1}. Thus, classical trajectories do not capture anharmonicity correctly.

Hence, we need a model that describes the deformation of the OH stretch potential upon external forces. To illustrate the methodology, we choose a model introduced by Oxtoby [149], which gives very accurate results for systems as different as neat N_2 [149], CN$^-$ in H_2O [155] and isotope-diluted water [138, 169], while being quite simple to understand and implement. To that end, we expand the interaction potential between the solute and the solvent in powers of the normal coordinate, defined with respect to its equilibrium value

$$V = \left[\frac{dV}{dQ}\right]_{Q=0} Q + \frac{1}{2}\left[\frac{d^2V}{dQ^2}\right]_{Q=0} Q^2 + \cdots \equiv F_1 Q + F_2 Q^2 + \cdots \quad (10.1)$$

which defines the "forces" F_1 and F_2. In particular, F_1 is (minus) the force exerted by the solvent on the fixed oscillator coordinate, while F_2 is (one-half) the derivative of the solvent force with respect to Q. It has been found that this term dominates the frequency shift [138], and so we neglect F_2. For a localized mode, such as the OD vibration of HOD in H_2O, F_1 is evaluated as the mass-weighted intermolecular force projected onto the OH bond axis:

$$F_1 = \mu \left(\frac{\vec{F}_O}{m_O} - \frac{\vec{F}_H}{m_H}\right) \cdot \vec{r}_{OH} \quad (10.2)$$

where $\vec{F}_O$ and $\vec{F}_H$ denote the total forces on the O and H centers, respectively, μ is the reduced mass, and $\vec{r}_{OH}$ is the unit vector from O to H.

The term related to F_1 vanishes for a harmonic oscillator. This fact is illustrated in Fig. 10.2(a), which shows that a linear force $F_1 Q$ acting on a harmonic oscillator $V(Q)$ shifts the potential but does not alter the curvature. In contrast, adding a linear force $F_1 Q$ to an anharmonic oscillator shifts the minimum and changes the curvature of the potential around its minimum, thereby altering the vibrational frequency.

Using perturbation theory, one can express the frequency shift as [149]:

$$\delta\omega(t) = \frac{3f}{\mu^2 \omega_0^3} F_1(t) \quad (10.3)$$

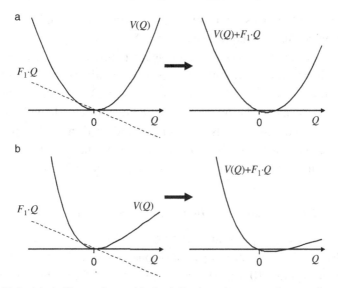

Figure 10.2 (a) A linear force (dashed line) acting on a harmonic oscillator (solid line) reveals a shifted but still harmonic oscillator with identical curvature. (b) A linear force (dashed line) acting on a Morse oscillator (solid line) changes the curvature around the minimum of the potential.

where ω_0 is the unperturbed gas-phase vibrational frequency and f is the third-order expansion term of the potential energy surface, which could be obtained from a gas-phase potential energy surface calculation. Instead, we estimate it from the anharmonic shift $\Delta = 240$ cm^{-1} of the OH vibration. That is, assuming a Morse potential for the OH stretch coordinate, we can relate the third-order expansion term f to the anharmonic shift Δ using perturbation theory [177], and plug it into Eq. 10.3 for which we then obtain:

$$\delta\omega(t) = \frac{3}{2\omega_0}\sqrt{\frac{\Delta}{\hbar\mu}} F_1(t). \tag{10.4}$$

In essence, according to this approach, the frequency shift $\delta\omega(t)$ depends linearly on the force F_1 exerted by the environment on the OH bond. The proportionality factor, in turn, scales with the square-root of the anharmonic shift Δ. The anharmonic shift of a hydrogen-bonded OH vibration is exceptionally high, which is one of the reasons why the linewidth of the OH vibration of water is so large (>100 cm^{-1}).

The force F_1 is mass weighted (Eq. 10.2), hence it is dominated almost exclusively by the force on the proton. Furthermore, since common water models assign a charge to the hydrogens, but no Lennard-Jones term, the force is exclusively the

Coulomb force.[1] Thus, it is the electrostatic interaction of the hydrogen with the environment which causes the frequency shift $\delta\omega(t)$.

This simple approach has been verified against more computationally expensive methods which are nonperturbative [48, 55, 138]. Similar results are obtained. An alternative and very successful method of mapping MD simulations into frequency trajectories is to use a correlation based on *ab initio* cluster calculations [32, 89]. In these methods, electrostatic fields or potentials are used instead of forces, but the idea is quite similar. These or other approaches are also being applied to the solvent interactions of the C=O frequency of amide I vibrations [90, 101, 118, 154, 162].

10.1.3 Frequency fluctuation correlation function

Using the *Mathematica* program 2DwaterCumulant.nb from the book webpage (http://www.2d-ir-spectroscopy.com), one can evaluate the outcome of the MD simulation. Figure 10.3(a) shows a short piece of the frequency trajectory, Fig. 10.3(b) the distribution of frequencies, and Fig. 10.3(c) the resulting frequency

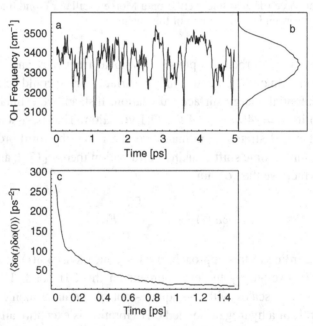

Figure 10.3 (a) A short piece of the frequency trajectory deduced from an MD simulation of water, with (b) its frequency distribution and (c) the resulting frequency fluctuation correlation function.

---

[1] We use the SPC water model; a very nice summary of water models is found in http://www1.lsbu.ac.uk/water/models.html.

10.1 2D lineshapes: Spectral diffusion of water

fluctuation correlation function, $c(t) = \langle \delta\omega(t)\delta\omega(0)\rangle$. There is a fast initial drop of the FFCF within the first 100 fs (the so-called *inertial component*), a slightly underdamped oscillatory component which reflects the intermolecular hydrogen bond vibration, and a slower decay on a 1 ps time-scale (often called the *diffusive component*). Following a proposal from Ref. [138], we fit this frequency fluctuation correlation function with a function of the form:

$$c(t) = a_1 \cos(\omega_{00}t)e^{-t/\tau_1} + a_2 e^{-t/\tau_2} + a_3 e^{-t/\tau_3} \tag{10.5}$$

which allows us to calculate the lineshape function $g(t)$ analytically by double integration (see Eq. 7.19):

$$g(t) = \int_0^t d\tau' \int_0^{\tau'} d\tau'' \langle \delta\omega_{01}(\tau'')\delta\omega_{01}(0)\rangle. \tag{10.6}$$

The two exponential terms give just Kubo-like terms (see Eq. 7.25), whereas integration of the term with the cosine function produces a lengthy expression that is not repeated here (*Mathematica* does it for us).

10.1.4 2D IR spectra

The lineshape function $g(t)$ is what connects the MD simulation to the formalism of nonlinear spectroscopy. That is, the lineshape function $g(t)$ enters into the response function (see Eq. 7.40):

$$R_{1,2,3} = i\mu_{01}^4 \left(e^{-i\omega_{01}(t_3-t_1)} - e^{-i((\omega_{01}-\Delta)t_3-\omega_{01}t_1)}\right) \cdot$$
$$\cdot e^{-g(t_1)+g(t_2)-g(t_3)-g(t_1+t_2)-g(t_2+t_3)+g(t_1+t_2+t_3)}$$

$$R_{4,5,6} = i\mu_{01}^4 \left(e^{-i\omega_{01}(t_3+t_1)} - e^{-i((\omega_{01}-\Delta)t_3+\omega_{01}t_1)}\right) \cdot$$
$$\cdot e^{-g(t_1)-g(t_2)-g(t_3)+g(t_1+t_2)+g(t_2+t_3)-g(t_1+t_2+t_3)}. \tag{10.7}$$

The resulting rephasing and non-rephasing signals in the time domain are shown in Fig. 10.4. 2D IR spectra are obtained through a 2D Fourier transform (see Eq. 4.31) after multiplying the points $R(t_1 = 0, t_2, t_3)$ and $R(t_1, t_2, t_3 = 0)$ by 0.5 (Section 9.5.3) and padding with zeros to twice the data size (Section 9.5.4)

$$R_{1,2,3}(\omega_3, t_2, \omega_1) = \int_0^\infty \int_0^\infty iR_{1,2,3}(t_3, t_2, t_1)e^{+i\omega_3 t_3}e^{+i\omega_1 t_1}dt_1 dt_3$$

$$R_{4,5,6}(\omega_3, t_2, \omega_1) = \int_0^\infty \int_0^\infty iR_{4,5,6}(t_3, t_2, t_1)e^{+i\omega_3 t_3}e^{+i\omega_1 t_1}dt_1 dt_3. \tag{10.8}$$

Finally, the rephasing and non-rephasing spectra are added after inverting the ω_1-coordinate of the rephasing diagram, to obtain purely absorptive spectra (see Eq. 4.36):

224   *Simple simulation strategies*

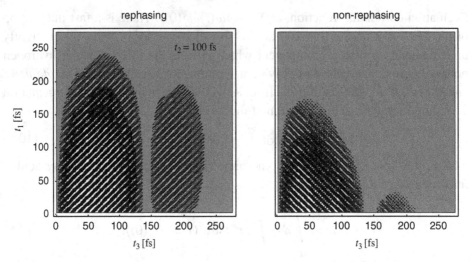

Figure 10.4 Rephasing and non-rephasing signal in the time domain. The population time was set to $t_2 = 100$ fs.

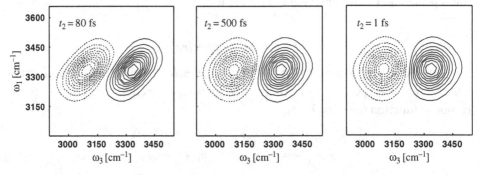

Figure 10.5 A series of purely absorptive 2D IR spectra of the OH vibration of HDO in D_2O at population times $t_2 = 80$ fs, $t_2 = 500$ fs, $t_2 = 1$ ps, respectively, calculated within the framework of the cumulant expansion.

$$R(\omega_1, \omega_3) = \Re\left(R_{1,2,3}(-\omega_1, \omega_3) + R_{4,5,6}(\omega_1, \omega_3)\right). \quad (10.9)$$

Figure 10.5 shows a series of simulated 2D IR spectra at different waiting times calculated using these programs. The 0–1 and 1–2 contributions to the spectra are symmetric along the diagonal due to the cumulant expansion.

10.1.5 Avoiding the cumulant expansion

It turns out that the approximation of the cumulant expansion truncated after second order is not very accurate for water. The distribution of frequencies (Fig. 10.3b) is clearly asymmetric, with a shoulder to the high-frequency side.

Hence it deviates quite significantly from Gaussian statistics. Consequently, rather than using Eqs. 10.5 through 10.7 to calculate the response function, we should use Eq. 7.35 directly:

$$R_{1,2,3} = i\mu_{01}^4 \left(e^{-i\omega_{01}(t_3-t_1)} - e^{-i((\omega_{01}-\Delta)t_3-\omega_{01}t_1)}\right) .$$

$$\cdot \left\langle \exp\left(+i\int_0^{t_1} \delta\omega_{01}(\tau)d\tau - i\int_{t_2+t_1}^{t_3+t_2+t_1} \delta\omega_{01}(\tau)d\tau\right)\right\rangle$$

$$R_{4,5,6} = i\mu_{01}^4 \left(e^{-i\omega_{01}(t_3+t_1)} - e^{-i((\omega_{01}-\Delta)t_3+\omega_{01}t_1)}\right) .$$

$$\cdot \left\langle \exp\left(-i\int_0^{t_1} \delta\omega_{01}(\tau)d\tau - i\int_{t_2+t_1}^{t_3+t_2+t_1} \delta\omega_{01}(\tau)d\tau\right)\right\rangle .$$

(10.10)

These equations were modified to a three-level system of an anharmonic oscillator, assuming that the 0–1 fluctuations are strictly correlated with the 1–2 fluctuations, and that we have for the transition dipoles $\mu_{12} = \sqrt{2}\mu_{01}$.

The *Mathematica* program 2DwaterWithoutCumulant.nb from the book webpage calculates the integrals of Eq. 10.10 on a grid of 40 fs. In contrast to the previous program 2DwaterCumulant.nb, the Fourier transform is done without the $e^{-i\omega_{01}(t_3 \pm t_1)}$ terms, hence the resulting 2D IR spectra will be centered at $\omega_1 = \omega_3 = 0$ (Fig. 10.6). Without the highly oscillatory term, it is sufficient to evaluate the response functions with 40 fs time steps. Effectively, this is the same as measuring 2D IR spectra in the rotating frame (Section 9.3.3). If needed, the spectra can be shifted to the correct frequency by just adding ω_{01} to the ω_1- and ω_3-axis *after* Fourier transformation.

Computation of Eq. 10.10 is much more tedious than Eqs. 10.5 through 10.7, because the ensemble average in the former converges very slowly. This is why one prefers to use the cumulant expression when possible. Assuming ergodicity (see Section 7.2), the ensemble average is actually done with a time average. The noise

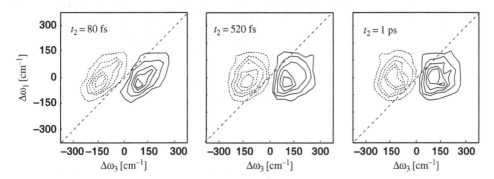

Figure 10.6 A series of purely absorptive 2D IR spectra of the OH vibration of HDO in D_2O at population times $t_2 = 80$ fs, $t_2 = 520$ fs, $t_2 = 1$ ps, respectively, calculated without the cumulant expansion.

level of Fig. 10.6 is what we get from a 1 ns trajectory. One could improve the noise level by running a longer MD trajectory, or, in this particular case, make use of the fact that there are ca. 1000 water molecules in the simulation box, each of which could be used as a test chromophore. Despite the noisy result of Fig. 10.6, one can nevertheless see the asymmetric lineshape and the spectral diffusion process.

Finally, we note that *non-Condon effects* are quite pronounced in water [161], which are not included in the present simulation. The non-Condon effect reduces the transition dipole strength for frequencies towards the blue side of the spectrum, which deforms the 2D IR lineshape of water even further.

10.2 Molecular couplings by *ab initio* calculations

In the last section, we focused on converting MD simulations to lineshapes. Here we address how to calculate couplings and hence the cross-peaks in 2D IR spectra. To that end, we calculate molecular couplings from an *ab initio* quantum chemistry calculations, using the *Gaussian09* program package [58]. In principle, the approach is applicable to any molecule and to any set of modes (to the extent that the quantum chemistry method is correct and applicable), even when the exciton model breaks down such as for modes that are separated by a large frequency range. *Gaussian09* starts from a structure, minimizes it, calculates normal modes from a Hessian matrix, for which it needs the second derivatives of the potential energy surface that are calculated analytically. In addition, anharmonic corrections are calculated perturbatively from third-order derivatives and some of the fourth-order derivatives of the potential energy surface. The latter have to be done numerically by deflecting the molecule along all atomic coordinates, which is why the calculation is extremely computer-time consuming, even for relatively small molecules. We demonstrate the approach for the two $-C\equiv O$ stretch vibrations of dicarbonylacetylacetonato rhodium ($Rh(CO)_2C_5H_7O_2$, RDC, see Fig. 10.7). The 2D IR spectroscopy of dicarbonylacetylacetonato rhodium and many other metal carbonyls have been studied extensively with 2D IR spectroscopy [104].

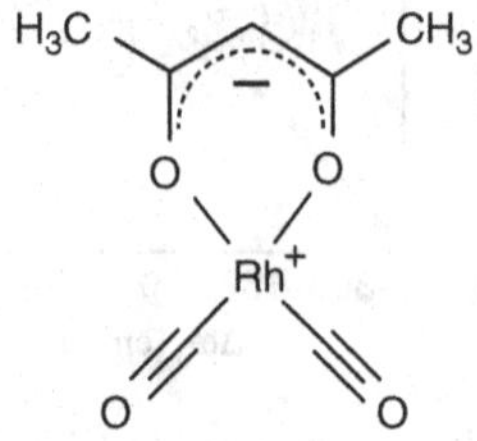

Figure 10.7 Chemical structure of dicarbonylacetylacetonato rhodium ($Rh(CO)_2C_5H_7O_2$, RDC).

10.2 *Molecular couplings by ab initio calculations*

Table 10.1 *Relevant results from the Gaussian09 calculation.*

	ω_A	ω_S	ω_{2A}	ω_{2S}	ω_{A+S}
Frequency (cm^{-1})	1889.184	1947.621	3767.517	3883.911	3813.698
Intensity (km/Mole)	908	671	–	–	–

To start the calculation, download the following file from the book webpage:

Rhcomplex.dat

specifies the *Gaussian09* job. The header:

B3LYP/CEP-121G Opt Freq=(Anharmonic,SelectAnharmonicModes)

tells *Gaussian09*, which method (B3LYP) and which basis set (CEP-121G) to use. It also tells it to begin by optimizing the geometry starting from the conformation given, and then make an anharmonic frequency calculation. The option SelectAnharmonicModes restricts the anharmonic frequency calculation to the modes specified in the last line (i.e. modes 43, 44, which are the two carbonyl modes).[2] Run the job with

g09 Rhcomplex.dat

which will take a few hours on a single processor. It will produce an output file Rhcomplex.log (also on the book webpage), which contains all the information we need (besides a lot of useless stuff).

What we need in this particular case are the anharmonic frequencies of the two fundamental –C≡O modes, i.e the asymmetric and symmetric stretch vibrations, their overtones and the mixed-combination mode. These are the eigenstates illustrated in Fig. 1.5 for a two-oscillator system. One finds them in the file Rhcomplex.log by searching for "Vibrational Energies" (modes 8 and 9), and a little below for "Overtones" and "Combination Bands"; the results are listed in Table 10.1. We also need the transition dipoles of the transitions. *Gaussian09* calculates the intensities of the fundamentals, which are proportional to the square of the transition dipoles. They can be found by searching the file for "Frequencies" in the list of all normal modes. However, *Gaussian09* does not calculate the intensities

---

[2] *Gaussian03* does not allow one to restrict the calculation of anharmonic constants to a certain subset of vibrational modes, which causes the calculation to take many days.

of the overtones, hence we estimate them from the fundamental using the usual harmonic approximation (Eqs. 6.13 and 6.14). We also need the directions of the transition dipoles. *Gaussian09* calculates them, and outputs them in the cryptic output at the end of the log file in a format. If needed, the directions can be extracted using *GaussView*, the graphical frontend of *Gaussian09*. For our purpose, we know that the transitions dipoles of the two –C≡O vibrations are perpendicular to each other, for symmetry reasons, and we assume that the overtones have the same directions.

The *Mathematica* program Rhcomplex.nb uses these results to calculate purely absorptive 2D IR spectra. We follow the nomenclature of Figs. 4.10 and 4.12. Calculating the response functions requires three nested loops with indexes $i = \{A, S\}$ and $j = \{A, S\}$ running over the single excited symmetric (S) and antisymmetric (A) states and $k = \{2A, 2S, A + S\}$ running over the double excited states. The rephasing diagrams for the $\langle ZZZZ \rangle$ polarization condition are:

$$R_1 = i\langle(\hat{\mu}_{0i} \cdot \hat{Z})^2(\hat{\mu}_{0j} \cdot \hat{Z})^2\rangle e^{+i\omega_j t_1 + i(\omega_j - \omega_i)t_2 - i\omega_i t_3} e^{-(t_1+t_3)/T_2}$$
$$R_2 = i\langle(\hat{\mu}_{0i} \cdot \hat{Z})^2(\hat{\mu}_{0j} \cdot \hat{Z})^2\rangle e^{+i\omega_j t_1 - i\omega_i t_3} e^{-(t_1+t_3)/T_2}$$
$$R_3 = i\langle(\hat{\mu}_{0i} \cdot \hat{Z})(\hat{\mu}_{0j} \cdot \hat{Z})(\hat{\mu}_{ik} \cdot \hat{Z})(\hat{\mu}_{jk} \cdot \hat{Z})\rangle \cdot$$
$$\cdot e^{+i\omega_j t_1 + i(\omega_j - \omega_i)t_2 - i(\omega_k - \omega_j)t_3} e^{-(t_1+t_3)/T_2} \quad (10.11)$$

and for the non-rephasing diagrams:

$$R_4 = i\langle(\hat{\mu}_{0i} \cdot \hat{Z})^2(\hat{\mu}_{0j} \cdot \hat{Z})^2\rangle e^{-i\omega_j t_1 - i(\omega_j - \omega_i)t_2 - i\omega_i t_3} e^{-(t_1+t_3)/T_2}$$
$$R_5 = i\langle(\hat{\mu}_{0i} \cdot \hat{Z})^2(\hat{\mu}_{0j} \cdot \hat{Z})^2\rangle e^{-i\omega_j t_1 - i\omega_i t_3} e^{-(t_1+t_3)/T_2}$$
$$R_6 = i\langle(\hat{\mu}_{0i} \cdot \hat{Z})(\hat{\mu}_{0j} \cdot \hat{Z})(\hat{\mu}_{ik} \cdot \hat{Z})(\hat{\mu}_{jk} \cdot \hat{Z})\rangle \cdot$$
$$\cdot e^{-i\omega_j t_1 - i(\omega_j - \omega_i)t_2 - i(\omega_k - \omega_i)t_3} e^{-(t_1+t_3)/T_2}. \quad (10.12)$$

We assume $T_2 = 2$ ps for the homogeneous dephasing time. The dipole terms read (see Eq. 5.28):

$$\langle(\hat{\mu}_{0i} \cdot \hat{Z})^2(\hat{\mu}_{0j} \cdot \hat{Z})^2\rangle = \frac{1}{15}|\hat{\mu}_{0i}|^2|\hat{\mu}_{0j}|^2\left(1 + 2\cos\theta_{0i,0j}\right) \quad (10.13)$$

$$\langle(\hat{\mu}_{0i} \cdot \hat{Z})(\hat{\mu}_{0j} \cdot \hat{Z})(\hat{\mu}_{ik} \cdot \hat{Z})(\hat{\mu}_{jk} \cdot \hat{Z})\rangle = \frac{1}{15}|\hat{\mu}_{0i}||\hat{\mu}_{0j}||\hat{\mu}_{ik}||\hat{\mu}_{jk}| \cdot$$
$$\cdot \left(\cos\theta_{0i,0j}\cos\theta_{ik,jk} + \cos\theta_{0i,ik}\cos\theta_{0j,jk} + \cos\theta_{0i,jk}\cos\theta_{0j,ik}\right).$$

Purely absorptive 2D IR spectra are calculated by adding up the 2D Fourier transforms of rephasing and non-rephasing diagrams after inverting the ω_1-axis of the former. The resulting purely absorptive 2D IR spectrum is shown in Fig. 10.8.

10.3 2D spectra using an exciton approach

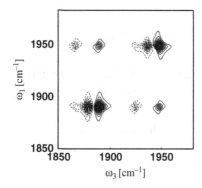

Figure 10.8 Simulated purely absorptive 2D IR spectrum of RDC for population time $t_2=0$ fs.

Figure 10.9 The tryptophan zipper 2 from the Protein Data Bank entry 1LE1.pdb [30].

10.3 2D spectra using an exciton approach

The exciton model is commonly used to compute 2D IR spectra of the amide I band of peptides and proteins (see Chapter 6). To illustrate this approach, we simulate the 2D IR spectrum of a tryptophan zipper (Fig. 10.9), whose NMR structure has been solved [30], and which forms a stable β-hairpin structure. β-Hairpins, β-sheets and also aggregated proteins exhibit a characteristic 2D IR spectrum with two excitonic bands, the modes parallel and perpendicular to the sheet structure, that are well separated (see Section 6.3.4). 2D IR spectra of β-hairpins have been measured experimentally [168, 189] and modeled on essentially the same level as we use below. More elaborate simulation protocols to calculate 2D IR spectra have been worked out [209].

To start the simulation, download the following files from the book webpage:

- 1LE1.pdb
  structure file of the tryptophan zipper 2 from the Protein Data Bank [30]. As the structure has been determined by NMR spectroscopy, it contains in total 20 individual structures that mimic the structural disorder of the system.

- peptide.c
  the actual simulation program written in C.
- coupling.par
  parameter file that contains the parameterized nearest neighbor coupling calculated from a full *ab initio* calculation (Fig. 6.12d) [73], stored as a Fourier series.
- nma.par
  parameter file that contains the amide I normal mode and transition charges of an isolated NMA molecule (Fig. 6.5); needed for the calculation of couplings other than nearest neighbors [84].

Compile the C-code with (under Linux)

```
gcc -O3 -o peptide peptide.c -lm
```

and run it by typing:

```
./peptide 1LE1
```

It will read the subsequent structures from the pdb file, and generate two output files, 1LE1_lin.dat and 1LE1_2D.dat, which contain the linear and 2D IR spectra, respectively (the latter for the $\langle ZZZZ \rangle$ polarization). We use a pdb file from the Protein Data Bank, but one could also generate a sequence of structures from an MD simulation.

The header of the C code contains simulation parameters, which are set to typical values for soluble peptides. The parameters are: inhomogeneous width $\Delta\omega_{inhom} = 10$ cm^{-1}, homogeneous dephasing time $T_2 = 1$ ps, local mode frequency of a non-hydrogen-bonded amide I vibration $\omega_0 = 1660$ cm^{-1}, anharmonic shift $\Delta = 16$ cm^{-1}, and a hydrogen-bond induced shift of $\Delta\omega_{hbond} = 30$ cm^{-1}/Å. In addition, parameters for the Fourier transform are given: number of time steps $= 32$ and step size $= 0.2$ ps. To save computation time, the Fourier transform is performed in a rotating frame (Section 9.3.3) and the amide I frequency is added to the frequency axis only after the Fourier transform. The parameter for the offset frequency ω_{off} allows one to center the spectrum in the spectral window.

The program uses the following ingredients:

- It reads only the backbone atoms, i.e. the CO–NH–C$_\alpha$ repeat units. Side-chains are disregarded in the program although some side-chains absorb in the amide I frequency range (Section 6.4 and Ref. [10]).

10.3 2D spectra using an exciton approach

- From the position of the backbone atoms, it calculates the one-exciton Hamiltonian. To that end, nearest neighbor couplings are calculated using the dihedral angles and the precomputed coupling map in the file coupling.par [73]. For other couplings, the transition charge model is used [84], from which also the local mode transition dipoles are determined.
- The diagonal elements of the one-exciton Hamiltonian are modulated by hydrogen bonding, an effect which we model by an empirical expression and gives rise to diagonal disorder. That is, when a C=O group is hydrogen-bonded to one –NH group of the peptide backbone (with an acceptance angle smaller than 60°), then the corresponding local mode frequency is downshifted by [83]:

$$\delta\omega = -\Delta\omega_{\text{hbond}}(2.6 \text{ Å} - r_{\text{O}\cdots\text{H}}). \tag{10.14}$$

- The structural distribution in the pdb file will cause inhomogeneous broadening through the variation in hydrogen bond lengths and coupling constants. In addition, in order to mimic the influence of the fluctuating solvent, a random frequency shift is added to the diagonal elements of the one-exciton Hamiltonian selected from a Gaussian distribution with width $\Delta\omega_{\text{inhom}}$.
- Once the one-exciton Hamiltonian is ready, the two-exciton Hamiltonian can be determined along the lines of Eq. 6.11. The only additional parameter is the anharmonic shift Δ.
- Both the one- and two-exciton Hamiltonians are diagonalized, and the transition dipoles μ_{0i} from the ground state to the one-exciton states, and μ_{ik} from the one-exciton states to the two-exciton states are calculated by a proper unitary transformation.
- Linear and nonlinear response functions are calculated as before (Eqs. 10.11–10.13), except that it is programmed in a numerically more efficient manner.
- They are averaged over all structures, and finally Fourier transformed to reveal linear and purely absorptive 2D IR spectra.

Figure 10.10(a) shows the simulated purely absorptive 2D IR spectrum. One can clearly identify the two exciton bands expected for a β-hairpin (Section 6.3.4) and a cross-peak between them. The diagonal bands are elongated along the diagonal, reflecting the inhomogeneous broadening. The spectra contain the major features of the experimental 2D IR spectra shown in Fig. 10.10(b) [168]. In this simulation, we purposely used an inhomogeneous distribution that is smaller than occurs naturally in order to emphasize the cross-peaks.

Diagonalization of the two-exciton Hamiltonian scales with the sixth power of the number of amino acids, while the calculation of the response function scales with the fourth power of the number of amino acids times the second power of the

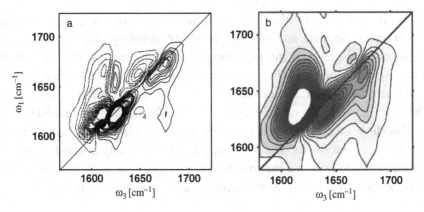

Figure 10.10 (a) Simulated purely absorptive 2D IR spectrum of the tryptophan zipper 2 from Protein Data Bank entry 1LE1. (b) Corresponding experimental spectrum, adapted from Ref. [168] with permission. Note that the ω_1- and ω_3-axes have been flipped in the experimental spectrum to facilitate comparison.

number of time points. In other words, while it is easy to simulate 2D IR spectra of small peptides (the current example takes only a few seconds on a laptop), it quickly becomes relatively computer expensive for larger proteins. One time-saving measure is to calculate stick spectra and then convolute them with a pre-simulated lineshape using a fast Fourier transform routine [114]. In this approach, one can use pre-calculated coupling maps and Bloch dynamics for the lineshape theory. Under these approximations, it is quite straightforward to simulate rather large systems of coupled oscillators.

Exercises

10.1 For Fig. 10.5, explain why the 2D lineshapes change with time. Is the nodal slope consistent with the frequency correlation function from which the spectra were calculated?

10.2 For Fig. 10.8, explain why the cross-peaks are further separated than the diagonal peaks.

10.3 Recalculate Fig. 10.5 with improved signal-to-noise ratio.

10.4 Simulate the 2D IR spectrum of the antibiotic ovispirin which has been studied with 2D IR spectroscopy [194] and has a solution NMR structure in the Protein Data Bank (1HU5). Do two simulations. In one simulation, isotope-label an end residue; in the other a middle residue and see if they have different lineshapes and cross-peaks.

11
Pulse sequence design: Some examples

The concepts outlined in the previous chapters lay a foundation from which new pulse sequences can be designed. We have covered two types of 2D IR spectra that can be generated with third-order pulse sequences, the so-called rephasing and non-rephasing pulse sequences. We start this chapter by presenting the third type of third-order 2D IR pulse sequence, which we term the two-quantum (2Q) pulse sequence, for reasons that will become apparent. We then improve upon this two-quantum pulse sequence by adding two more laser pulses to generate a fifth-order two-quantum coherence, which also enables 3D IR experiments. Purely absorptive 3D IR experiments are described as are transient 2D IR spectroscopies which are also fifth-order nonlinear experiments. Currently, 3D IR pulse sequences are largely unexplored, and the ones that have been implemented have only been applied to a few molecules. This chapter is written to illustrate some of the basic concepts to serve as a platform for more sophisticated experiments in the future.

11.1 Two-quantum pulse sequence

Third-order 2D IR spectra are generated from pulse sequences that interact three times with the sample. The electric field for one of the pulse interactions must be the complex conjugate of the other two, which gives rise to the third basic types of third-order 2D IR spectra: $E_1^* E_2 E_3$, $E_1 E_2^* E_3$, $E_1 E_2 E_3^*$.[1] If the pulses interact with the sample in the same time ordering as written here, then $E_1^* E_2 E_3$ is the rephasing, $E_1 E_2^* E_3$ is the non-rephasing and $E_1 E_2 E_3^*$ is the 2Q pulse sequence. There are only two Feynman diagrams that survive the rotating wave approximation for the 2Q pulse sequence, which are shown in Fig. 11.1 (i can equal j). Unlike the rephasing and non-rephasing spectra whose diagonal peaks evolve as a population during t_2 ($i = j$ in Figs. 4.10 and 4.12), the 2Q pulse sequence evolves as a coherence.

---

[1] In a third-harmonic experiment, one could also have $E_1 E_2 E_3$.

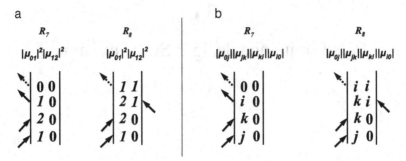

Figure 11.1 Feynman diagrams for the two-quantum response functions for (a) just the "diagonal" peaks and (b) for any number of coupled eigenstates. i and j are fundamentals (1Q states) and k are the overtone and combination band states (2Q states).

The coherence is a continuation of the coherence created during t_1. That is, the first pulse creates a coherence between the $\nu = 0$ and 1 states, $|j\rangle\langle 0|$, which the second pulse converts into a coherence between $\nu = 0$ and 2, $|k\rangle\langle 0|$, where k is either an overtone or a combination band. Thus, during t_2, the macroscopic polarization is oscillating at the frequencies of the overtones and combination bands (in the case of a multi-oscillator molecule), which is what we will refer to as a 2Q coherence. The third pulse creates a 1Q coherence from which the sample emits.

Following the rules from Chapter 2, we can write the response functions for these two Feynman diagrams, which for the diagonal peaks are (setting $i = 1$, $j = 1$, $k = 2$):

$$R_7(t_1, t_2, t_3) = i\mu_{01}^2\mu_{12}^2 e^{-i\omega_{01}t_1 - t_1/T_2} e^{-i\omega_{02}t_2 - t_2/T_2} e^{-i\omega_{01}t_3 - t_3/T_2}$$
$$R_8(t_1, t_2, t_3) = -i\mu_{01}^2\mu_{12}^2 e^{-i\omega_{01}t_1 - t_1/T_2} e^{-i\omega_{02}t_2 - t_2/T_2} e^{-i\omega_{12}t_3 - t_3/T_2}. \quad (11.1)$$

In these equations, the 2Q coherence decays as $1/2T_2$ because the energy gap is twice as large. One could collect a 2D IR spectrum by incrementing t_1 and t_3 as is usually done, but this would not be very useful because it would correlate the 1Q states with each other (both axes are ω_{01}), just as the other two types of third-order 2D IR pulse sequences already do. But, incrementing t_2 and t_3 is very different, because the 2Q frequencies would be correlated along one axis to the 1Q fundamentals on the other (ω_{02} versus ω_{01}) [60, 205]. A 2Q–1Q correlation is not possible with the other two pulse sequences using R_1 through R_6. The resulting 2D IR spectrum for a coupled two-oscillator system is shown schematically in Fig. 11.2 (the peaks from "forbidden" pathways are not drawn). Along the 1Q-axis, the spectra appear as pairs of out-of-phase peaks, just like for the typical rephasing and non-rephasing spectra. But along the 2Q-axis (ω_3), there are three lines of peaks that lie at the overtone and combination band frequencies.

11.1 Two-quantum pulse sequence

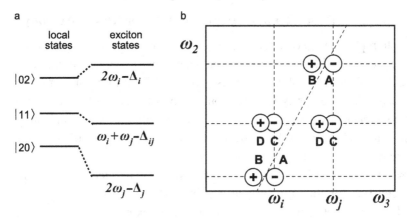

Figure 11.2 2Q 2D IR spectroscopy. (a) 2Q eigenstates as described in Chapter 6, Fig. 6.3, and (b) schematic of the 2Q 2D IR spectrum generated from the response function R_7 and R_8. The weak peaks associated with forbidden pathways are not shown. Notice that the "diagonal" peaks are actually offset from the diagonal (defined as $\omega_2 = 2\omega_3$) by the anharmonicities.

The "diagonal" peaks lie slightly below the overtone frequency along the 2Q-axis and the fundamental frequency along the 1Q-axis. Thus, the diagonal is defined as $\omega_3 = 2\omega_1$. The cross-peaks from allowed transitions all lie at the combination band frequency.

Since the 2Q pulse sequence probes the same eigenstates as the other third-order pulse sequences, there is no new information about the eigenstates in the 2Q spectra. However, the way in which the eigenstates are exhibited is very useful from a practical standpoint. In the typical 2D rephasing and non-rephasing spectra, one must carefully fit the pairs of peaks in order to extract the anharmonicities that give the overtone and combination band frequency, which is usually difficult because the anharmonicities are smaller than the linewidths (see Chapter 9, Section 9.5.8). But in a 2Q 2D IR spectrum, all one needs are the frequencies along the ω_2-axis (Fig. 11.2a), and the peak pairs themselves do not need to be fit. As a result, the overtone and combination band frequencies are easier to obtain [60], leading to better couplings.

There is one drawback to the 2Q pulse sequence. It is a non-rephasing pulse sequence and so the peaks are inhomogeneously broadened. Furthermore, absorptive spectra cannot be generated because there is no complementary rephasing pathway, at least for third-order pulse sequences. Thus, in practice, the 2Q pulse sequence is not very useful because typical spectra are congested and the phase twist causes difficult-to-interpret interferences, although in some cases phase twist is beneficial [175]. An improvement is to generate rephased 2Q spectra using a fifth-order pulse sequence, which we outline in the next section.

11.2 Rephased 2Q pulse sequence: Fifth-order spectroscopy

In order to rephase a 2Q coherence, a sequence of laser pulses must be used to generate a coherence with the opposite sign. That way the dephasing that occurs during the 2Q coherence evolution time is reversed. The minimum number of pulses required for this process is five: two pulses to generate a $|0\rangle\langle 2|$ coherence (or a $|2\rangle\langle 0|$ coherence), two pulses to stop it via a population state, and one pulse to reverse it to $|1\rangle\langle 0|$ (or $|0\rangle\langle 1|$, respectively). One possible Feynman diagram that can accomplish this feat is shown in Fig. 11.3(a). Others are possible as well.

Shown in Fig. 11.3(b) is the photon echo signal generated from this fifth-order pulse sequence for the antisymmetric stretch of the azide ion (N_3^-) in an ionic glass. The forces between the glass and the ion are extremely strong and yet static, which creates a 41 cm^{-1} linewidth that is almost solely due to inhomogeneous broadening. As a result, the photon echo is well localized in time (often in vibrational systems the echo is not well formed because the homogeneous linewidth is comparable to the inhomogeneity). Notice that the echo appears at $t_5 = 2t_2$. This is because there is a $|2\rangle\langle 0|$ coherence during t_2 and a $|1\rangle\langle 2|$ coherence during t_5, which evolves at half the frequency. Thus, the photon echo takes twice as long to form.

For comparison, the third- and fifth-order 2Q 2D IR spectra of azide in the ionic glass is shown in Fig. 11.4 [59, 61]. In the third-order spectrum, it is clear from their phase difference that there are two peaks (labeled 1–0 and 2–1 according to their coherences during t_3), but the spectra are so broad that they are severally overlapped (the actual anharmonic shift is 26 cm^{-1}). It appears as if the two peaks have different ω_2 frequencies, but we know that this is an illusion caused

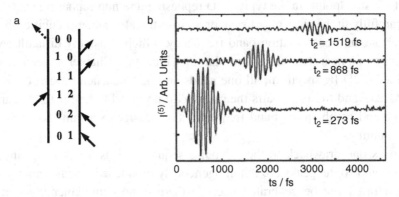

Figure 11.3 Fifth-order pulse sequence and echo. (a) One possible Feynman diagram that rephases a 2Q coherence. (b) Measured photon echo of the azide ion in an ionic glass for a rephased 2Q coherence using the pulse sequence that corresponds to the Feynman diagram in (a). Notice that the echo rephases at $t_5 = 2t_2$, because the 2Q coherence is being rephased by a 1Q coherence. The echo is very well formed because the azide vibrational mode is extremely inhomogeneously broadened. Adapted from Ref. [59] with permission.

11.2 Rephased 2Q pulse sequence

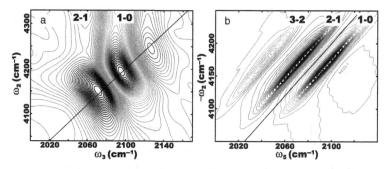

Figure 11.4 Comparison of (left) third- and (right) fifth-order 2Q 2D IR spectra of the azide ion in an ionic glass (real part spectra). The coherences observed during t_3 or t_5 are labeled at the top, respectively. Adapted from Ref. [59] with permission.

by overlap of the phase twisted peaks, since both Feynman pathways evolve with the same coherence during t_2. In fact, when these two peaks are well resolved in the fifth-order 2Q 2D IR spectra, which are extremely narrow due to the nearly perfect inhomogeneous broadening, they appear with the same ω_2 center frequency. The fifth-order peaks are still phase twisted, since only a rephasing spectrum was collected, but the phase twist is minimal since the corresponding non-rephasing spectrum would be very weak. A third peak is also observed in the fifth-order spectrum (3–2 coherence). This peak arises from a Feynman diagram that accesses the $\nu = 3$ quantum state (see Problem 11.1). The Feynman pathways for this pulse sequence for a coupled oscillator system have also been worked out, as have the fifth-order tensor elements from which the polarization dependence of the signal can be calculated [44].

We also note that the fifth-order signal reported here was not collected using five independent mid-IR pulses. Instead, three pulses were used with three different wavevectors, and the signal measured in the $\vec{k}_s = -2\vec{k}_1 + \vec{k}_2 + 2\vec{k}_3$ phase matching direction. As a result, the signal appearing in this direction necessarily originates when the first and third pulse each interact twice with the sample.

11.2.1 Cascaded signals

In the first fifth-order 2D IR studies, one of the largest uncertainties was whether or not cascaded signals contributed to the spectra. These concerns originate from problems researchers experienced in fifth-order Raman spectroscopy, where cascaded signals are much more intense than the true higher-order signals [14, 70, 115, 126]. Cascaded signals occur when a lower-order polarization emits a field that is then reabsorbed by the sample to generate a new coherence that radiates a second lower-order field. For example, consider a generic fifth-order pulse sequence that has five pulses. Any three of these pulses will generate a standard third-order electric field. This emitted field has some probability of being reabsorbed while it is exiting the sample cell. If it is, then it will generate a second third-order polarization in conjunction with the remaining two laser pulses. As a result, one gets a signal that is the convolution of two third-order signals. The second third-order signal, often called a *cascaded* signal, contains no new information, but has the same phase matching conditions as the desired fifth-order signal, so if it exists, it pollutes the spectra. One can always perform a concentration study to ascertain whether a signal is a cascade (which scales quadratically with the concentration, while the desired fifth-order signal scales linearly). Furthermore, cascades can be recognized by their spectra, which can be simulated using coupled Feynman diagrams just like one simulates other response functions [44, 59]. The cascading signal is opposite in sign relative to the corresponding fifth-order signal (originating from the $(-i)^n$-factor in the response function, see Eq. 3.61) and often has different spectral patterns. It turns out that in the mid-IR we do not need to be concerned about cascaded signals, because the long mid-IR wavelengths and the low concentrations of most samples cause cascaded signals to contribute less than a few percent to higher-order IR spectra. In fact, IR cascades have yet to be observed [43, 44, 59, 66].

To understand why cascading is small in IR spectroscopy, one can derive from Eq. C.1 that the ratio of the cascaded to the desired fifth-order signal scales as:

$$\frac{E^{(\text{cas})}}{E^{(\text{5th})}} = -\frac{\omega_0 l N \mu_{01}^2}{2nc\epsilon_0 \hbar} \frac{R^{(3)} R^{(3)}}{R^{(5)}} \qquad (11.2)$$

where $\omega_0 = 2\pi \nu_0$ is the center frequency, l the sample thickness, N the particle density, μ_{01} the transition dipole and n the index of refraction [59]. The response functions are already convoluted over the laser pulses, but dimensionless otherwise (we pulled out the transition dipole moments), hence the fifth-order response functions scale as:

$$R^{(5)} \propto \Delta t_{\text{pulse}}^5 \qquad (11.3)$$

with t_{pulse} the laser pulse duration. The cascading signal scales as:

$$R^{(3)} R^{(3)} \propto \Delta t_{\text{pulse}}^5 \Delta t_{\text{tot}}. \qquad (11.4)$$

Here, Δt_{tot} is the total free induction decay time including inhomogeneous dephasing, which reflects the duration of the light pulse emitted by the first third-order process and reabsorbed by the second third-order process. Plugging Eq. 6.50 into this relationship, we obtain, together with $\Delta \nu_{\text{tot}} = 1/\pi \Delta t_{\text{tot}}$ (see Eq. 4.10):

$$\frac{E^{(\text{cas})}}{E^{(\text{5th})}} = -\frac{3 \ln 10}{2} \frac{\Delta \nu_{\text{hom}}}{\Delta \nu_{\text{tot}}} \frac{R^{(3)} R^{(3)}}{R^{(5)}} \Delta A \qquad (11.5)$$

where $\Delta\nu_{\text{hom}}$ is the homogeneous width of the line originating from integration in Eq. 6.49 (the integration over a homogeneous Lorentzian line reveals $\frac{\pi}{2}\Delta\nu_{\text{hom}}\epsilon(\nu_0)$). In this step, we assumed that only molecules within the homogeneous linewidth share a common transition dipole moment. The last term, ΔA, is the optical density of the vibrational transition in OD. The response functions also contain orientational factors, which amount to 7/25 for a diagonal peak, if all pulses are polarized in parallel (Chapter 5). For an optical density of 0.2 OD, cascading will then contribute <10% to the total signal, which matches estimates based on simulated spectra [59, 66]. Thus, it will be a small contribution if indeed it is ever observed.

11.3 3D IR spectroscopy

Implementing a fifth-order pulse sequence is the first true step towards collecting 3D IR spectra. Of course, one can generate a 3D IR spectrum from the 2Q pulse sequence in Section 11.1 above by collecting data for all three delay times t_1, t_2 and t_3 (which may be a useful approach for studying some types of vibrational dynamics), but it is the only 3D IR spectrum that one can obtain using a third-order pulse sequence [182]. One can also generate "3D" spectra by collecting 2D spectra as a function of a third variable, such as the waiting time, which can either evolve as a coherence or a population depending on the phase matching conditions and the time ordering of the pulses [98]. This type of 3D plot cannot probe higher-lying vibrational states nor non-Gaussian frequency fluctuations because there are still only three interactions between the laser pulses and the sample. In the applications described here, we define 3D IR spectroscopy as the result of a 3D Fourier transform of a signal arising from at least a fifth-order resonant interaction between the infrared electric fields and the sample. In the case where each pulse interacts with the sample once, this requires that at least five field interactions generate the 3D IR signal. With five pulse interactions, higher vibrational states can be accessed than with three pulse interactions and for some Feynman pathways there exist up to two waiting times (t_2 and t_5). The signal strength scales with $|\mu_{ab}|^2|\mu_{bc}|^2|\mu_{cd}|^2 E^5$, where a, b, c and d are the eigenstates accessed by the laser pulses. With five pulses, all sorts of 3D IR experiments are possible. In this section, we present the only two sets of experiments that have been carried out so far.

11.3.1 Two-quantum 3D IR spectroscopy

The first 3D IR experiment used the fifth-order pulse sequence discussed in Section 11.2 to study the vibrational dynamics and couplings of several model systems: $W(CO)_6$ in hexane, azide in an ionic glass, and a metal dicarbonyl that

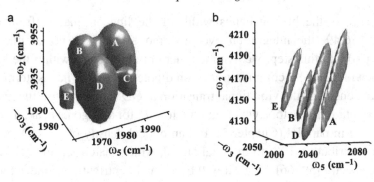

Figure 11.5 Absolute-value 3D IR spectra of (left) W(CO)$_6$ and (right) azide in an ionic glass using the pulse sequence corresponding to the fifth-order 2Q Feynman pathway shown in Fig. 11.3. W(CO)$_6$ is homogeneously broadened while azide is inhomogeneously broadened. Adapted from Ref. [43] with permission.

we refer to as Ir(CO)$_2$ [43]. These three systems were chosen as the first systems to study using 3D IR spectroscopy because their carbonyl vibrational eigenstates, vibrational dynamics and couplings are well understood, and thus the 3D IR spectra could be simulated from known parameters.

The 3D IR spectra of W(CO)$_6$ and azide are shown in Fig. 11.5. The spectra were collected using the pulse sequence shown in Fig. 11.3, in which the t_2, t_3 and t_5 times were measured and a 3D Fourier transform performed to give the final spectrum ($t_1 = t_4 = 0$). Thus, the 2Q coherences during t_2 are correlated to 1Q coherences during t_1 and t_5. For a single-oscillator system in which only one vibrational mode can be excited, there are seven Feynman pathways that contribute to the signal, giving rise to five "diagonal" peaks in the spectrum. All the peaks lie in a plane at the overtone frequency along t_2. The peak spacings in t_3 and t_5 are given by the differences in energy level spacing between $\nu = 0$–1, 1–2, and 2–3. For W(CO)$_6$, the peaks appear as spheres, because the vibrational mode has homogeneous dynamics. As a result, the response functions decay with the homogeneous lifetime T_2 along all three delays. For azide, the peaks are extremely elongated along the 3D diagonal defined as $\omega_2 = 2\omega_3 = 2\omega_5$. The "cigar" shape is how an inhomogeneously broadened lineshape appears in this style of 3D IR spectrum.

The 3D IR spectrum of Ir(CO)$_2$ is shown in Fig. 11.6. Since it consists of two coupled oscillators, there are two overtone states and one combination band. As a result, the 3D IR spectrum consists of three planes of peaks along the ω_2-axis (Fig. 11.3b). The "diagonal" peaks are a set of five peaks grouped together, just like for W(CO)$_6$ and azide (see Problem 11.1). They appear in the planes at the overtone frequencies, similar to the lines of peaks found in third-order 2Q spectroscopy (Fig. 11.2). The majority of the cross-peaks appear in the middle plane at the combination band frequency. Some cross-peaks also appear in the overtone

11.3 3D IR spectroscopy

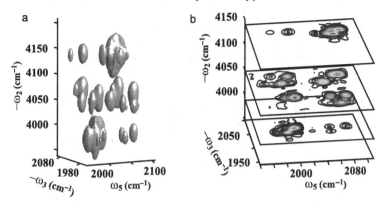

Figure 11.6 (a) Absolute-value 3D IR spectrum of an Ir(CO)$_2$ compound that exhibits two coupled carbonyl stretch modes using the pulse sequence corresponding to the fifth-order 2Q Feynman pathway shown in Fig. 11.3. (b) Also shown are three slices through the 3D spectrum at the overtone and combination band frequencies. Adapted from Ref. [43] with permission.

planes, which are a result of reaching vibrational states beyond those accessible to third-order pulse sequences. This 3D IR pulse sequence allowed the 3Q eigenstates to be correlated, which enabled the potential energy surface to be characterized with a higher level of accuracy [44].

11.3.2 Purely absorptive 3D IR spectroscopy

In 3D IR spectroscopy, there are many more possible ways to manipulate the vibrational motions of the molecule to extract the desired information. In previous chapters we have seen that it is advantageous to have absorptive peaks rather than the broad and phase twisted peaks caused by measuring a single response function. The same is true for 3D IR spectroscopy. Figure 11.7(a,b) shows the pulse sequence and Feynman diagrams, respectively, for a purely absorptive 3D IR spectrum. By a different choice of phase matching, this pulse sequence is designed to switch between coherence and population states with each pulse in the pulse sequence, in contrast to the 2Q pulse sequence discussed in the previous section. Times t_1, t_3 and t_5 are now coherence times with respect to which a 3D Fourier transformation is taken. As long as one is considering only one of these diagrams, the 3D lineshape is phase twisted (Fig. 11.7c), in analogy to rephasing and non-rephasing diagrams in 2D IR spectroscopy (e.g. Fig. 4.7). Summing together the four spectra collected using the four diagrams gives a purely absorptive 3D IR spectrum (Fig. 11.7d), which exhibits the narrowest possible lineshapes and retains the signs of the peaks.

Figure 11.8 shows two examples of purely absorptive 3D IR spectra [66]. Like the 2Q 3D IR spectra (Fig. 11.6), five peaks are observed because the pulse

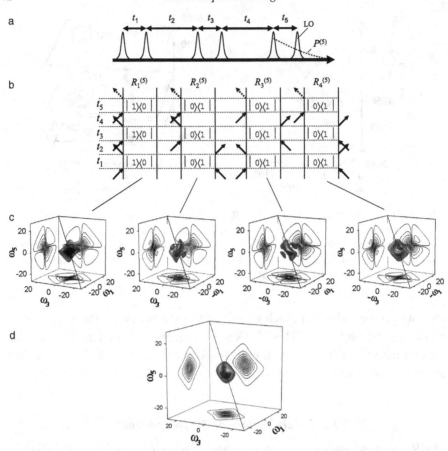

Figure 11.7 Purely absorptive 3D IR spectroscopy. (a) Pulse sequence, (b) fundamental Feynman diagrams for a two-level system, and (c) 3D lineshapes of the individual contributions, which include a phase twist in analogy to 2D IR spectroscopy (Fig. 4.7). (d) Sum over the four contributions, revealing a purely absorptive spectrum. Adapted from Ref. [66] with permission.

sequence reaches as high as the second overtone (see Problem 11.1). The 3D IR lineshapes also contain new information on the spectral diffusion process. In Chapter 7, we often approximated the distribution of the vibrational frequencies as Gaussian, which was a result of truncating the Cumulant expansion after second order. However, non-Gaussian lineshapes are often observed in infrared spectroscopy, and 3D IR spectroscopy provides a means of quantifying their dynamics [65, 79]. For instance, the lineshape of water is quite non-Gaussian. 3D IR spectroscopy can time-resolve the dynamics of the hydrogen-bond making and breaking that is the source of the asymmetry, and thus provide insight into the origin of the non-Gaussian frequency fluctuations (Fig. 11.8b).

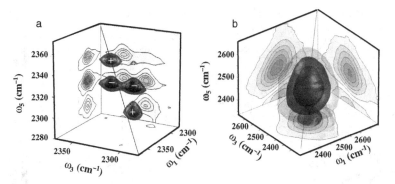

Figure 11.8 Purely absorptive 3D IR spectra of (a) the asymmetric stretch vibration of CO_2 dissolved in water and (b) the OD stretch vibration of HDO in H_2O. Adapted from Refs. [64, 66] with permission.

11.4 Transient 2D IR spectroscopy

A subclass of fifth-order nonlinear experiments is what has become known as transient 2D IR spectroscopy. That is, one disturbs a molecular system by initiating a photochemical or photophysical process using an actinic pump, typically with a short UV or Vis laser pulse, and then measures the 2D IR spectrum of the generated transient state after some delay time t_d. If the molecular system is undergoing a structural or electronic transition, then transient 2D IR spectroscopy reveals information about the structure and dynamics of the time-evolving species. Transient 2D IR spectroscopy is currently being applied to study protein dynamics [29, 110], chemical dynamics [8] and electron transfer at interfaces [196].

For example, consider the –C≡O stretch modes of Re(CO)$_3$(dmbpy)Cl. It has three antisymmetric stretch frequencies, which we label as a'(2), a'', and a'(1) in Fig. 11.9(a). A transient pump–probe (transient 1D IR) experiment is shown in Fig. 11.9(b). Upon electronic excitation with a 400 nm pulse, the compound undergoes a metal-to-ligand charge transfer (MLCT), which in turn changes the bond order of the –C≡O groups at the metal center and hence their vibrational frequency. The three modes appear negative, because they have bleached due to the electron transfer out of the ground state. Three positive bands appear, which are the new absorptions of the carbonyl stretches in the electronic excited state.

Figure 11.9(c) shows the corresponding transient 2D IR spectrum. Just as a transient 1D spectrum consists of ground state bleaches and new excited state absorptions, the transient 2D IR spectrum is composed of a bleach of the original ground state 2D spectrum (solid lines) overlaid with a new 2D IR spectrum of the excited state (dashed lines). In this molecule, the product peaks are weaker since the transitions are broader, which depends on the vibrational dynamics of the system and the

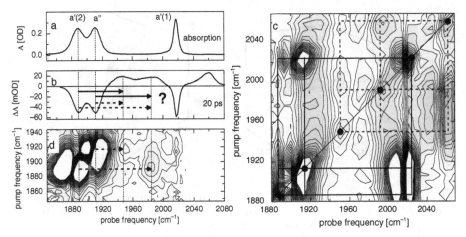

Figure 11.9 Modes of transient 2D IR spectroscopy. (a) Absorption spectrum in the electronic ground state. (b) Transient pump–probe (transient 1D-IR) spectrum 20 ps after UV excitation. It cannot be deduced if the bands shift according to the solid arrows or according to the dotted ones. (c) Purely absorptive transient 2D IR spectrum, showing a completely resolved set of diagonal and cross-peaks in the electronic ground state (solid grid) and the MLCT state (dashed grid). (d) A triggered exchange experiment determines that the shift occurs according to the dotted arrows. Adapted from Refs. [17, 18] with permission.

nature of the transient complex (electronic excitation versus conformation change, for example). The reactant and product 2D IR spectra overlap if the ground and excited state frequencies are similar, but the resolution is improved over standard transient IR spectroscopy. Moreover, we can implement other pulse sequences, such as the one described below.

One might assume that the three carbonyl modes retain the same energy ordering in the electronically excited state as they do in the ground state, so that upon visible excitation, the ground state absorption bands shift to those indicated by the solid arrows in Fig. 11.9(b). This question can be addressed by a variant of transient 2D IR spectroscopy, in which the time ordering of the UV pump pulse and the IR pump process is interchanged. This pulse sequence correlates the vibrational states in the electronic ground with those in the excited state, clearly showing that the two lower vibrational modes switch their ordering (Fig. 11.9d, dashed arrows). We call this pulse sequence triggered exchange 2D IR spectroscopy.

The time resolution of transient 2D IR spectroscopy depends upon the duration of the pulse sequence (which is set by the vibrational dynamics) and the time-scale of initiating the reaction. If the actinic pump is a femtosecond pulse, then it will not degrade the inherent time resolution of the spectrum. If a nanosecond pump is used, then it will set the time resolution of the experiment. Of course, signal averaging is required for each time delay associated with the 2D IR spectrum (or series of 2D IR spectra), which requires the experiment to be repeated many times over. Thus, replenishable samples are most useful, such as photoswitchable chromophores or temperature jump experiments, although irreversible processes can also be studied if enough starting material is available.

11.4.1 Transient 2D IR via rapid scanning

Another mode of collecting transient 2D IR spectra is what is being called rapid-scanning 2D IR, in which kinetics are monitored by continuously recording 2D IR spectra. In rapid-scanning mode, some sort of kinetic event is initiated by an actinic pump or some other means and then a series of 2D IR spectra are collected during the course of the kinetics. In this data collection mode, the time resolution will be set by how long it takes to collect a 2D IR spectrum. Thus, one wants to collect 2D IR spectra as quickly as possible. One way to do this is to use the pulse shaping method for data collection, so that a new pulse delay is measured with each laser shot with no time lost because of moving translation stages. In this manner, a typical 2D IR spectrum will be collected in about 0.4 s (assuming 1 kHz laser repetition rate, 100 time delays and four phase cycles). Of course, one laser shot per data point is insufficient signal-to-noise and so a running average is performed to smooth the data. The running average decreases the time resolution, but because it is done after data collection is complete, one can choose the best compromise between the two. The method is especially useful on systems which are not replenishable and/or cannot be triggered many times in succession with an actinic pump. Thus, it is currently finding use in processes like amyloid fiber kinetics [167, 172].

11.5 Enhancement of 2D IR spectra through coherent control

This entire textbook presents the formalism behind 2D IR spectroscopy using a perturbative approach. In perturbation theory, the change in the wavefunction created by each laser pulse is small enough that the Schrödinger equation does not need to be solved self-consistently. For example, when a laser pulse excites a molecule from $\nu = 0$ to 1, we calculate the change in the $\nu = 1$ coefficient c_1, but do not alter $\nu = 0$ coefficient of the wavefunction, c_0. Thus, we use Feynman diagrams to understand multidimensional infrared spectroscopies because they are a graphical representation of a perturbative expansion of the time-evolving density matrix.

It is also possible that the fields are strong enough and/or the molecules absorb enough light that the change in the wavefunction is so large that we cannot solve the Schrödinger equation perturbatively. NMR spectroscopy uses radio frequency pulses with variable lengths to actively control the populations and coherences of nuclear spins. In fact, the radio frequency pulses in NMR can saturate the transitions so that the wavefunction has equal coefficients $c_0 = c_1$, which is called a $\pi/2$ pulse. As a result, perturbation theory is no longer useful in NMR, and the product operator formalism is used instead. In either case, in the perturbative limit or with strong $\pi/2$ pulses, simple bookkeeping of the various coherence pathways is possible with the help of either Feynman diagrams or the product operator formalism, respectively. In fact, both formalisms lead to an almost isomorphous description despite the fact they work in opposite limits [160].

The regime in between these two limits is much less explored [130], but as the energy of IR OPAs increases, it will become relevant. There is experimental evidence that phase shaped mid-IR pulses can be used to control ground state vibrations and optimize 2D IR spectra [171]. Shown in Fig. 11.10 are the probe spectra following a pump that is (a) transform-limited (roughly Gaussian) and (b) shaped. The Gaussian pump creates a typical transient absorption spectrum that exhibits a bleach and a series of higher lying vibrational bands that come from transition to higher-lying vibrational states. In the case of the shaped pump, its envelope has been altered to optimize the intensity of the $v = 2$–3 over the $v = 1$–2 band. As a result, the 2–3 band is enhanced and the 1–2 band is depleted, indicating that the population of $v = 2$ is now larger than the population of $v = 1$. Population inversion indicates that the perturbative description of the light–matter interaction has broken down. Pulses like these might be used to enhance spectral features in 2D IR spectroscopy.

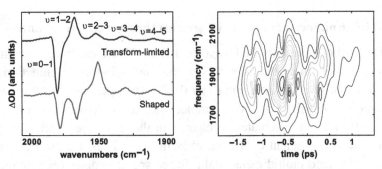

Figure 11.10 An example of how shaped mid-IR pulses can be used to selectively populate vibrational levels. A transient absorption spectrum of $W(CO)_6$ using (a) a transform limited and (b) a shaped pump pulse. The second bleach indicates that the population of $v = 2$ is now larger than the population of $v = 1$. Adapted from Ref. [171] with permission.

11.6 Mixed IR–Vis spectroscopies[2]

This book has focused mostly on 2D IR spectroscopy and to some extent on 2D Vis spectroscopy, but there exists the possibility of combining visible and infrared pulses into a single pulse sequence to generate 2D IR spectra. In fact, one of the first methods for measuring 2D IR spectroscopy used two infrared pulses and one visible Raman pulse [206]. In the literature, this method appears under various names: Doubly Vibrationally Enhanced (DOVE) 2D IR [206], Electron-Vibration-Vibration (EVV) coupling 2D IR [76] and IR-IR-Vis difference frequency generation (DFG) [25].

Consider again the energy levels of Fig. 4.9, but this time we make one of the time-ordered pulses of the 2D experiment interact resonantly or non-resonantly with the electronic states of the molecule (Fig. 11.11a). Out of the many Feynman diagrams possible, we focus on three pairs of diagrams (Fig. 11.11b), where two infrared (vibrational) interactions are followed by one electronic interaction. One of the vibrational transitions is resonant with the fundamentals of the vibrational spectrum (i or j), the other with overtones and combination modes (k). To describe a non-resonant electronic interaction, the antiresonant electronic terms normally neglected in the RWA have to be retained. As including both terms does not greatly change matters conceptually, we denote each resonant/antiresonant pair as diagram 1, 2 and 3. The transition dipole products $\mu_{ke}\mu_{ei}$ and $\mu_{je}\mu_{e0}$ are proportional to the Raman transition polarizabilities α_{ki} and α_{j0}. Diagrams involving forbidden Raman transitions are not shown here. The usual selection rules apply to each mode perturbed by the IR and Raman steps.

An experimentally straightforward method for measuring diagrams 1–3 as 2D IR spectra is achieved with two tuneable narrowband picosecond infrared pulses of frequencies ω_α (resonant with fundamental transitions i or j) and ω_β (resonant with an overtone $2i$ or $2j$ or a combination mode k) and a narrowband picosecond visible pulse ω_γ. The beams are focused into the sample in a box-CARS arrangement and generate a third-order signal of frequency $\omega_\delta = -\omega_\alpha + \omega_\beta + \omega_\gamma$ which can be spectrally isolated and measured as an intensity on a single-channel detector [206]. Measuring the signal as a function of infrared frequencies ω_α and ω_β generates a two-dimensional spectrum. Cross-peaks appear only when both ω_α is resonant with a fundamental band i and ω_β is resonant with the same fundamental's overtone or combination band k. The infrared pulses leave the system in the states $|k\rangle\langle i|$ (diagrams 1 and 2) or $|j\rangle\langle 0|$ (diagram 3), the Raman step acts on these states to remove a quantum of j (for an overtone we take $j = i$), returning the system to a population of i (diagrams 1 and 2) or to the ground state (diagram 3).

---

[2] We are very grateful to Paul Donaldson for writing this section.

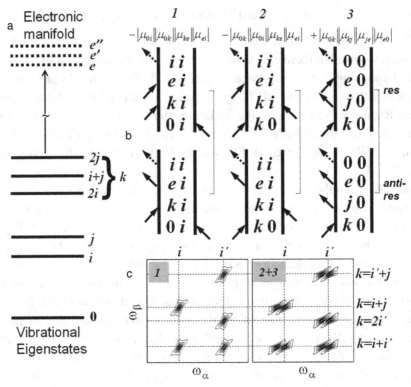

Figure 11.11 (a) Eigenstates for IR-Vis 2D IR experiments. (b) Three pairs of Feynman diagrams describing a DOVE/EVV 2D IR measurement. *res* and *anti-res* denote two visible frequency terms. The *anti-res* diagrams can be neglected when the visible pulse is resonant with an electronic state (due to the RWA). For an overtone signal, substitute j for i in diagram 3. Shown in (c) are schematic depictions of diagram 1 and 2+3 2D IR spectra, with two fundamentals i and i' making cross-peaks with three combination bands and an overtone.

Thus, the 2D spectrum maps overtone and combination bands to their contributing fundamental modes. Note that as of yet all experiments of this sort used homodyne detection, and the information on the phase of the emitted field is lost. This implies that a Fourier transform cannot be used to obtain spectral resolution, hence the method works with spectrally narrowband pulses in a quasi-frequency domain.

Although probing forbidden infrared transitions (overtones or combination modes) reduces the size of the nonlinear response, this is compensated by the fact that for $\omega_\alpha \neq \omega_\beta$ the signal ω_δ can be measured completely background-free and in the visible part of the spectrum. Detectors in the visible spectral range are many orders of magnitude more sensitive than mid-IR detectors. In a typical experiment, the time ordering of the infrared pulses can be used to uniquely select diagram 1

(ω_α = pulse 1) or diagrams 2+3 (ω_β=pulse 1). Signals from diagrams 2+3 destructively interfere, however the cross-peak positions for diagram 3 are sensitive along the ω_α coordinate to the mechanical anharmonic shift of the combination band (see Problem 11.10), resulting in the diagram 2+3 cross-peaks being split by the k band's mechanical anharmonic shift, as depicted in Fig. 11.11(c). In contrast, a diagram 1 2D IR spectrum shows single cross-peaks, simplifying the 2D spectrum and allowing for the appearance of cross-peaks from combination bands with negligible mechanical anharmonic shift. Such peaks arise from nonlinear dependence of the transition dipole with normal mode coordinates (the so-called electrical anharmonicity) and can be identified by their absence in the diagram 2+3 2D IR spectrum.

Applications of IR-IR-Vis 2D IR spectroscopy to aromatic and aliphatic compounds demonstrate that like the fully infrared methods discussed in this book, the method provides information about combination bands, overtones and Fermi resonance bands that is difficult and time-consuming to deduce using linear IR absorption or Raman spectroscopy [46, 47]. Electronically resonant experiments add further information and improve sensitivity; studies on retinal in bacteriorhodopsin show a 10^4 increase in 2D IR signal size compared with off-resonance conditions. The immunity of the method to pump beam scatter also makes imaging applications possible [57]. The electrical anharmonicity measurements possible through diagram 1 spectra provide a useful direct probe of nonbonded electrostatic interactions. Molecular complexes exhibit "interaction" cross-peaks in diagram 1 2D IR spectra which report on the geometry of the complex [76].

11.7 Some of our dream experiments

As we said in the Introduction, we have written this book with the expectation that 2D IR spectroscopy will grow in its range of applications and dimensionality. To end our book, we include a few briefly sketched ideas that we hope multidimensional IR spectroscopy will someday be able to accomplish.

11.7.1 Study of molecular interfaces

Molecular interfaces are a fascinating subject. In biology, solvation is responsible for the low-frequency fluctuations of proteins that allow large-scale structural motions responsible for function. Chemical denaturation is a solvation effect. In electron transfer, solvation is critical to reducing the energy gap between donor and acceptor. The list is nearly endless. Currently, the only optical spectroscopy that is applicable to interfaces is sum-frequency generation (SFG), which can be extended to surface-selective 2D IR spectroscopy [20]. However, SFG spectroscopy can really only be applied to bulk interfaces, and not to microscopic interfaces such as the protein–water interface on the inside of an ion channel.[3] One dream experiment is to utilize multidimensional IR spectroscopy to study interfaces like these. Consider an ion channel. Are the structure and dynamics of water on the inside of an ion channel the same as in the bulk? In gramicidin, an antibiotic proton channel, the answer is certainly "no," since the channel is so narrow that only a single water wire can form. So, how does one study water like this? The water inside the channel will have about the same frequency as water outside the channel, and so be swamped by signal from the bulk. We believe that the way to get at this water is to perform a 3D IR experiment in which the protein vibrational modes are correlated to the water modes. Such a spectrum could be generated with a fifth-order pulse sequence in which two pulses are resonant with the isotope label and the other four are resonant with the water. The result would be a 2D IR spectrum of just the water in the channel. The inhomogeneity and frequency fluctuations could then be measured and compared to simulations. To probe a specific region of the channel, one could use a protein with an isotope label that was located on the inside of the channel. Is such a signal measurable? The signal strength will depend on the transition dipole strength and the coupling. The transition dipole strength dependence is $|\mu_{protein}|^2|\mu_{water}|^4$. Because the protein mode is a strong absorber, this term is several orders of magnitude larger than that of water, and it is possible to measure the 3D IR spectrum of water (see Fig. 11.8b) [64]. Thus, it will be the coupling strength between the water and the protein that dictates the difficulty of this experiment. Water alters the amide I lineshapes of proteins and ions alter the linear spectrum of water, which are good indicators that the two are coupled enough to see an effect, and so the coupling might be appreciable. This type of 3D experiment is routinely performed in NMR spectroscopy to isolate particular signals, but is difficult to apply to solvation because the dynamics of water are too fast for them to catch. And, of course, NMR is even more difficult to apply to membrane proteins. Thus, if possible, a 2D IR experiment like this would have a tremendous impact throughout a range of scientific topics.

11.7.2 Beyond pair potentials

Much of our understanding of the structure and dynamics of condensed-phase molecular systems originates from MD computer simulations. These simulations are based on empirical force fields, which in almost all cases are constructed as pairwise. That is, the interaction strength of molecules A and B is independent of

---

[3] The interfaces have to be about as big as the laser spot size.

whether or not molecule A is in contact also with another molecule C. But pairwise potentials cannot always be a good approximation since the interaction of molecules is mediated by complicated correlations of their electronic structure. For example, the affinity of a water molecule to accept a hydrogen bond is reduced upon donating a hydrogen bond, since the latter changes the electron density at the oxygen. While such effects can be studied quite accurately in the gas phase with the help of high-level *ab initio* calculations, their impact in the complex environment of the condensed phase is largely unknown.

This subject is another that might be improved by 3D IR spectroscopy. We envision measuring a 3D cross-peak between molecules A, B, and C (which could for example be HOD and HOT in H_2O) and compare its strength with the two 2D IR cross-peaks of the $A-B$ and $A-C$ pairs. The theoretical background of this idea has not been worked out, yet one might expect that the 3D cross-peak strength is just a simple product of the 2D cross-peak strengths, if pair potentials dominate, whereas it deviates from such a simple relationship if multi-body terms become strong.

11.7.3 Molecular movies

One of the greatest potentials of 2D IR spectroscopy is its capability to time-resolve structural change over a wide range of time-scales. A dream of ours is that the time resolution of 2D IR spectroscopy will be routinely used in the future to make movies of chemical reactions, folding proteins, and moving electrons. The major roadblocks to achieving this dream are signal strength and spectral congestion. Signal strengths are certain to improve in the future as new technologies for generating intense and high-repetition-rate mid-IR lasers are developed. As for congestion, we already know how to isolate specific vibrational modes using isotope labeling. For example, one can use pairs of isotope labels to precisely monitor secondary structures. So, to make a movie of an α-helix folding, one could isotope-label residues i and $i+3$, which are strongly coupled when the helix forms $\beta_{i,i+3} \approx -5$ cm^{-1} [53], and then initiate folding with a temperature jump or some other actinic pump. By investigating a series of pairwise combinations of double isotope labels, one could compile a picture of the folding process with unprecedented detail. Movies like this would be difficult to attain with any other technique for systems that exhibit sub-millisecond dynamics, form aggregates, or are involved with membranes.

But even isotope labeling is sometimes not enough to overcome spectral congestion, particularly in transient 2D IR spectroscopy because the spectra before and after initiation often overlaps. However, spectral congestion could be reduced even further with the appropriate choice of pulse sequence. The two-quantum pulse sequence discussed in Sect. 11.2 has better spectral resolution than conventional

2D IR spectra, at least it would if it were possible to collect absorptive 2Q spectra. Thus, our dream is to implement fifth-order pulse sequences that create line-narrowed 2Q spectra (Section 11.2) and to do so in a transient experiment. Such an experiment would help alleviate much of the overlap between reactants and products. It would be technically demanding, since the experiments would be seventh-order, but we do not believe that these are out of reach. Compared to a standard third-order experiment, fifth-order spectra are typically 10 times smaller and the actinic pump will reduce the signal further (although some initiation methods alter the entire ensemble and so will not decrease the signal strength). Thus, our dream experiment is currently feasible on many systems and will in the future become possible on others as well.

Exercises

11.1 In the 3D spectra presented above, five "diagonal" peaks are observed both for the fifth-order 2Q and the purely absorptive pulse sequences. Use Feynman diagrams to explain the differences.

11.2 Write the fifth-order response function for the Feynman diagram in Fig. 11.3.

11.3 The 2Q 2D IR spectra in Section 11.2 are not absorptive. Describe how one could generate an absorptive 2Q spectrum in order to implement our dream experiment in Section 11.7.3.

11.4 Draw the Feynman diagrams and predict the spectrum for a transient 2D IR experiment in which the actinic pump excites the molecule to a stable electronic state. Do the same for a pulse sequence in which the actinic pump appears after k_1 and k_2, but before k_3.

11.5 Draw all possible Feynman diagrams for sequential and parallel third-order cascading signals for (a) two-quantum 3D IR spectroscopy and (b) purely absorptive 3D IR spectroscopy. A sequential cascade is one in which the first three pulses emit a third-order field that is reabsorbed by the sample. A parallel cascade is one in which both third-order signals evolve simultaneously.

11.6 Draw the Feynman diagrams for a fifth-order Raman experiment, as well as those of two cascaded third-order Raman processes. Verify that a forbidden two-quantum transition is necessary for the fifth-order Raman process, but not for the third-order cascading process, which is why it is very difficult to measure the fifth-order signal in the Raman case. Contrast this to the IR case.

11.7 Visible dyes absorbed to semiconductor surfaces are often used in solar cells because when they absorb light they inject an electron into the semiconductor that is collected to make electricity. The efficiency at which they inject electrons is related to the coupling of the electronic orbitals of the dye to the semiconductor, among other factors. Strong coupling sometimes manifests itself as a shift in the electronic absorption spectrum. In a recent 2D IR experiment [196], it was found that one particular dye appears to bind to the surface in three different conformations because it exhibits three infrared bands when there is normally just one. Moreover, each band has a different electron transfer rate, as measured with transient 2D IR spectroscopy that excites the dye with 400 nm light. Design a 2D Vis/IR experiment that measures the electronic absorption bands of these three conformations that might reveal why the electron transfer rates differ.

11.8 The shaped pump pulse in Fig. 11.10 optimizes the population of the $\nu = 2$ state as compared to $\nu = 1$. Explain how a shaped pulse might be used to optimize the signal strength of a 2Q experiment.

11.9 It is more common in the EVV/DOVE 2D IR community to use so-called Wave Mixing Energy Level (WMEL) diagrams instead of Feynman diagrams to visualize various pathways [57]. WMEL diagrams depict the field interactions with energy levels of the system in a similar (but not identical) manner to the arrows used in Fig. 4.9. For a WMEL diagram, the interaction ordering goes from left to right. Bra side transitions have solid arrows and ket side transitions have dotted arrows. The inward (outward) arrows of a Feynman diagram are represented as upward (downward) arrows on a WMEL diagram. Using these rules, translate the Feynman diagrams of Fig. 11.11.

11.10 Demonstrate that compared with pathways 1 and 2, pathway 3 cross-peaks are lower in frequency along ω_α by the mechanical anharmonic shift of the state k.

Appendix A
Fourier transformation

In the following, we will summarize some of the properties of the Fourier transformation which will be needed explicitly for the purpose of 2D IR spectroscopy. A more comprehensive summary is given, e.g. in Ref. [152]. The Fourier transformation of a time domain function $f(t)$ is defined as:

$$f(\omega) = \mathrm{FT}(f(t)) \equiv \frac{1}{\sqrt{2\pi}} \int_{-\infty}^{\infty} f(t) e^{i\omega t} dt \qquad (A.1)$$

and its inverse:

$$f(t) = \mathrm{FT}^{-1}(f(\omega)) \equiv \frac{1}{\sqrt{2\pi}} \int_{-\infty}^{\infty} f(\omega) e^{-i\omega t} d\omega. \qquad (A.2)$$

The Fourier transformation is bijective:

$$f(t) = \mathrm{FT}^{-1}(\mathrm{FT}(f(\omega))) \qquad (A.3)$$

i.e. the principal information content of both the time domain $f(t)$ and the frequency domain $f(\omega)$ version of a function are the same. Even though $f(t)$ and $f(\omega)$ are of course different functions, one commonly writes the same letter f and specifies implicitly by using t or ω as an argument whether one is referring to the time domain or frequency domain version of the function. We will in general omit the prefactor $1/\sqrt{2\pi}$ since we use proportionalities in essentially all equations anyway.

A 2D Fourier transformation is defined as:

$$f(\omega_3, \omega_1) \equiv \int_{-\infty}^{\infty} \int_{-\infty}^{\infty} f(t_3, t_1) e^{i\omega_1 t_1} e^{i\omega_3 t_3} dt_1 dt_3 \qquad (A.4)$$

and its inverse:

$$f(t_3, t_1) \equiv \int_{-\infty}^{\infty} \int_{-\infty}^{\infty} f(\omega_3, \omega_1) e^{-i\omega_1 t_1} e^{-i\omega_3 t_3} d\omega_1 d\omega_3. \qquad (A.5)$$

Fourier transformation

These are just two subsequent loops of 1D Fourier transformations for frequencies ω_1 and ω_3, respectively (the order is irrelevant), hence it is sufficient to discuss the properties of 1D Fourier transformations only.

For most purposes of this book, the time domain functions are single sided and nonzero for $t \geq 0$ only, which is a consequence of causality. Furthermore, experimental data are real-valued in the time domain, in which case we obtain as a symmetry for the frequency domain data:

$$f(-\omega) = f^*(\omega), \tag{A.6}$$

i.e. it is sufficient to consider the positive frequency range only.

Some important properties of the Fourier transformation are:

Linearity:	$f(t) + g(t)$	$\Leftrightarrow$ $f(\omega) + g(\omega)$	contributions are additive in both the frequency and time domains.
Time shifting:	$f(t - t_0)$	$\Leftrightarrow$ $f(\omega)e^{i\omega t_0}$	a shift in time-zero causes a frequency-dependent phase (Section 9.5.6).
Frequency shifting:	$f(t)e^{-i\omega_0 t}$	$\Leftrightarrow$ $f(\omega - \omega_0)$	multiplying a response function by $e^{-i\omega_0 t}$ (such as used in the rotating frame), causes a frequency shift (Section 9.3.3).
Convolution theorem:	$f(t) \otimes g(t)$	$\Leftrightarrow$ $f(\omega) \cdot g(\omega)$	see, e.g. inhomogeneous broadening (Eq. 2.72).
	$f(t) \cdot g(t)$	$\Leftrightarrow$ $f(\omega) \otimes g(\omega)$	see, e.g. window functions (Sect. 9.5.7).

Some Fourier transformations of important functions are:

$$f(t) \Leftrightarrow f(\omega) \tag{A.7}$$

$$e^{-i\omega_0 t} \Leftrightarrow \delta(\omega - \omega_0) \tag{A.8}$$

$$\cos(\omega_0 t) \Leftrightarrow \frac{1}{2\sqrt{2\pi}} (\delta(\omega + \omega_0) + \delta(\omega - \omega_0)) \tag{A.9}$$

$$\delta(t - t_0) \Leftrightarrow \frac{1}{\sqrt{2\pi}} e^{i\omega t_0} \tag{A.10}$$

$$\delta(t - t_0) + \delta(t + t_0) \Leftrightarrow \sqrt{\frac{2}{\pi}} \cos(\omega t_0) \tag{A.11}$$

$$e^{-t^2/2\sigma^2} \Leftrightarrow \sigma^2 e^{-\sigma^2 \omega^2/2} \tag{A.12}$$

$$e^{-t/\tau} \Theta(t) \Leftrightarrow \frac{1}{\sqrt{2\pi}} \frac{1}{i\omega - 1/\tau} \tag{A.13}$$

$$e^{-t/\tau}e^{-i\omega_0 t}\Theta(t) \Leftrightarrow \frac{1}{\sqrt{2\pi}}\frac{1}{i(\omega-\omega_0)-1/\tau} \quad (A.14)$$

$$\Pi(t/t_0) \Leftrightarrow \sqrt{\frac{2}{\pi}}\frac{\sin(t_0\omega/2)}{\omega} \quad (A.15)$$

where $\delta(\omega)$ is the δ-function (i.e. $\delta(\omega) = 0$ for $\omega \neq 0$, $\delta(\omega) = \infty$ for $\omega = 0$ with $\int_{-\infty}^{\infty}\delta(\omega)d\omega = 1$), $\Theta(t)$ the Heaviside step function (i.e. $\Theta(t) = 0$ for $t < 0$ and $\Theta(t) = 1$ for $t \geq 0$), and $\Pi(t)$ the rectangle function (i.e. $\Pi(t) = 1$ for $|t| \leq 1/2$ and $\Pi(t) = 0$ for $|t| > 1/2$).

A.1 Sampling theorem, aliasing and under-sampling

As long as we deal with analytical expressions for the response functions, the integrals in Eqs. A.1 and A.2 are the method of choice to calculate the Fourier transformation (in case they can be calculated). However, in a real experiment, and also for any numerical simulation, one will obtain the data on some finite time grid with N data points and with equidistant step size Δt:[1]

$$t_n = n \cdot \Delta t \quad \text{with } n = 0, 1, 2, \ldots, N-1. \quad (A.16)$$

The maximum frequency one may detect with a given step size Δt is:

$$\frac{\omega_N}{2\pi} = \frac{1}{2\Delta t} \quad (A.17)$$

which is called the Nyquist critical frequency. For a sine wave, this would be one point at, e.g. the peak, and one point at the bottom. If the signal contains frequency components above this cutoff frequency, it cannot be distinguished from one below (Fig. A.1a). In other words, any frequency component above the cutoff frequency will be folded back into the frequency window of interest (Fig. A.1b), an effect called *aliasing*. Once aliasing has occurred, there is no way to distinguish the aliased frequency components from real frequency components. When choosing a step size Δt, one must know for sure that the corresponding frequency spectrum is below the Nyquist critical frequency. This can be achieved by, e.g. knowing the frequency spectrum of the laser, or by explicitly putting a cutoff filter into the laser beam. One should recognize that it is not the center frequency of the laser spectrum or the signal spectrum which determines the maximum step-size, but the most-blue frequency.

---

[1] We discourage using non-equidistant time steps resulting, e.g. from a poor delay stage. None of the discussion below, in particular the Nyquist sampling theorem, would apply for non-equidistant time steps.

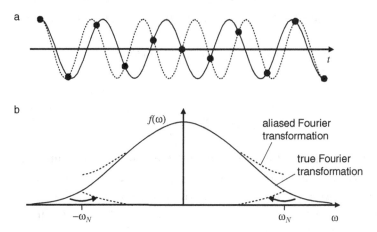

Figure A.1 Aliasing: (a) two sine waves with frequencies slightly below the Nyquist critical frequency (i.e. $0.9\omega_N$, solid line) and slightly above the Nyquist critical frequency (i.e. $1.1\omega_N$, dotted line). For the sampling points (dots), these two sine waves are indistinguishable. (b) Any frequency component above the Nyquist critical frequency is folded back into the frequency window of interest.

A.2 Discrete Fourier transformation

As we put only the knowledge of N time points into the Fourier transformation, it is reasonable to assume that we will produce N frequency points as an output. For the purpose of our discussion, these frequency points will be:

$$\frac{\omega_n}{2\pi} = \frac{n}{N\Delta t} \quad \text{with } n = -\frac{N}{2}+1, \ldots, \frac{N}{2}. \tag{A.18}$$

i.e. an equidistant frequency grid with spacing $\Delta\omega/2\pi = 1/(N\Delta t)$. We assume for the following that N is an even number. The larger the total scanned time, $N\Delta t$, the larger the frequency resolution of the result.

With that in mind, we can approximate the integral in Eq. A.1 as a sum:

$$f(\omega_n) = \frac{1}{\sqrt{2\pi}} \int_{-\infty}^{\infty} f(t)e^{i\omega_n t}dt \approx \frac{\Delta t}{\sqrt{2\pi}} \sum_{k=0}^{N-1} f(t_k)e^{i\omega_n t_k}$$

$$= \frac{\Delta t}{\sqrt{2\pi}} \sum_{k=0}^{N-1} f(t_k)e^{2\pi i kn/N}. \tag{A.19}$$

We write the inverse Fourier transform in a symmetric manner:

$$f(t_k) = \frac{\Delta\omega}{\sqrt{2\pi}} \sum_{n=0}^{N-1} f(\omega_n)e^{-2\pi i kn/N}. \tag{A.20}$$

One easily sees that both Eqs. A.19 and A.20 are periodic, i.e.

$$f(t_k + jN\Delta t) = f(t_k) \tag{A.21}$$

and

$$f(\omega_n + jN\Delta\omega) = f(\omega_n) \tag{A.22}$$

with an integer j. Interestingly, discretizing the Fourier transformation renders the resulting spectrum, and implicitly also the input time data, periodic. The periodicity explains the missing point in Eq. A.18 with $n = -N/2$, which is just identical to the one with $n = +N/2$. It also justifies *a posteriori* why the sum in Eq. A.20 runs from 0 to $N - 1$ rather than from $-N/2 + 1$ to $N/2$.

The implicit periodicity has important consequences for the interpretation of discrete Fourier transformation data. For the purpose of 2D IR spectroscopy, the natural way to arrange the input and output data would be:

| 0 | Δt | ... | ... | ... | ... | $(N-2)\Delta t$ | $(N-1)\Delta t$ |

$$\downarrow FT$$

| $(-N/2+1)\Delta\omega$ | $(-N/2+2)\Delta\omega$ | ... | ... | ... | ... | $(N/2-1)\Delta\omega$ | $N/2\cdot\Delta\omega$ |

Unfortunately, most FT algorithm (*Numerical Recipes* [152], *Mathematica*) arrange the data like:

| 0 | Δt | ... | ... | ... | ... | $(N-2)\Delta t$ | $(N-1)\Delta t$ |

$$\downarrow FT$$

| 0 | $\Delta\omega$ | ... | ... | ... | ... | $(N-2)\Delta\omega$ | $(N-1)\Delta\omega$ |

which, owing to the periodicity, is the same as:

| 0 | Δt | ... | ... | ... | ... | $(N-2)\Delta t$ | $(N-1)\Delta t$ |

$$\downarrow FT$$

| 0 | $\Delta\omega$ | ... | $N/2\cdot\Delta\omega$ | $(-N/2+1)\Delta\omega$ | ... | $-2\Delta\omega$ | $-\Delta\omega$ |

Hence, one will have to rotate the output vector by $N/2$ to get the desired form of the spectrum with the frequency origin $\omega = 0$ being in the middle (*Matlab* provides a function *fftshift* to rearrange the data). Also, it is a good idea to symmetrize the

A.2 Discrete Fourier transformation

output vector by either deleting the $+N/2$ point, or by adding a $-N/2$ point that is identical to the $+N/2$ point.

Note that the time data are implicitly periodic as well. If negative times are relevant, the time vector might be reinterpreted as:

| 0 | Δt | ... | $N/2 \cdot \Delta t$ | $(-N/2+1) \cdot \Delta t$ | ... | $-2\Delta t$ | $-\Delta t$ |

$$\downarrow FT$$

| $(-N/2+1)\Delta\omega$ | $(-N/2+2)\Delta\omega$ | ... | ... | ... | ... | $(N/2-1)\Delta\omega$ | $N/2\cdot\Delta\omega$ |

Direct computation of Eq. A.19 scales quadratically with the number of time points. It turns out that the computational effort can be improved significantly by intelligently rearranging the terms in the sum, which leads to the Fast Fourier Transform (FFT) algorithm. The most elementary versions of the FFT algorithm (see, e.g. *Numerical Recipes* [152]) require the number of time points to be a power of 2 ($N = 2^m$), in which case the computational cost scales like $N \ln_2 N$, but more advanced algorithms (*Mathematica, Matlab*) accept any number for N.

Appendix B
The ladder operator formalism

If we write the Hamiltonian of a harmonic oscillator in dimensionless momentum and position coordinates:

$$H = \hbar\omega \left(\frac{\hat{p}^2}{2} + \frac{\hat{q}^2}{2} \right) \quad \text{(B.1)}$$

and define the operators

$$b = \frac{1}{\sqrt{2}} (\hat{q} + i\hat{p})$$

$$b^\dagger = \frac{1}{\sqrt{2}} (\hat{q} - i\hat{p}) \quad \text{(B.2)}$$

with the commutator:

$$[b, b^\dagger] = 1 \quad \text{(B.3)}$$

then one can easily show that the Hamiltonian can be rewritten as:

$$H = \hbar\omega \left(b^\dagger b + \frac{1}{2} \right). \quad \text{(B.4)}$$

The operators $b^\dagger$ and b are called *creation* and *annihilation* operators, respectively, since they climb up or down the harmonic ladder:

$$b^\dagger \varphi_n = \sqrt{n+1}\, \varphi_{n+1}$$
$$b \varphi_n = \sqrt{n}\, \varphi_{n-1}. \quad \text{(B.5)}$$

Here, φ_n is the nth eigenstate of the harmonic oscillator. The operator $N \equiv b^\dagger b$ is called the *number operator* since its eigenvalues are the number of quanta:

$$b^\dagger b \varphi_n = n \varphi_n. \quad \text{(B.6)}$$

This leads to the common expression for the eigenvalues of the harmonic oscillator:

$$H\varphi_n = \hbar\omega\left(b^\dagger b + \frac{1}{2}\right)\varphi_n = \hbar\omega\left(n + \frac{1}{2}\right)\varphi_n. \tag{B.7}$$

The ladder operator formalism is extremely useful since one can directly translate a Taylor-expanded potential energy surface into an intuitive physical picture (see Sections 6.1 and 6.8 for examples).

We often need to use position ($\hat{q}$) or momentum ($\hat{p}$) operators when evaluating integrals. By inverting Eq. B.2, these are

$$\begin{aligned}\hat{p} &= \frac{1}{\sqrt{2}}(b + b^\dagger) \\ \hat{q} &= \frac{i}{\sqrt{2}}(b - b^\dagger).\end{aligned} \tag{B.8}$$

Appendix C
Units and physical constants

In this appendix we list some of the commonly needed physical constants in spectroscopy. We also calculate the emitted field intensity directly from the third-order response functions for standard experimental conditions and molecular properties. The purpose of this calculation is not to obtain a highly accurate result, but to serve as a (somewhat interesting) exercise of units. In the following, we use units of meters-kilograms-seconds (SI units) unless otherwise noted.

C.1 Physical constants

Constant	Symbol	Value
Avogadro's number	N_A	6.022×10^{23} mol^{-1}
Planck's constant	h	6.626×10^{-34} J s
	$\hbar$	1.055×10^{-34} J s
Speed of light in vacuum	c	2.998×10^8 m s^{-1}
Permittivity of a vacuum	ε_0	8.854×10^{-12} C^2 s^2 kg^{-1} m^{-3} (C^2 N^{-1} m^2)

C.2 Units of common physical quantities

Physical quantity	Unit
Force	N = kg m s^{-2}
Energy	J = N m = kg m^2 s^{-2}
Power	W = J s^{-1} = kg m^2 s^{-3}
Electric charge	C = A s
Electric potential	V = J A^{-1} s^{-1}
Electric field	N C^{-1} = V m^{-1}
Macroscopic polarization	C m^{-2}
Intensity	J s^{-2} m^{-1}
Frequency	Hz = s^{-1} = 3.336×10^{-11} cm^{-1} (non-SI unit)
Dipole	D (non-SI unit) = 3.356×10^{-30} C m

C.3 Emitted field $E^{(3)}_{sig}$

To calculate the emitted electric field, we need the polarization. Previously we wrote $E^{(3)}_{sig} \propto i P^{(3)}$ (Eq. 2.63). With units, this proportionality becomes the equality

$$E^{(3)}_{sig} = \frac{i\omega\ell}{2nc\epsilon_0} P^{(3)} \tag{C.1}$$

where $\omega(s^{-1})$ is the frequency of the polarization, ℓ(m) the thickness of the sample, n the index of refraction, $c = 2.99 \times 10^8$ (m/s) the speed of light, and $\epsilon_0 = 8.85 \times 10^{-12}$ C^2/Nm^2 the permittivity of free space (1 N = 1 kg m/s^2). The units of $E^{(3)}_{sig}$, or any electric field for that matter, are N/C or equivalently V/m and the units for $P^{(3)}$ are C/m^2.

The polarization is calculated from the response functions, which we have written previously as $P^{(3)} \propto i^3 R^{(3)}$ (Eq. 2.58), which becomes the equality

$$P^{(3)} = N \left(\frac{i}{\hbar}\right)^3 \mu^4 R^{(3)} E^3 \tag{C.2}$$

where N (molecules m^{-3}) is the number density, μ(C m) is the transition dipole strength, $R^{(3)}$(s^3) is the molecular response function with units of time to the third power because it is a third-order response function, and E is the electric field strength of each of the three laser pulses.

To determine $E^{(3)}_{sig}$, one of the quantities we need to know is E. Let us assume that each of the three laser pulses is 100 fs and 0.5 μJ, and that they are focused into a spot size of 100 μm diameter at the sample. We need to calculate the intensity, I (J/sm^2), which is:

$$I = \frac{0.5 \times 10^{-6} \text{ J}}{(100 \times 10^{-15} \text{ s})\pi(50 \times 10^{-6} \text{ m})^2} = 6.3 \times 10^{14} \frac{\text{J}}{\text{sm}^2}. \tag{C.3}$$

From the intensity we get the electric field strength using

$$I = \frac{1}{2} n\epsilon_0 c E^2 \tag{C.4}$$

which gives $E = 6.0 \times 10^8$ V/m or N/C.

We also need to know some properties of the sample, like N at the laser focus. Using a typical sample concentration of 50 mM, we get

$$N = 50 \times 10^{-6} \frac{\text{mol}}{\text{cm}^3} \times \left(\frac{100 \text{ cm}}{\text{m}}\right)^3 \times \frac{6.022 \times 10^{23}}{\text{mol}} = 3.0 \times 10^{25} \text{ m}^{-3}. \tag{C.5}$$

A typical transition dipole strength might be $\mu = 0.3$ D. As for $R^{(3)}$, the molecule might have a homogeneous dephasing time of 1 ps, but the generated polarization is the convolution of the response with the pulses (Eq. 2.74). So, we approximate each dimension of $R^{(3)}$ as the pulse duration, which gives $R^{(3)}=(10^{-13}$ s$)^3$. Finally, let us consider the fundamental frequency to be $\omega = 2050$ cm$^{-1} = 6.15 \times 10^{13}$ s^{-1} and the path length to be $\ell = 50$ µm. Plugging all this into Eq. C.2 and then Eq. C.1, along with $\hbar = 1.05 \times 10^{-34}$ Js, we get $E_{sig}^{(3)} = 1.9 \times 10^7$ N/C. Thus, the emitted signal electric field is 3% that of the input electric field strength of the laser pulses. Using Eq. C.4 (which gives 6.5×10^{11} J/ms^2), multiplying by the 100 fs duration of the emitted field and the area of the focal spot, we calculate that the energy of the emitted field is 0.5 nJ, as compared to the incident pulses which were each 500 nJ.

Is this number accurate? In a four-wave mixing geometry, researchers usually do not bother to report signal strengths because one has to calibrate the detectors to get an absolute intensity. In a pump–probe geometry, it is quite common to report signal strengths in terms of changes in optical density of the probe upon excitation with a pump, often called ΔOD. It is easy to calculate because the probe provides a convenient reference. A typical sample might have an absorbance of 0.5 OD that transmits half of the probe intensity while it is quite common in the mid-IR to measure ΔOD of 0.1–1 mOD. Thus, the signal strength is on the order of 0.1% relative to the probe intensity, which is about what we estimated above for the pulse energies.

Appendix D
Legendre polynomials and spherical harmonics

It is quite common when working in spherical coordinates to need the associated Legendre polynomials, $P_{\ell,m}(x)$. The first few are

$P_{0,0}(\cos\theta) = 1$
$P_{1,0}(\cos\theta) = \cos\theta$
$P_{1,1}(\cos\theta) = \sin\theta$
$P_{2,0}(\cos\theta) = \frac{1}{2}(3\cos^2\theta - 1)$
$P_{2,1}(\cos\theta) = 3\cos\theta\sin\theta$
$P_{2,2}(\cos\theta) = 3\sin^2\theta.$

$P_{3,0}(\cos\theta) = \frac{1}{2}(5\cos^3\theta - 3\cos\theta)$
$P_{3,1}(\cos\theta) = \frac{3}{2}(5\cos^2\theta - 1)\sin\theta$
$P_{3,2}(\cos\theta) = 15\cos\theta\sin^2\theta$
$P_{3,3}(\theta) = 15\sin^3\theta$

If m is not specified, then only the Legendre polynomials are needed, which are just the subset above with $m = 0$.

The spherical harmonics, $Y_{\ell,m}$ come from the Legendre polynomials

$$Y_{\ell,m}(\theta,\phi) = \left[\frac{(2\ell+1)(\ell-|m|)!}{4\pi(\ell+|m|)!}\right]^{1/2} P_{\ell,|m|}(\cos\theta)e^{im\phi}. \tag{D.1}$$

Some of the lower spherical harmonics are

$Y_{0,0} = \frac{1}{(4\pi)^{1/2}}$

$Y_{1,0} = \left(\frac{3}{4\pi}\right)^{1/2}\cos\theta$

$Y_{1,1} = -\left(\frac{3}{8\pi}\right)^{1/2}\sin\theta e^{i\phi}$

$Y_{1,-1} = \left(\frac{3}{8\pi}\right)^{1/2}\sin\theta e^{-i\phi}$

$Y_{2,0} = \left(\frac{5}{16\pi}\right)^{1/2}(3\cos^2\theta - 1)$

$Y_{2,1} = -\left(\frac{15}{8\pi}\right)^{1/2}\sin\theta\cos\theta e^{i\phi}$

$Y_{2,-1} = \left(\frac{15}{8\pi}\right)^{1/2}\sin\theta\cos\theta e^{-i\phi}$

$Y_{2,2} = \left(\frac{15}{32\pi}\right)^{1/2}\sin^2\theta e^{2i\phi}$

$Y_{2,-2} = \left(\frac{15}{32\pi}\right)^{1/2}\sin^2\theta e^{-2i\phi}.$

Legendre polynomials

The spherical harmonics are orthogonal

$$\int Y^*_{\ell',m'}(\Omega)Y_{\ell,m}(\Omega)d\Omega = \delta_{\ell',\ell}\delta_{m',m}. \tag{D.2}$$

When computing the orientational response functions, one often runs into integrals over three spherical harmonic functions, such as $\int Y^*_{10}(\Omega)Y_{1,-1}(\Omega) \cdot Y_{\ell,m}(\Omega)d\Omega$. To solve an integral like this one, one uses Clebsch–Gordan coefficients [31, 158], although nowadays one can use *Mathematica* or other software to easily compute the integrals. A few integrals that one encounters when solving the orientational response function for $\langle (Z \cdot \alpha)(Z \cdot \alpha)(Z \cdot \alpha)(Z \cdot \alpha) \rangle$ are

$$\int Y^*_{00}(\Omega)Y_{1,0}(\Omega)Y_{\ell,m}(\Omega)d\Omega = \left(\frac{1}{4\pi}\right)^{1/2}\delta_{\ell,1}\delta_{m,0}$$

$$\int Y^*_{10}(\Omega)Y_{1,0}(\Omega)Y_{\ell,m}(\Omega)d\Omega = \left(\frac{1}{4\pi}\right)^{1/2}\delta_{\ell,0}\delta_{m,0} + \left(\frac{1}{5\pi}\right)^{1/2}\delta_{\ell,2}\delta_{m,0}$$

$$\int Y^*_{20}(\Omega)Y_{1,0}(\Omega)Y_{\ell,m}(\Omega)d\Omega = \left(\frac{1}{5\pi}\right)^{1/2}\delta_{\ell,1}\delta_{m,0} + \frac{3}{2}\left(\frac{3}{35\pi}\right)^{1/2}\delta_{\ell,3}\delta_{m,0}$$

$$\tag{D.3}$$

and when solving $\langle (Z \cdot \alpha)(Z \cdot \alpha)(X \cdot \alpha)(X \cdot \alpha) \rangle$ one also needs

$$\int Y^*_{00}(\Omega)Y_{1,\pm 1}(\Omega)Y_{\ell,m}(\Omega)d\Omega = -\left(\frac{1}{4\pi}\right)^{1/2}\delta_{\ell,1}\delta_{m,\mp 1}$$

$$\int Y^*_{20}(\Omega)Y_{1,\pm 1}(\Omega)Y_{\ell,m}(\Omega)d\Omega = \frac{1}{2}\left(\frac{1}{5\pi}\right)^{1/2}\delta_{\ell,1}\delta_{m,\mp 1} - 3\left(\frac{1}{70\pi}\right)^{1/2}\delta_{\ell,3}\delta_{m,\mp 1}.$$

$$\tag{D.4}$$

Appendix E
Recommended reading

In this appendix we list some textbooks that we found helpful in learning the subjects contained in this book.

Quantum mechanics and/or density matrices

C. Cohen-Tannoudji, B. Diu, and F. Laloë, *Quantum Mechanics*, Volume II, John Wiley, New York, 1977.

G. C. Schatz and M. A. Ratner, *Quantum Mechanics in Chemistry*, Prentice Hall, New Jersey, 1993.

I. N. Levine, *Quantum Chemistry*, Pearson Prentice Hall, New Jersey, 2009.

M. D. Fayer, *Elements of Quantum Mechanics*, Oxford University Press, Oxford, 2001.

Electrodynamics

D. J. Griffiths, *Introduction to Electrodynamics*, Addison Wesley, New York, 1999.

J. D. Jackson, *Classical Electrodynamics*, John Wiley, New York, 1999.

Nonlinear optical spectroscopy

S. Mukamel, *Principles of Nonlinear Optical Spectroscopy*, Oxford University Press, Oxford, 1995.

M. Cho, *Two-Dimensional Optical Spectroscopy*, CRC Press, Boca Raton, 2009.

R. W. Boyd, *Nonlinear Optics*, Academic Press, Amsterdam, 2008.

General spectroscopy

J. L. McHale, *Molecular Spectroscopy*, Pearson Education, New Jersey, 1999.

W. S. Struve, *Fundamentals of Molecular Spectroscopy*, Wiley Inter-Science, New York, 1989.

D. A. McQuarrie, *Quantum Chemistry*, University Science Books, California, 1983.

References

[1] Andrews, D. L. and Ghoul, W. A. 1981. Eighth rank isotropic tensors and rotational averages. *J. Phys. A: Math. Gen.*, **14**, 1281–1290.

[2] Asbury, J. B., Steinel, T., Kwak, K., Corcelli, S. A., Lawrence, C. P., Skinner, J. L., and Fayer, M. D. 2004a. Dynamics of water probed with vibrational echo correlation spectroscopy. *J. Chem. Phys.*, **121**, 12431–12446.

[3] Asbury, J. B., Steinel, T., and Fayer, M. D. 2004b. Vibrational echo correlation spectroscopy probes of hydrogen bond dynamics in water and methanol. *J. Lumin.*, **107**, 271–286.

[4] Ashcroft, N. W. and Mermin, N. D. 1976. *Solid State Physics*. Philadephia: Sounders College Publishing.

[5] Asplund, M. C., Zanni, M. T., and Hochstrasser, R. M. 2000. Two-dimensional infrared spectroscopy of peptides by phase-controlled femtosecond vibrational photon echoes. *Proc. Natl. Acad. Sci. USA*, **97**, 8219–8224.

[6] Atkins, P. W. 2004. *Physical Chemistry*. New York: W. H. Freeman and Company.

[7] Backus, E. H. G., Garrett-Roe, S., and Hamm, P. 2008. Phasing problem of heterodyne-detected two-dimensional infrared spectroscopy. *Opt. Lett.*, **33**, 2665–2667.

[8] Baiz, C. R., McRobbie, P. L., Anna, J. M., Geva, E., and Kubarych, K. J. 2009. Two-dimensional infrared spectroscopy of metal carbonyls. *Acc. Chem. Res.*, **42**, 1395–1404.

[9] Barber-Armstrong, W., Donaldson, T., Wijesooriya, H., Silva, R. A. G. D., and Decatur, S. M. 2004. Empirical relationships between isotope-edited IR spectra and helix geometry in model peptides. *J. Am. Chem. Soc.*, **126**, 2339–2345.

[10] Barth, A. 2000. The infrared absorption of amino acid side chains. *Prog. Biopyhys. Mol. Biol.*, **74**, 141–173.

[11] Bartholdi, E. and Ernst, R. R. 1973. Fourier spectroscopy and the causality principle. *J. Magn. Resonance*, **11**, 9–19.

[12] Berne, B. J. and Pecora, R. 2000. *Dynamic Light Scattering*. New York: Dover.

[13] Blanco, E. A. and Zanni, M. T. 2010. Unpublished.

[14] Blank, D. A., Kaufman, L. J., and Fleming, G. R. 2000. Direct fifth-order electronically nonresonant Raman scattering from CS_2 at room temperature. *J. Chem. Phys.*, **113**, 771–778.

[15] Bloem, R., Garret-Roe, S., Strzalka, H., Hamm, P. and Donaldson, P. 2011. Enhancing signal detection and completely eliminating scattering using quasi-phase-cycling in 2D IR experiments. Submitted.

[16] Boyd, R. W. 2008. *Nonlinear Optics*. Amsterdam: Academic Press.

[17] Bredenbeck, J., Helbing, J., and Hamm, P. 2004a. Labeling vibrations by light: Ultrafast transient 2D-IR spectroscopy tracks vibrational modes during a photoreaction. *J. Am. Chem. Soc.*, **126**, 990–991.

[18] Bredenbeck, J., Helbing, J., and Hamm, P. 2004b. Transient two-dimensional IR spectroscopy – exploring the polarization dependence. *J. Chem. Phys.*, **121**, 5943–5957.

[19] Bredenbeck, J., Helbing, J., Kolano, C., and Hamm, P. 2007. Ultrafast 2D-IR spectroscopy of transient species. *Chem. Phys. Chem.*, **8**, 1747–1756.

[20] Bredenbeck, J., Ghosh, A., Smits, M., and Bonn, M. 2008. Ultrafast two dimensional-infrared spectroscopy of a molecular monolayer. *J. Am. Chem. Soc.*, **130**, 2152–2153.

[21] Bristow, A. D., Karaiskaj, D., Dai, X., and Cundiff, S. T. 2008. All-optical retrieval of the global phase for two-dimensional Fourier-transform spectroscopy. *Opt. Express*, **16**, 18017–18027.

[22] Brixner, T., Stenger, J., Vaswani, H. M., Cho, M., Blankenship, R. E., and Fleming, G. R. 2005. Two-dimensional spectroscopy of electronic couplings in photosynthesis. *Nature*, **434**, 625–628.

[23] Bulova Watch Company, Inc. (New York, NY). 1971. United States Patent 3609485.

[24] Cervetto, V., Helbing, J., Bredenbeck, J., and Hamm, P. 2004. Double-resonance versus pulsed Fourier transform 2D-IR spectroscopy: An experimental and theoretical comparison. *J. Chem. Phys.*, **121**, 5935–5942.

[25] Cho, M. 1999. Theoretical description of the vibrational echo spectroscopy by time-resolved infrared-infrared-visible difference-frequency generation. *J. Chem. Phys.*, **111**, 10587–10594.

[26] Cho, M. 2006. Coherent two-dimensional optical spectroscopy. *Bull. Korean Chem. Soc.*, **27**, 1940–1960.

[27] Cho, M. 2009. *Two-Dimensional Optical Spectroscopy*. Boca Raton: CRC Press.

[28] Cho, M., Yu, J. Y., Joo, T., Nagasawa, Y., Passino, S. A., and Fleming, G. R. 1996. The integrated photon echo and solvation dynamics. *J. Phys. Chem.*, **100**, 11944–11953.

[29] Chung, H. S., Ganim, Z., Jones, K. C., and Tokmakoff, A. 2007. Transient 2D IR spectroscopy of ubiquitin unfolding dynamics. *Proc. Natl. Acad. Sci. USA*, **104**, 14237–14242.

[30] Cochran, A. G., Skelton, N. J., and Starovasnik, M. A. 2001. Tryptophan zippers: Stable, monomeric β-hairpins. *Proc. Natl. Acad. Sci. USA*, **98**, 5578–5583.

[31] Cohen-Tannoudji, C., Diu, B., and Laloë, F. 1977. *Quantum Mechanics*, Volume II. New York: John Wiley.

[32] Corcelli, S. A., Lawrence, C. P., Asbury, J. B., Steinel, T., Fayer, M. D., and Skinner, J. L. 2004. Spectral diffusion in a fluctuating charge model of water. *J. Chem. Phys*, **121**, 8897–8900.

[33] Cowan, M. L., Bruner, B. D., Huse, N., Dwyer, J. R., Chugh, B., Nibbering, E. T. J., Elsaesser, T., and Miller, R. J. D. 2005. Ultrafast memory loss and energy redistribution in the hydrogen bond network of liquid H_2O. *Nature*, **434**, 199–202.

[34] Craig, D. P. and Thirunamachandran, T. 1984. *Molecular Quantum Electrodynamics*. London: Academic Press.

[35] Darling, B. T. and Dennison, D. M. 1940. The water vapor molecule. *Phys. Rev.*, **57**, 128–139.

[36] Davydov, A. S. 1971. *Theory of Molecular Excitons*. New York: Plenum.

[37] de Boeij, W. P., Pshenichnikov, M. S., and Wiersma, D. A. 1996. On the relation between the echo-peak shift and Brownian oscillator correlation function. *Chem. Phys. Lett.*, **253**, 53–60.

[38] de Boeij, W. P., Pshenichnikov, M. S., and Wiersma, D. A. 1998. Heterodyne-detected stimulated photon echo: Applications to optical dynamics in solution. *Chem. Phys.*, **233**, 287–309.

[39] Decatur, S. M. and Antonic, J. 1999. Isotope-edited infrared spectroscopy of helical peptides. *J. Am. Chem. Soc.*, **121**, 11914–11915.

[40] DeFlores, L. P., Nicodemus, R. A., and Tokmakoff, A. 2007. Two dimensonal Fourier transform spectroscopy in the pump-probe geometry. *Opt. Lett.*, **32**, 2966–2968.

[41] Demirdöven, N., Khalil, M., and Tokmakoff, A. 2002. Correlated vibrational dynamics revealed by two-dimensional infrared spectroscopy. *Phys. Rev. Lett.*, **89**, 237401.

[42] Dijkstra, A. G. and Jansen, T. L. C. 2008. Localization and coherent dynamics of excitons in the two-dimensional optical spectrum of molecular J-aggregates. *J. Chem. Phys.*, **128**, 164511.

[43] Ding, F. and Zanni, M. T. 2007. Heterodyned 3D-IR spectroscopy. *Chem. Phys.*, **341**, 95–105.

[44] Ding, F., Fulmer, E. C., and Zanni, M. T. 2005. Heterodyned fifth-order 2D-IR spectroscopy: Third-quantum states and polarization selectivity. *J. Chem. Phys.*, **123**, 094502.

[45] Ding, F., Mukherjee, P., and Zanni, M. T. 2006. Passively correcting phase drift in two-dimensional infrared spectroscopy. *Opt. Lett.*, **31**, 2918–2920.

[46] Donaldson, P. M., Guo, R., Fournier, F., Gardner, E. M., Barter, L. M. C., Barnett, C. J., Gould, I. R., Klug, D. R., Palmer, D. J., and Willison, K. R. 2007. Direct identification and decongestion of Fermi resonances by control of pulse time-ordering in 2D-IR spectroscopy. *J. Chem. Phys.*, **127**, 114513.

[47] Donaldson, P. M., Guo, R., Fournier, F., Gardner, E. M., Gould, I. R., and Klug, D. R. 2008. Decongestion of methylene spectra in biological and non-biological systems using picosecond 2DIR spectroscopy measuring electron-vibration–vibration coupling. *Chem. Phys.*, **350**, 201–211.

[48] Eaves, J. D., Tokmakoff, A., and Geissler, P. L. 2005a. Electric field fluctuations drive vibrational dephasing in water. *J. Phys. Chem. A*, **109**, 9424–9436.

[49] Eaves, J. D., Loparo, J. J., Fecko, C. J., Roberts, S. T., Tokmakoff, A., and Geissler, P. L. 2005b. Hydrogen bonds in liquid water are broken only fleetingly. *Proc. Natl. Acad. Sci. USA*, **102**, 13019–13022.

[50] Edler, J. and Hamm, P. 2002. Self-trapping of the amide I band in a peptide model crystal. *J. Chem. Phys.*, **117**, 2415–2424.

[51] Edler, J. and Hamm, P. 2003. Two-dimensional vibrational spectroscopy of the amide I band of crystalline acetanilide: Fermi resonance, conformational sub-states or vibrational self-trapping? *J. Chem. Phys.*, **119**, 2709–2715.

[52] Engel, G. S., Calhoun, T. R., Read, E. L., Ahn, T.-K., Mancal, T., Cheng, Y.-C., Blankenship, R. E., and Fleming, G. R. 2007. Evidence for wavelike energy transfer through quantum coherence in photosynthetic systems. *Nature*, **446**, 782–786.

[53] Fang, C., Wang, J., Kim, Y. S., Charnley, A. K., Barber-Armstrong, W., Smith III, A. B., Decatur, S. M., and Hochstrasser, R. M. 2004. Two-dimensional infrared spectroscopy of isotopomers of an alanine rich α-helix. *J. Phys. Chem. B*, **108**, 10415–10427.

[54] Fang, C., Senes, A., Cristian, L., DeGrado, W. F., and Hochstrasser, R. M. 2006. Amide vibrations are delocalized across the hydrophobic interface of a transmembrane helix dimer. *Proc. Natl. Acad. Sci. USA*, **103**, 16740–16745.

[55] Fecko, C. J., Eaves, J. D., Loparo, J. J., Tokmakoff, A., and Geissler, P. L. 2003. Ultrafast hydrogen-bond dynamics in the infrared spectroscopy of water. *Science*, **301**, 1698–1702.

[56] Finkelstein, I. J., Zheng, J., Ishikawa, H., Kim, S., Kwak, K., and Fayer, M. D. 2007. Probing dynamics of complex molecular systems with ultrafast 2D IR vibrational echo spectroscopy. *Phys. Chem. Chem. Phys.*, **9**, 1533–1549.

[57] Fournier, F., Guo, R., Gardner, E. M., Donaldson, P. M., Loeffeld, C., Gould, I. R., Willison, K. R., and Klug, D. R. 2009. Biological and biomedical applications of two-dimensional vibrational spectroscopy: Proteomics, imaging, and structural analysis. *Acc. Chem. Res.*, **42**, 1322–1331.

[58] Frisch, M. J., Trucks, G. W., Schlegel, H. B., Scuseria, G. E., Robb, M. A., Cheeseman, J. R., Scalmani, G., Barone, V., Mennucci, B., Petersson, G. A., Nakatsuji, H., Caricato, M., Li, X., Hratchian, H. P., Izmaylov, A. F., Bloino, J., Zheng, G., Sonnenberg, J. L., Hada, M., Ehara, M., Toyota, K., Fukuda, R., Hasegawa, J., Ishida, M., Nakajima, T., Honda, Y., Kitao, O., Nakai, H., Vreven, T., Montgomery, Jr., J. A., Peralta, J. E., Ogliaro, F., Bearpark, M., Heyd, J. J., Brothers, E., Kudin, K. N., Staroverov, V. N., Kobayashi, R., Normand, J., Raghavachari, K., Rendell, A., Burant, J. C., Iyengar, S. S., Tomasi, J., Cossi, M., Rega, N., Millam, J. M., Klene, M., Knox, J. E., Cross, J. B., Bakken, V., Adamo, C., Jaramillo, J., Gomperts, R., Stratmann, R. E., Yazyev, O., Austin, A. J., Cammi, R., Pomelli, C., Ochterski, J. W., Martin, R. L., Morokuma, K., Zakrzewski, V. G., Voth, G. A., Salvador, P., Dannenberg, J. J., Dapprich, S., Daniels, A. D., Farkas, O., Foresman, J. B., Ortiz, J. V., Cioslowski, J., and Fox, D. J. 2009. *Gaussian 09 Revision A.1*. Wallingford CT: Gaussian Inc.

[59] Fulmer, E. C., Ding, F., and Zanni, M. T. 2005a. Heterodyned fifth-order 2D-IR spectroscopy of the azide ion in an ionic glass. *J. Chem. Phys.*, **122**, 034302.

[60] Fulmer, E. C., Mukherjee, P., Krummel, A. T., and Zanni, M. T. 2005b. A pulse sequence for directly measuring the anharmonicities of coupled vibrations: Two-quantum two-dimensional infrared spectroscopy. *J. Chem. Phys.*, **120**, 8067–8078.

[61] Fulmer, E. C., Ding, F., Mukherjee, P., and Zanni, M. T. 2005c. Vibrational dynamics of ions in glass from fifth-order two-dimensional infrared spectroscopy. *Phys. Rev. Lett.*, **94**, 067402.

[62] Gallagher Faeder, S. M. and Jonas, D. M. 1999. Two-dimensional electronic correlation and relaxation spectra: Theory and model calculations. *J. Phys. Chem. A*, **103**, 10489–10505.

[63] Ganim, Z., Chung, H. S., Smith, A. W., DeFlores, L. P., Jones, K. C., and Tokmakoff, A. 2008. Amide I two-dimensional infrared spectroscopy of proteins. *Acc. Chem. Res.*, **41**, 432–441.

[64] Garrett-Roe, S. and Hamm, P. The dynamical heterogeneity of water revealed by 3D-IR spectroscopy. Submitted.

[65] Garrett-Roe, S. and Hamm, P. 2008. Three-point frequency fluctuation correlation functions of the OH stretch vibration in liquid water. *J. Chem. Phys.*, **128**, 104507.

[66] Garrett-Roe, S. and Hamm, P. 2009. Purely absorptive three-dimensional infrared spectroscopy. *J. Chem. Phys.*, **130**, 164510.

[67] Ge, N. H. and Hochstrasser, R. M. 2002. Femtosecond two-dimensional infrared spectroscopy: IR-COSY and THIRSTY. *Phys. Chem. Comm.*, **5**, 17–26.

[68] Ge, N. H., Zanni, M. T., and Hochstrasser, R. M. 2002. Effects of vibrational frequency correlations on two-dimensional infrared spectra. *J. Phys. Chem. A*, **106**, 962–972.

[69] Golonzka, O. and Tokmakoff, A. 2001. Polarization-selective third-order spectroscopy of coupled vibronic states. *J. Chem. Phys.*, **115**, 297–309.

[70] Golonzka, O., Demirdöven, N., Khalili, M., and Tokmakoff, A. 2000. Separation of cascaded and direct fifth-order Raman signals using phase-sensitive intrinsic heterodyne detection. *J. Chem. Phys.*, **113**, 9893–9896.

[71] Golonzka, O., Khalil, M., Demirdöven, N., and Tokmakoff, A. 2001. Coupling and orientation between anharmonic vibrations characterized with two-dimensional infrared vibrational echo spectroscopy. *J. Chem. Phys.*, **115**, 10814–10828.

[72] Goodno, G. D., Dadusc, G., and Miller, R. J. D. 1998. Ultrafast heterodyne-detected transient-grating spectroscopy using diffractive optics. *J. Opt. Soc. Am. B*, **15**, 1791–1794.

[73] Gorbunov, R. D., Kosov, D. S., and Stock, G. 2005. Ab initio-based exciton model of amide I vibrations in peptides: Definition, conformational dependence, and transferability. *J. Chem. Phys.*, **122**, 224904.

[74] Griffiths, D. J. 1999. *Introduction to Electrodynamics*, 3rd edn. New York: Addison Wesley.

[75] Grumstrup, E. M., Shim, S.-H., Montgomery, M. A., Damrauer, N. H., and Zanni, M. T. 2007. Facile collection of two-dimensional electronic spectra using femtosecond pulse-shaping technology. *Optics Express*, **15**, 16681.

[76] Guo, R., Fournier, F., Donaldson, P. M., Gardner, E. M., Gould, I. R., and Klug, D. R. 2009. Detection of complex formation and determination of intermolecular geometry through electrical anharmonic coupling of molecular vibrations using electron-vibration-vibration two-dimensional infrared spectroscopy. *Phys. Chem. Chem. Phys.*, **11**, 8417–8421.

[77] Hahn, S., Kim, S.-S., Lee, C., and Cho, M. 2005. Characteristic two-dimensional IR spectroscopic features of antiparallel and parallel β-sheet polypeptides: Simulation studies. *J. Chem. Phys.*, **123**, 084905.

[78] Ham, S. and Cho, M. 2003. Amide I modes in the N-methylacetamide dimer and glycine dipeptide analog: Diagonal force constants. *J. Chem. Phys.*, **118**, 6915.

[79] Hamm, P. 2006. 3D-IR spectroscopy: Beyond the two-point frequency fluctuation correlation function. *J. Chem. Phys.*, **124**, 124506.

[80] Hamm, P. and Hochstrasser, R. M. 2001. Structure and dynamics of proteins and peptides: Femtosecond two-dimensional infrared spectroscopy. In: Fayer, M. D. (ed.), *Ultrafast Infrared and Raman Spectroscopy*. New York: Marcel Dekker, pp. 273–347.

[81] Hamm, P. and Woutersen, S. 2002. Coupling of the amide I modes of the glycine di-peptide. *Bull. Chem. Soc. Jpn.*, **75**, 985–988.

[82] Hamm, P., Lim, M. H., and Hochstrasser, R. M. 1998a. Non-Markovian dynamics of the vibrations of ions in water from femtosecond infrared three-pulse photon echoes. *Phys. Rev. Lett.*, **81**, 5326–5329.

[83] Hamm, P., Lim, M. H., and Hochstrasser, R. M. 1998b. Structure of the amide I band of peptides measured by femtosecond nonlinear-infrared spectroscopy. *J. Phys. Chem. B*, **102**, 6123–6138.

[84] Hamm, P., Lim, M. H., DeGrado, W. F., and Hochstrasser, R. M. 1999. The two-dimensional IR nonlinear spectroscopy of a cyclic penta-peptide in relation to its three-dimensional structure. *Proc. Natl. Acad. Sci. USA*, **96**, 2036–2041.

[85] Hamm, P., Kaindl, R. A., and Stenger, J. 2000a. Noise suppression in femtosecond mid-infrared light sources. *Opt. Lett.*, **25**, 1798–1800.

[86] Hamm, P., Lim, M. H., DeGrado, W. F., and Hochstrasser, R. M. 2000b. Pump/probe self heterodyned 2D spectroscopy of vibrational transitions of a small globular peptide. *J. Chem. Phys.*, **112**, 1907–1916.

[87] Hamm, P., Helbing, J., and Bredenbeck, J. 2008. Two-dimensional infrared spectroscopy of photoswitchable peptides. *Annu. Rev. Phys. Chem.*, **59**, 291–317.

[88] Harel, E., Fidler, A. and Engel, G. 2010. Real-time mapping of electronic structure with single-shot tow-dimensional electronic spectroscopy. *Proc. Natl. Acad. Sci. USA*, DOI:10.1073/pnas.1007579107.

[89] Hayashi, T., la Cour Jansen, T., Zhuang, W., and Mukamel, S. 2005a. Collective solvent coordinates for the infrared spectrum of HOD in D_2O based on an *ab initio* electrostatic map. *J. Phys. Chem. A*, **109**, 64–82.

[90] Hayashi, T., Zhuang, W., and Mukamel, S. 2005b. Electrostatic DFT map for the complete vibrational amide band of NMA. *J. Phys. Chem. A*, **109**, 9747.

[91] Helbing, J. and Bonmarin, M. 2009. Vibrational circular dichroism signal enhancement using self-heterodyning with elliptically polarized laser pulses. *J. Chem. Phys.*, **131**, 174507.

[92] Helbing, J., Nienhaus, K., Nienhaus, G. U., and Hamm, P. 2005. Restricted rotational motion of CO in a protein internal cavity: Evidence for nonseparating correlation functions from IR pump-probe spectroscopy. *J. Chem. Phys.*, **122**, 124505.

[93] Herzberg, G. 1939. *Molecular Spectra and Molecular Structure*. New York: Prentice-Hall.

[94] Herzberg, G. 1945. *Infrared and Raman Spectra of Polyatomic Molecules*. New York: Van Nostrand.

[95] Hochstrasser, R. M. 2001. Two-dimensional IR-spectroscopy: Polarization anisotropy effects. *Chem. Phys.*, **266**, 273–284.

[96] Hochstrasser, R. M. 2007. Multidimensional ultrafast spectroscopy. *Proc. Natl. Acad. Sci. USA, Special Issue*, **104**, 14189 and all articles in that issue.

[97] Horng, M. L., Gardecki, J. A., Papazyan, A., and Maroncelli, M. 1995. Subpicosecond measurements of polar solvation dynamics: Coumarin 153 revisited. *J. Phys. Chem.*, **99**, 17311–17337.

[98] Iwaki, L. K. and Dlott, D. D. 2000. Three-dimensional spectroscopy of vibrational energy relaxation in liquid methanol. *J. Phys. Chem. A*, **104**, 9101–9112.

[99] Jackson, J. D. 1999. *Classical Electrodynamics*, 3rd edn. New York: John Wiley.

[100] Jang, S., Dempster, S. E., and Silbey, R. J. 2001. Characterization of the static disorder in the B850 band of LH2. *J. Phys. Chem. B*, **105**, 6655–6665.

[101] Jansen, T. L. and Knoester, J. 2006. A transferable electrostatic map for solvation effects on amide I vibrations and its application to linear and two-dimensional spectroscopy. *J. Chem. Phys.*, **124**, 044502.

[102] Jansen, T. L. C., Auer, B. M., Yang, M., and Skinner, J. L. 2010. Two-dimensional infrared spectroscopy and ultrafast anisotropy decay of water. *J. Chem. Phys.*, **132**, 224503.

[103] Jonas, D. M. 2003. Two-dimensional femtosecond spectroscopy. *Ann. Rev. Phys. Chem.*, **54**, 425–463.

[104] Khalil, M., Demirdöven, N., and Tokmakoff, A. 2003a. Coherent 2D IR spectroscopy: Molecular structure and dynamics in solution. *J. Phys. Chem. A*, **107**, 5258–5279.

[105] Khalil, M., Demirdöven, N., and Tokmakoff, A. 2003b. Obtaining absorptive line shapes in two-dimensional infrared vibrational correlation spectra. *Phys. Rev. Lett.*, **90**, 047401.

[106] Khalil, M., Demirdöven, N., and Tokmakoff, A. 2004. Vibrational coherence transfer characterized with Fourier-transform 2D IR spectroscopy. *J. Chem. Phys.*, **121**, 362–373.

[107] Kim, Y. S. and Hochstrasser, R. M. 2005. Chemical exchange 2D IR of hydrogen-bond making and breaking. *Proc. Natl. Acad. Sci. USA*, **102**, 11185–11190.

[108] Kim, Y. S. and Hochstrasser, R. M. 2009. Applications of 2D IR spectroscopy to peptides, proteins, and hydrogen-bond dynamics. *J. Phys. Chem. B*, **113**, 8231–8251.

[109] Kim, Y. S., Liu, L., Axelsen, P. H., and Hochstrasser, R. M. 2008. Two-dimensional infrared spectra of isotopically diluted amyloid fibrils from A beta 40. *Proc. Natl. Acad. Sci. USA*, **105**, 7720–7725.

[110] Kolano, C., Helbing, J., Kozinski, M., Sander, W., and Hamm, P. 2006. Watching hydrogen bond dynamics in a β-turn by transient two-dimensional infrared spectroscopy. *Nature*, **444**, 469–472.

[111] Koziński, M., Garrett-Roe, S., and Hamm, P. 2007. Vibrational spectral diffusion of CN^- in water. *Chem. Phys.*, **341**, 5–10.

[112] Krimm, S. and Bandekar, J. 1986. Vibrational spectroscopy and conformation of peptides, polypeptides, and proteins. *Adv. Protein Chem.*, **38**, 181–364.

[113] Krummel, A. T., Mukherjee, P., and Zanni, M. T. 2003. Inter and intrastrand vibrational coupling in DNA studied with heterodyned 2D-IR spectroscopy. *J. Phys. Chem. B*, **107**, 9165–9169.

[114] Krummel, A. T. and Zanni, M. T. 2006. DNA vibrational coupling revealed with two-dimensional infrared spectroscopy: Why vibrational spectroscopy is sensitive to DNA structure. *J. Phys. Chem. B*, **110**, 13991–14000.

[115] Kubarych, K. J., Milne, C. J., and Miller, R. J. D. 2003. Fifth-order two-dimensional Raman spectroscopy: A new direct probe of the liquid state. *Int. Rev. Phys. Chem.*, **22**, 497–532.

[116] Kubo, R. 1969. A stochastic theory of line shape. *Adv. Chem. Phys.*, **15**, 101–127.

[117] Kurochkin, D. V., Naraharisetty, S. R. G., and Rubtsov, I. V. 2005. Dual-frequency 2D IR on interaction of weak and strong IR modes. *J. Phys. Chem. A*, **109**, 10799–10802.

[118] Kwac, K. and Cho, M. 2003. Molecular dynamics simulation study of N-methylacetamide in water. I. Amide I mode frequency fluctuation. *J. Chem. Phys.*, **119**, 2247–2255.

[119] Laage, D. and Hynes, J. T. 2006. A molecular jump mechanism of water reorientation. *Science*, **311**, 832–835.

[120] Lawrence, C. P. and Skinner, J. L. 2005. Quantum corrections in vibrational and electronic condensed phase spectroscopy: Line shapes and echoes. *Proc. Natl. Acad. Sci. USA*, **102**, 6720–6725.

[121] Lazonder, K., Pshenichnikov, M. S., and Wiersma, D. A. 2007. Two-dimensional optical correlation spectroscopy applied to liquid/glass dynamics. In: Miller, R. J. D., Weiner, A. M., Corkum, P., and Jonas, D. M. (eds), *Ultrafast Phenomena XV*. New York: Springer, pp. 356–358.

[122] Lee, C. and Cho, M. H. 2007. Vibrational dynamics of DNA: IV. Vibrational spectroscopic characteristics of A-, B-, and Z-form DNA's. *J. Chem. Phys.*, **126**, 145102.

[123] Lehmann, K. K. 1983. On the relation of Child and Lawton's harmonically coupled anharmonic-oscillator model and Darling–Dennison coupling. *J. Chem. Phys.*, **79**, 1098.

[124] Lepetit, L., Chériaux, G., and Joffre, M. 1995. Linear techniques of phase measurement by femtosecond spectral interferometry for applications in spectroscopy. *J. Opt. Soc. Am. B*, **12**, 2467–2474.

[125] Levine, I. N. 2009. *Quantum Chemistry*, 6th edn. New Jersey: Pearson Prentice Hall.

[126] Li, Y. L., Huang, L., Miller, R. J. D., Hasegawa, T., and Tanimura, Y. 2008. Two-dimensional fifth-order Raman spectroscopy of liquid formamide: Experiment and theory. *J. Chem. Phys.*, **128**, 234507.

[127] Ling, Y. L., Strasfeld, D. B., Shim, S.-H., Raleigh, D. P., and Zanni, M. T. 2009. Two-dimensional infrared spectroscopy provides evidence of an intermediate in the membrane-catalyzed assembly of diabetic amyloid. *J. Phys. Chem. B*, **113**, 2498–2505.

[128] Maekawa, H., Toniolo, C., Broxterman, Q. B., and Ge, N. H. 2002. Two-dimensional infrared spectral signatures of 3(10)- and α-helical peptides. *J. Phys. Chem. B*, **111**, 3222–3235.

[129] Marion, J. B. and Thornton, S. T. 2004. *Classical Dynamics of Particles and Systems*. Fort Worth: Saunders College Publishing.

[130] Mathew, N. A., Yurs, L. A., Block, S. B., Pakoulev, A. V., Kornau, K. M., Sibert III, E. L., and Wright, J. C. 2010. Fully and partially coherent pathways in multiply enhanced odd-order wave-mixing spectroscopy. *J. Phys. Chem. A*, **114**, 817–832.

[131] McHale, J. L. 1999. *Molecular Spectroscopy*. New Jersey: Pearson Education.

[132] McQuarrie, D. A. 1983. *Quantum Chemistry*. California: University Science Books.

[133] Middleton, C. T., Strasfeld, D. B., and Zanni, M. T. 2009. Polarization shaping in the mid-IR and polarization-based balanced heterodyne detection with application to 2D IR spectroscopy. *Optics Express*, **17**, 14526–14533.

[134] Middleton, C. T., Woys, A. M., Mukherjee, S. S., and Zanni, M. T. 2010. Residue-specific structural kinetics of proteins through the union of isotope labeling, Mid-IR pulse shaping, and coherent 2D IR spectroscopy. *Methods*, **52**, 12–22.

[135] Mills, I. M. and Robiette, A. G. 1985. On the relationship of normal modes to local modes in molecular vibrations. *Mol. Phys.*, **56**, 743–765.

[136] Miyazawa, T. 1960. Perturbation treatment of the characteristic vibrations of polypeptide chains in various configurations. *J. Chem. Phys.*, **32**, 1647–1652.

[137] Moffitt, W. 1956. Optical rotatory dispersion of helical polymers. *J. Chem. Phys.*, **25**, 467–478.

[138] Moller, K. B., Rey, R., and Hynes, J. T. 2004. Hydrogen bond dynamics in water and ultrafast infrared spectroscopy: A theoretical study. *J. Phys. Chem. A*, **108**, 1275–1289.

[139] Moran, A. and Mukamel, S. 2004. The origin of vibrational mode couplings in various secondary structural motifs of polypeptides. *Proc. Natl. Acad. Sci. USA*, **101**, 506–510.

[140] Moran, S. D. and Zanni, M. T. 2010. Unpublished.

[141] Mukamel, S. 1995. *Principles of Nonlinear Optical Spectroscopy*. Oxford: Oxford University Press.

[142] Mukamel, S. 2000. Multidimensional femtosecond correlation spectroscopies of electronic and vibrational excitations. *Annu. Rev. Phys. Chem.*, **51**, 691–729.

[143] Mukamel, S. and Hochstrasser, R. M. 2001. 2D Spectroscopy. *Chem. Phys., Special Issue*, **266**, 135–136 and all articles in that issue.

[144] Mukamel, S., Tanimura, Y., and Hamm, P. 2009. Coherent multidimensional optical spectroscopy. *Acc. Chem. Research, Special Issue*, **42**, 1207–1209 and all articles in that issue.

[145] Mukherjee, S. and Zanni, M. T. 2010. Unpublished.

[146] Myers, J. A., Lewis, K. L. M., Tekavec, P. F., and Ogilvie, J. P. 2008. Two-color two-dimensional Fourier transform electronic spectroscopy with a pulse-shaper. *Optics Express*, **16**, 17420–17428.

[147] Nee, M. J., McCanne, R., Kubarych, K. J., and Joffre, M. 2007. Two-dimensional infrared spectroscopy detected by chirped pulse upconversion. *Opt. Lett.*, **32**, 713–715.
[148] Otting, G., Widmer, H., Wagner, G., and Wüthrich, K. 1986. Origin of T_1 and T_2 ridges in 2D *NMR*-spectra and procedures for supression. *J. Magn. Reson.*, **66**, 187–193.
[149] Oxtoby, D. W., Levesque, D., and Weis, J. J. 1978. A molecular dynamics simulation of dephasing in liquid nitrogen. *J. Chem. Phys.*, **68**, 5528–5533.
[150] Painter, P. C., Coleman, M. M., and Koenig, J. L. 1982. *The Theory of Vibrational Spectroscopy and its Application to Polymeric Materials*. New York: John Wiley.
[151] Petty, S. A. 2005. Intersheet rearrangement of polypeptides during nucleation of beta-sheet aggregates. *Proc. Natl. Acad. Sci. USA*, **102**, 14272–14277.
[152] Press, W. H., Teukolsky, S. A., Vetterling, W. T., and Flannery, B. P. 1992. *Numerical Recipes in C*. Cambridge: Cambridge University Press.
[153] Read, E. L., Schlau-Cohen, G. S., Engel, G. S., Wen, J., Blankenship, R. E., and Fleming, G. R. 2008. Visualization of excitonic structure in the Fenna–Matthews–Olson photosynthetic complex by polarization-dependent two-dimensional electronic spectroscopy. *Biophys. J.*, **95**, 847–856.
[154] Reddy, A. S., Wang, L., Singh, S., Ling, Y. L., Buchanan, L., Zanni, M. T., Skinner, J. L., and de Publo, J. J. 2010. Folded and misfolded states of human amylin in solution. *Biophys. J*, submitted.
[155] Rey, R. and Hynes, J. T. 1998. Vibrational phase and energy relaxation of CN^- in water. *J. Chem. Phys.*, **108**, 142.
[156] Rhee, H., June, Y.-G., Lee, J.-S., Lee, K.-K., Ha, J.-H., Kim, Z. H., Jeon, S.-J., and Cho, M. 2009. Femtosecond characterization of vibrational optical activity of chiral molecules. *Nature*, **458**, 310–313.
[157] Roberts, S. T., Loparo, J. J., and Tokmakoff, A. 2006. Characterization of spectral diffusion from two-dimensional line shapes. *J. Chem. Phys.*, **125**, 084502.
[158] Rose, M. E. 1957. *Elementary Theory of Angular Momentum*. New York: John Wiley.
[159] Rubtsov, I. V., Wang, J., and Hochstrasser, R. M. 2003. Dual frequency 2D-IR of peptide amide-A and amide-I modes. *J. Chem. Phys.*, **118**, 7733–7736.
[160] Scheurer, C. and Mukamel, S. 2001. Design strategies of pulse sequences in multidimensional optical spectroscopy. *J. Chem. Phys.*, **115**, 4989–5004.
[161] Schmid, J. R., Corcelli, S. A., and Skinner, J. L. 2005. Pronounced non-Condon effects in the ultrafast infrared spectroscopy of water. *J. Chem. Phys.*, **123**, 044513.
[162] Schmidt, J. R., Corcelli, S. A., and Skinner, J. L. 2004. Ultrafast vibrational spectroscopy of water and aqueous N-methylacetamide: Comparison of different electronic structure/molecular dynamics approaches. *J. Chem. Phys.*, **121**, 8887–8896.
[163] Selig, U., Langhojer, F., Dimler, F., Löhrig, T., Schwarz, C., Gieseking, B., and Brixner, T. 2008. Inherently phase-stable coherent two-dimensional spectroscopy using only conventional optics. *Opt. Lett.*, **33**, 2851–2853.
[164] Seyfried, M. S., Lauber, B. S., and Luedtke, N. W. 2009. Multiple-turnover isotopic labeling of Fmoc- and Boc-protected amino acids with oxygen isotopes. *Organic Lett.*, **12**, 104–106.
[165] Shim, S. H. and Zanni, M. T. 2009. How to turn your pump-probe instrument into a multidimensional spectrometer: 2D IR and Vis spectroscopies via pulse shaping. *Phys. Chem. Chem. Phys.*, **11**, 748–761.

[166] Shim, S. H., Strasfeld, D. B., Ling, Y. L., and Zanni, M. T. 2007. Automated 2D IR spectroscopy using a mid-IR pulse shaper and application of this technology to the human islet amyloid polypeptide. *Proc. Natl. Acad. Sci. USA*, **104**, 14197–14202.

[167] Shim, S.-H., Gupta, R., Ling, Y.L., Strasfeld, D.B., Raleigh, D.P., and Zanni, M. T. 2009. 2D IR spectroscopy and isotope labeling defines the pathway of amyloid formation with residue specific resolution. *Proc. Natl. Acad. Sci. USA*, **106**, 6614–6619.

[168] Smith, A. W. and Tokmakoff, A. 2007. Amide I two-dimensional infrared spectroscopy of β-hairpin peptides. *J. Chem. Phys.*, **126**, 045109.

[169] Smith, J. D., Saykally, R. J., and Geissler, P. L. 2007. The effects of dissolved halide ions on hydrogen bonding in liquid water. *J. Am. Chem. Soc.*, **129**, 13847–13856.

[170] Steinel, T., Asbury, J. B., Zheng, J., and Fayer, M. D. 2004. Watching hydrogen bonds break: A transient absorption study of water. *J. Phys. Chem. A*, **108**, 10957–10964.

[171] Strasfeld, D. B., Shim, S.-H., and Zanni, M. T. 2007. Controlling vibrational excitation with shaped mid-IR pulses. *Phys. Rev. Lett.*, **99**, 038102.

[172] Strasfeld, D. B., Ling, Y. L., Shim, S.-H., and Zanni, M. T. 2008. Tracking fibril formation in human islet amyloid polypeptide with automated 2D-IR spectroscopy. *J. Am. Chem. Soc.*, **130**, 6698–6699.

[173] Strasfeld, D. B., Ling, Y. L., Gupta, R., Raleigh, D. P., and Zanni, M. T. 2009. Strategies for extracting structural information from 2D IR spectroscopy of amyloid: Application to islet amyloid polypeptide. *J. Phys. Chem. B*, **113**, 15679.

[174] Struve, W. S. 1989. *Fundamentals of Molecular Spectroscopy*. New York: Wiley Inter-Science.

[175] Sul, S., Karaiskaj, D., Jiang, Y., and Ge, N.-H. 2006. Conformations of N-acetyl-L-prolinamide by two-dimensional infrared spectroscopy. *J. Phys. Chem. B*, **110**, 19891.

[176] Tian, P. F., Keusters, D., Suzaki, Y., and Warren, W. S. 2003. Femtosecond phase-coherent two-dimensional spectroscopy. *Science*, **300**, 1553–1555.

[177] Tipping, R. H. and Ogilvie, J. F. 1983. Expectation values for Morse oscillators. *J. Chem. Phys.*, **79**, 2537–2540.

[178] Torii, H. and Tasumi, M. 1992. Model calculations on the amide I infrared bands of globular proteins. *J. Chem. Phys.*, **96**, 3379–3387.

[179] Torii, H. and Tasumi, M. 1998. *Ab initio* molecular orbital study of the amide I vibrational interactions between the peptide groups in di- and tripeptides and considerations on the conformation of the extended helix. *J. Raman Spectrosc.*, **29**, 81–86.

[180] Torres, J., Kukol, A., Goodman, J. N., and Arkin, I. T. 2001. Site-specific examination of secondary structure and orientation determination in membrane proteins: The peptidic 13C=18O group as a novel infrared probe. *Biopolymers*, **59**, 396–401.

[181] Tseng, C.-H., Matsika, S., and Weinacht, T. C. 2009. Two-dimensional ultrafast Fourier transform spectroscopy in the deep ultraviolet. *Opt. Express*, **17**, 18788–18793.

[182] Turner, D. B., Stone, K. W., Gundogdu, K., and Nelson, K. A. 2009. Three-dimensional electronic spectroscopy of excitons in GaAs quantum wells. *J. Chem. Phys.*, **131**, 144510.

[183] van der Spoel, D., Lindahl, E., Hess, B., Groenhof, G., Mark, A. E., and Berendsen, H. J. C. 2005. Gromacs; Fast, flexible and free. *J. Comput. Chem.*, **26**, 1701–1718.

[184] Van Kampen, N. G. 1992. *Stochastic Processes in Physics and Chemistry*. Amsterdam: Elsevier.
[185] Vaughan, J. C., Hornung, T., Stone, K. W., and Nelson, K. A. 2007. Coherently controlled ultrafast four-wave mixing spectroscopy. *J. Phys. Chem. A*, **111**, 4873–4883.
[186] Volkov, V., Schanz, R., and Hamm, P. 2005. Active phase stabilization in Fourier-transform two-dimensional infrared spectroscopy. *Opt. Lett.*, **30**, 2010–2012.
[187] Voronine, D., Abramavicius, D., and Mukamel, S. 2006. Coherent control of pump-probe signals of helical structures by adaptive pulse polarizations. *J. Chem. Phys.*, **124**, 034104.
[188] Wang, J. and Hochstrasser, R. M. 2004. Characteristics of the two-dimensional infrared spectroscopy of helices from approximate simulations and analytic models. *Chem. Phys.*, **297**, 195–219.
[189] Wang, J., Chen, J., and Hochstrasser, R. M. 2006. Local structure of β-hairpin isotopomers by FTIR, 2D IR, and ab initio theory. *J. Phys. Chem. B*, **110**, 7545–7555.
[190] Woutersen, S. and Hamm, P. 2000. Structure determination of trialanine in water using polarization sensitive two-dimensional vibrational spectroscopy. *J. Phys. Chem. B*, **104**, 11316–11320.
[191] Woutersen, S. and Hamm, P. 2002. Nonlinear 2D vibrational spectroscopy of peptides. *J. Phys.: Condens. Matter*, **14**, R1035–R1062.
[192] Woutersen, S., Mu, Y., Stock, G., and Hamm, P. 2001a. Hydrogen-bond lifetime measured by time-resolved 2D-IR spectroscopy: N-methylacetamide in methanol. *Chem. Phys.*, **266**, 137–147.
[193] Woutersen, S., Mu, Y., Stock, G., and Hamm, P. 2001b. Subpicosecond conformational dynamics of small peptides probed by two-dimensional vibrational spectroscopy. *Proc. Natl. Acad. Sci. USA*, **98**, 11254–11258.
[194] Woys, A. M., Lin, Y.-S., Reddy, A. S., Xiong, W., de Pablo, J. J., Skinner, J. L., and Zanni, M. T. 2010. 2D IR line shapes probe ovispirin peptide conformation and depth in lipid bilayers. *J. Am. Chem. Soc.*, **132**, 2832–2838.
[195] Xiong, W. and Zanni, M. T. 2008. Signal enhancement and background cancellation in collinear two-dimensional spectroscopies. *Optics Lett.*, **33**, 1371–1373.
[196] Xiong, W., Laaser, J. E., Paoprasert, P., Franking, R. A., Hamers, R. J., Gopalan, P., and Zanni, M. T. 2009a. Transient 2D IR spectroscopy of charge injection in dye-sensitized nanocrystalline thin films. *J. Am. Chem. Soc*, **131**, 18040–18041.
[197] Xiong, W., Strasfeld, D. B., Shim, S.-H., and Zanni, M. T. 2009b. Automated 2D IR spectrometer mitigates the influence of high optical densities. *Vibrational Spec.*, **50**, 136–142.
[198] Yeremenko, S., Pshenichnikov, M. S., and Wiersma, D. A. 2003. Hydrogen-bond dynamics in water explored by heterodyne-detected photon echo. *Chem. Phys. Lett.*, **369**, 107–113.
[199] Yetzbacher, M. K., Belabas, N., Kitney, K. A., and Jonas, D. M. 2007. Propagation, beam geometry, and detection distortions of peak shapes in two-dimensional Fourier transform spectra. *J. Chem. Phys.*, **126**, 044511.
[200] Zanni, M. T. and Hochstrasser, R. M. 2001. Two-dimensional infrared spectroscopy: A promising new method for the time resolution of structures. *Curr. Opin. Struct. Biol.*, **11**, 516–522.
[201] Zanni, M. T., Ge, N. H., Kim, Y. S., and Hochstrasser, R. M. 2001a. 2D-IR can be designed to eliminate the diagonal peaks and expose only the crosspeaks needed for structure determination. *Proc. Natl. Acad. Sci. USA*, **98**, 11265–11270.

[202] Zanni, M. T., Gnanakaran, S., Stenger, J., and Hochstrasser, R. M. 2001b. Heterodyned two-dimensional infrared spectroscopy of solvent-dependent conformations of acetylproline-NH_2. *J. Phys. Chem. B*, **105**, 6520–6535.

[203] Zanni, M. T., Asplund, M. C., and Hochstrasser, R. M. 2001c. Two-dimensional heterodyned and stimulated infrared photon echoes of N-methylacetamide-D. *J. Chem. Phys.*, **114**, 4579–4590.

[204] Zhang, W. M., Meier, T., Chernyak, V., and Mukamel, S. 1998a. Exciton-migration and three-pulse femtosecond optical spectroscopies of photosynthetic antenna complexes. *J. Chem. Phys.*, **108**, 7763–7774.

[205] Zhang, W. M., Chernyak, V., and Mukamel, S. 1998b. Multidimensional femtosecond correlation spectroscopies of electronic and vibrational excitons. *J. Chem. Phys.*, **110**, 5011–5028.

[206] Zhao, W. and Wright, J. C. 2000. Doubly vibrationally enhanced four wave mixing: The optical analog to 2D-NMR. *Phys. Rev. Lett.*, **84**, 1411–1414.

[207] Zheng, J., Kwak, K., Asbury, J., Chen, X., Piletic, I. R., and Fayer, M. D. 2005. Ultrafast dynamics of solute-solvent complexation observed at thermal equilibrium in real time. *Science*, **309**, 1338–1343.

[208] Zheng, J., Kwak, K., and Fayer, M. D. 2007. Ultrafast 2D IR vibrational echo spectroscopy. *Acc. Chem. Research*, **40**, 75–83.

[209] Zhuang, W., Abramavicius, D., Hayashi, T., and Mukamel, S. 2006. Simulation protocols for coherent femtosecond vibrational spectra of peptides. *J. Phys. Chem. B*, **110**, 3362–3374.

Index

α-helix, 125
   3_{10}, 142
   Exciton, 120, 123–126
   Isotope labeling, 131–133
β-sheet, 127
   Exciton, 120, 126–128
   Hairpin simulation, 229–232
   Isotope labeling, 131
$\pi/2$-pulse, 24
2Q
   Dream experiment, 252
   Pulse sequence, 233–237, 239–240, 252
   see also Exciton two-quantum Hamiltonian, 233
3_{10}-helix, 142
3D IR spectroscopy, 2–3, 239–242

Absorption, 37–38
Absorption coefficient, 143
Absorption spectrum, *see* Linear spectrum
Absorptive lineshape, 65
Absorptive spectrum, *see* Lineshape, Phase twist
   2D, 75
   3D, 241
Aliasing, 183, 202, 203, 256
Amide I mode, 120
   Charge density, 134
   Coupling constants, 134–136
   Coupling maps, 136–137
   Isotope labeling, 128–133
   Lineshapes, 222
   Simulation, 229–232
Angular momentum operator, 102
Anharmonicity, 7
   Anharmonicity for coupled dimer, 117–119
   Calculation of, 226–228
   Diagonal anharmonic shift, 5, 7–8
   Extracting accurate values, 213–214
   Influenced by exciton delocalization, 120, 123, 128
   Intermode, 138

Local mode, 112, 117
Measuring with 2Q pulse sequence, 235
Molecular dynamics, 220
Normal mode, 137–140
Off-diagonal anharmonic shift, 5, 8

Balanced heterodyne detection, *see* Spectrometer design, 186–188, 206
Bilinear coupling, 110, 111
Bleach, 7
Bloch
   Dynamics, 154
   Theorem, 122
   Vector, 24
      Tilt angle, 25
box-CARS, 70, 184

Carrier envelope phase, 197
Cascaded signals, 237–239
   Parallel vs. sequential, 252
Causality, 59, 161, 194, 204, 255
Central limit theorem, 150
Chemical exchange, 13, 174–175
Clebsch–Gordan coefficients, 266
Coherence, 22, 28
   Interstate, 79
Coherence artifact, 87
Coherence spike, 87
Coherence transfer, 166
Coherent control, 245–246
Coherent state, 30
Condon approximation, 148
   Breakdown, 163
Correlation function
   Diffusive component, 223
   Ergodic hypothesis, 151
   Four-point, 90
   Frequency–frequency or frequency fluctuation (FFCF), 150

281

Six-point, 106
Stationary, 150
Coupling, 3–4, 8, 10, 109–137, 140, 144
  α-helix, 126, 132
  β-sheet, 127
  3_{10}-helix, 142
  Accuracy, 119–120
  Bilinear, 110, 111
  Calculations
    Excitons, 229–232
    Quantum, 226–227
  Darling–Dennison terms, 139
  Dimer, 114–116
  Fermi resonance, 140–142
  Isotope labeling, 130
  Map, 136–137
  Measurement, 213, 235
  Models, 134–136
  Off-diagonal disorder, see Off-diagonal disorder
  Population transfer, 167
  Quantum conserving terms, 110
  Strong vs. weak regimes, 117–119
  Transition dipole, 4, 114
  Units, 143
Cross-peaks, 9–10
  2D lineshapes, 11
  3D IR, 240
    Polarization dependence, 106–108
  Caused by chemical exchange, 13, 174–175
  Caused by kinetics, 14, 243–245
  Caused by population transfer, 166–172
  Extracting accurate anharmonicities, 213–214
  For typical pulse sequences, 77–84
  Forbidden, 79, 81
  Interstate coherences, 79
  On-diagonal, 81
  Polarization dependence, 88, 91, 93–100
  Spurious, 204, 212
  Strong vs. weak regimes, 117–119
  Transient pump–probe, 85
Cumulant expansion, 149–150, 153, 155, 157
  Avoiding it, 224–226
  Simulations, 224

Darling–Dennison coupling terms, 139
Data collection and processing, 176–215
Debye equation, 102
Density matrix, 27–31, 49
  Basis-free representation, 49
  Coherence, 28
  Coherent state, 30
  Incoherent state, 31
  Of a statistical average, 50–52
  Population, 28
  Pure state, 31
  Time evolution, Liouville–von Neumann equation, 49
Density operator, 31

Diagonal anharmonic shift, 5
Diagonal disorder, 124, 127, 131, 164, 167, 231
  Forbidden transitions, 124
Diagonal peak, 7
Diagonal peaks, 4
  2D spectrum, 72–76
  3D IR, 239–242
    Polarization dependence, 106–108
  Eliminating with polarization, 99–100
  Influenced by exciton delocalization, 120, 123, 128
  Lineshapes, 145–164
  Measuring population relaxation, 105–106
  Orientational contribution, 103–104
  Polarization dependence, 92–93
Diffusive component, 223
Dipole allowed transition, 9
Dipole operator, 19, 31, 32, 91
  Expectation value, 57
  In interaction picture, 57
Dispersion relation, 122, 123, 142
Dispersive lineshape, 65
  2D and phase twist, 74
Double sided Feynman diagrams, see Feynman diagram
Dunham expansion, 138

Ergodic hypothesis, 151
Etalon, see Spectrometer design
Euler angles, 94
Exciton, 109–113
  α-helix, 120, 123–126
  β-sheet, 120, 126–128
  Breakdown of model, 226
  Coupled dimer, 121
  Defect, 131
  Diagonal disorder, 124, 127, 131, 164, 167, 231
  Dispersion relation, 122, 123, 142
  Linear chain, 121–124
  Molecular, 109
  Off-diagonal disorder, 124, 127, 131, 164, 167
  One-quantum Hamiltonian, 111–114, 231
  Quantum conserving terms, 110
  Simulation, 229–232
  Tight binding model, 121
  Transition dipoles, 113, 114
  Two-quantum Hamiltonian, 111–113, 142
  Vibron, 109

Fabry–Perot, see Spectrometer design
Fast modulation limit, 152–153
Fermi resonance, 140–142
Feynman diagram
  2Q spectra, 234
  Chemical exchange, 173
  Double sided, 36
  Fifth-order or 3D IR, 242, 252
  For 2D narrow-band spectra, 83
  For 2D non-rephasing spectra, 80

For 2D rephasing spectra, 78
For pump–probe, 84
Linear response, 37
Population and coherence transfer, 166
Rules, 46–47
Single sided, 55, 56
Third-order, 43
Feynman pathway, 34
Fifth-order response function, *see* Response function, fifth-order
First-order response function, *see* Linear response function
Flipping angle, *see* Tilt angle
Forbidden transition, 9, 79, 81
  Coupling limits, 140
  Diagonal disorder, 124
  One vibrational quantum, 119
Fourier transform, 254–259
  Aliasing, 183, 202, 203, 256
  Causality, 194, 204
  Discrete, 204
  Double pulse, 15
  Initial value, 204
  Nyquist frequency, 202, 256
  Resolution, 202, 204
  Zero-padding, 204–206
Frame
  Laboratory, 92
  Molecular, 94
  Polarization, 106
  Rotating, 194–197
Free induction decay, *see* Linear response function
  Perturbed, 86
Frequency domain 2D IR spectroscopy, *see* Spectrometer design, Etalon, 82–84
  Polarization dependence, 98
  Pulse shaping, 176–180
  Spectrometer design, 176–180
Frequency resolution, 202, 204
Frequency–frequency correlation function, 150
FTIR Spectroscopy, *see* Linear (FTIR) Spectroscopy

Gaussian lineshape, *see* Lineshape, 150, 153, 155

Hamiltonian
  α-helix, 125
  β-sheet, 126
  Diagonal disorder, *see* Diagonal disorder
  Exciton, 110
  Fermi resonance, 141
  Harmonic oscillator, 260
  Interaction picture, 54
  Isotope labeling, 130
  Linear chain, 121
  Morse oscillator, 6
  Normal mode, 138
  Off-diagonal disorder, *see* Off-diagonal disorder
  One-quantum, 111–114, 231

  Quantum conserving terms, 110
  System, 52, 54, 55, 58
  Time dependence, 166
  Two-quantum, 111–113, 142
Heterodyne detection, 64, 70–71
  Balanced, 186–188, 206
  Local oscillator, 70
  Local oscillator intensity, 188
  Optical densities and spectral distortion, 188–189
  Scatter, 189
  Self-, 64, 185
  Signal strength, 163
  Versus homodyne, 70
Hole-burning, 11, *see* Frequency domain 2D IR spectroscopy and Spectrometer design, Etalon
Homodyne detection, 70
Homogeneous, *see* Lineshape
  0–1 vs. 1–2 transition, 68
  Dephasing, 28
  Limit, 152
  Lineshape, 10
  Rotational contribution to dephasing, 104

Incoherent state, 31
Inhomogeneous, *see* Lineshape
Inhomogeneous broadening, 10, 26
  Limit, 153, 157
Intermode anharmonicity, 138
Interstate coherence, 79
Isotope labeling, 128–133

Kubo model, 145, 152–153

Laboratory frame, 92
Ladder operator, 68, 110, 260–261
Legendre polynomials, 96, 265–266
Line-narrowed, 156
Linear
  (FTIR) Spectroscopy, 6, 184, 195, 213
  Lineshape, 27, 152–154, 214
  Response function, 32, 34, 37, 52, 59, 148, 150
  Response theory, 35
  Spectrum, 22, 86, 103
    CD3ζ peptide, 129
    Of a coupled dimer, 114–116
    Water, 250
Lineshape, 145–165
  2D, 156–157
  3D IR, 240–242
  Absorptive, 65, 72–75, 82–84
  Antidiagonal width, 11, 156, 179
  Condon approximation, 148
    Breakdown, 163
  Cross-peaks, 11
  Cumulant expansion, 149–150
  Diagonal width, 11, 156, 179
  Dispersive, 65
  Ellipticity, 160

Excitonically coupled systems, 164
Fast modulation limit, 152–153
Function, 149
Gaussian, 150, 153, 155
Homogeneous, 10
Homogeneous limit, 152
Inhomogeneous, 10, 26
Kubo model, 145, 152–153
Line-narrowed, 156
Lorentzian, 64, 152, 154
Nodal line, 159
Non-Gaussian, 157, 225, 239, 242
Phase twist, 72–75
Quasi-absorptive, 81
Simulations, 217–226
Spectral diffusion, 12, 154, 169
  Measuring, 156, 159–163
Voigt, 154
Wings, 65
Liouville pathway, 34
Liouville–von Neumann equation, 49, 51, 52, 56, 146
Local mode, 8, 109–144
  Anharmonic shift, 112
  Anharmonicity, 117
Local oscillator, *see* Heterodyne detection, 70
  Around or through the sample, 188–189
  Intensity, 188
Lorentzian lineshape, *see* Lineshape, 64, 152, 154

Macroscopic polarization, 21
  Bloch vector, 24
  Calculated from dipole operator, 57
  Convolution with laser pulse, 35, 62, 84
  Phase shift, 22, 37
Magic angle, 98, 105
Mass weighted coordinates, 144
Michelson interferometer, 75, 183, 186, 188
Molecular dynamics simulations, 217–232
Molecular exciton, 109
Molecular frame, 94
molecular response, 21
Momentum operator, 261
Morse oscillator, 6, 112, 140

Narrow band pump–probe, *see* Frequency domain 2D IR spectroscopy
Nodal line, 159
Non-Condon effects, 163, 226
Non-rephasing
  2D spectrum, 79–81
  Diagram, example, 41, 65, 158
Normal mode, 109, 137–140
  Amide I, *see* Amide I
  Dunham expansion, 138
  Hamiltonian, 138
Normal modes
  Gaussian calculations, 227
Number operator, 260

Nyquist frequency, 202, 256

Off-diagonal anharmonic shift, 5
Off-diagonal disorder, 124, 127, 131, 164, 167
On-diagonal cross-peaks, 81
Operator
  Angular momentum, 102
  Density, 31
  Dipole, 19, 31, 91
    Expectation value, 57
    In interaction picture, 57
  Ladder, 68, 110, 260–261
  Momentum, 261
  Number, 260
  Position, 261
  Time evolution, 55
  Transition dipole, 113
Optical density, 188–189
Orientational response, 88–108
  3D and higher-order pulse sequences, 106–108
  Cross-peaks, 93
  Diagonal peaks, 91–93, 100
  Dynamics, 100–106
  General expression, 96, 106

Peak shift measurement, 161–163
Perturbed free induction decay, 86
Phase cycling
  Frequency domain, 180
  Methods, 197–200
  Pathway selection, 44–46, 191–194
  Removing scatter, 191–193
  Removing transient absorption bkgd, 191–193
  Rotating frame, 194–197
Phase matching, 42
  box-CARS geometry, 70, 184
  Selecting pathways, 42–44
  Standard geometries, 184
  Thin sample limit, 185
Phase stability, 180
Phase twist, 72–76, 194
  Narrowband 2D spectra, 84
  Quasi-absorptive, 81
Photoelastic modulator, 198
Photon echo, 41
  Example of a heterodyned signal, 162, 236
  Fifth-order, 236
  Heterodyne, *see* Heterodyne detection
  Integrated, 70, 161
  Peak shift, 161–163
  Pulse sequence, 41
Physical constants, 262–264
Polarization
  $\langle -45°, +45°, 0°, 90°\rangle$, 99
  Combining pulse polarizations, 99–100
  Eliminating diagonal peaks, 99
  Frame, 106
  General expression, 96, 106

General expression for fifth-order and higher, 106–108
Magic angle, 98, 105
Response, *see* Orientational response
Table summarizing responses for XXXX,XXYY,XYXY,XYYX, 97
Population relaxation, 27
  0–1 vs. 1–2, 67
  Density matrix, 28, 66
  Feynman diagram, 167
  Measurement, 105–106
  Relation to homogeneous dephasing, 67
Position operator, 261
Product operator formalism, 246
Projection slice theorem, 85
Pulse sequence
  2Q, 233–237, 252
  2Q 3D IR, 239–240
  2Q dream experiment, 252
  Purely absorptive 3D IR, 241–242
  Transient 2D IR spectroscopy, 243–245, 252
Pulse shaping, 75, 82, 191, 204
  Data collection, 200–201
  Frequency domain, 176–180
  Phase stability, 181
  Phasing, 207
  Rapid scanning, 245
  Shaper design, 199–200
Pump–probe
  Background, 189
  Balanced heterodyne detection, 187
  Experimental setup, 83
  Extracting rephasing and non-rephasing signals, 193–194
  Feynman diagrams for narrowband, 83
  Frequency domain, *see* Frequency domain 2D IR spectroscopy
  Frequency domain spectrometer design, 176
  Hole-burning, 11
  Local oscillator intensity, 188
  Negative time delays, 86
  Phase matching geometry, 45, 75, 184, 185
  Spectrum for 2D IR calibration, 209
  Table of polarization conditions, 105
  Transient, 84–86
Pure dephasing, 29, 67
  Diagonal disorder, 169
  Fast modulation limit of Kubo, 152
Pure state, 31

Quantum conserving terms, 110
Quantum correction factor, 171
Quasi-absorptive 2D spectrum, 81

Rapid scanning 2D IR, 245
Rephasing
  2D spectrum, 77–79
  Diagram, example, 41, 66, 158

Response function
  Even-order (e.g. $R^{(2)}$), 59
  Fifth-order, 252
  Linear, 32, 34, 37, 52, 59, 148, 150
  nth-order, 59
  Single sided, 44, 59
  Third-order, 39, 59, 65–69
Rotating frame, 194–197, 202
  vs. undersampling, 204
Rotating wave approximation, 35–37, 43, 47, 53
Rotation matrices, 94
Rotational diffusion, 102–103
  Debye equation, 102
Rotational response, *see* Orientational response

Scatter, 189–193
Schrödinger equation, 19
Selection rules, 20
  Derivation, 133
  Harmonic oscillator, 8
Self-heterodyne, 64, 185
Semiclassical approximation, 19, 145
Semi-impulsive limit, 71, 72
Side-chain absorbances, 130
Simulation
  2D spectra using excitons, 229–232
  Lineshapes, 217–226
  Molecular couplings, 226–227
Single-sided Feynman diagram, 55, 56
Slow modulation limit, 153
Spectral density, 170
Spectral diffusion, 12, 154, 169
  Measuring, 156, 159–163
  Water, 217
Spectral interferometry, 206–207
Spectrometer design, 69–72, 176–215
  Alignment, 214–215
  Balanced heterodyne detection, 186–188, 206
    using polarization, 187, 188
  Etalon, 75, 82, 108, 176–179, 200, 201
  Fabry–Perot, 75
  Frequency domain, 176–180
  Homodyne detection, 70
  Michelson interferometer, 75, 183, 186, 188
  Optical density, 188–189
  Phase cycling, *see* Phase cycling
  Phase matching, 184–186
  Phase stability, 180–182
  Polarizaton control, 186
  Pulse intensities, 188
  Pulse shaping, *see* Pulse shaping
  Scatter, 189–193
  Time domain, 180–188
Spherical harmonics, 95, 265–266
  Addition theorem, 95
  Clebsch–Gordan coefficients, 266
  Closure property, 103
  Orthogonality, 102, 266

Strong coupling limit, 117–119
System Hamiltonian, 52, 54, 55, 58

Thin sample limit, 185
Third-order response function, *see* Response function, third-order
Tight binding model, 121
Tilt angle, 25
Time domain 2D IR spectroscopy, 14–16
Time-evolution operator, 55
Transient 2D IR spectroscopy, 14, 189, 243–245, 252
    Rapid scanning, 245
Transition charge density, 133, 134
Transition dipole, 133–134
    Coupling, 114
    Matrix, 113–114
    Moment, 20, 115
    Operator, *see* Operator, Dipole
    Point charge model, 135
    Strength, 133
        Amide I, 121
    Strength from absorption spectrum, 143
    Units, 143
Transition dipole operator, 113
Triggered exchange, 244
Two-quantum, *see* 2Q

Under-sampling, 202–204
Units, 262–264

Vibrational dynamics
    Bloch, 154
Vibrational exciton, *see* Exciton
Vibrational selection rules, 20
    Derivation, 133
Vibron, 109
Voigt lineshape, *see* Lineshape, 154

Wavepacket, 21
Wavevector, 19
    Definition, 185
Weak coupling limit, 117–119
Window functions, 211–212

Zero-padding, 204–206

Printed in the United States
By Bookmasters